1 Measuring Variation: How Values Differ

2 Correlation 29

3 Simple Regression 59

4 Using the LINEST() Function 103

5 Multiple Regression 151

6 Assumptions and Cautions Regarding Regression Analysis 199

7 Using Regression to Test Differences Between Group Means 245

8 The Analysis of Covariance 295

Index 337

Regression Analysis Microsoft® Excel®

Conrad Carlberg

800 East 96th Street,

Indianapolis, Indiana 46240 USA

Regression Analysis Microsoft® Excel®

ISBN-13: 978-0-7897-5655-8
ISBN-10: 0-7897-5655-2

Library of Congress Control Number: 2016930324

Printed in the United States of America

1 16

Trademarks

All terms mentioned in this book that are known to be trademarks or service marks have been appropriately capitalized. Que Publishing cannot attest to the accuracy of this information. Use of a term in this book should not be regarded as affecting the validity of any trademark or service mark.

Warning and Disclaimer

Every effort has been made to make this book as complete and as accurate as possible, but no warranty or fitness is implied. The information provided is on an "as is" basis. The author and the publisher shall have neither liability nor responsibility to any person or entity with respect to any loss or damages arising from the information contained in this book.

Special Sales

For information about buying this title in bulk quantities, or for special sales opportunities (which may include electronic versions; custom cover designs; and content particular to your business, training goals, marketing focus, or branding interests), please contact our corporate sales department at corpsales@pearsoned.com or (800) 382-3419.

For government sales inquiries, please contact governmentsales@pearsoned.com.

For questions about sales outside the U.S., please contact intlcs@pearson.com

Editor-in-Chief
Greg Wiegand

Acquisitions Editor
Joan Murray
Trina MacDonald

Development Editor
Mark Renfrow

Managing Editor
Sandra Schroeder

Project Editor
Seth Kerney
Mandie Frank

Copy Editor
Paula Lowell

Indexer
WordWise Publishing Services

Proofreader
Debbie Williams

Technical Editor
Bill Pate

Editorial Assistant
Cindy Teeters

Cover Designer
Chuti Prasertsith

Compositor
codeMantra

Contents

Introduction **1**

1 Measuring Variation: How Values Differ **5**

How Variation Is Measured 5

Sum of Deviations 6

Summing Squared Deviations 7

From the Sum of Squares to the Variance 10

Using the VAR.P( ) and VAR.S( ) Functions 11

The Standard Deviation 14

The Standard Error of the Mean 15

About z-Scores and z-Values 18

About t-Values 23

2 Correlation **29**

Measuring Correlation 29

Expressing the Strength of a Correlation 30

Determining a Correlation's Direction 32

Calculating Correlation 34

Step One: The Covariance 34

Watching for Signs 36

From the Covariance to the Correlation Coefficient 38

Using the CORREL( ) Function 41

Understanding Bias in the Correlation 41

Checking for Linearity and Outliers in the Correlation 44

Avoiding a Trap in Charting 48

Correlation and Causation 53

Direction of Cause 54

A Third Variable 55

Restriction of Range 55

3 Simple Regression **59**

Predicting with Correlation and Standard Scores 60

Calculating the Predictions 61

Returning to the Original Metric 63

Generalizing the Predictions 64

Predicting with Regression Coefficient and Intercept 65

The SLOPE( ) Function 65

The INTERCEPT( ) Function 69

Charting the Predictions 70

Shared Variance 71

The Standard Deviation, Reviewed 71

More About Sums of Squares 73

Sums of Squares Are Additive 74

R^2 in Simple Linear Regression 77

Sum of Squares Residual versus Sum of Squares Within 81

The TREND( ) Function....82
Array-entering TREND( )....84
TREND( )'s *new x's* Argument....85
TREND( )'s *const* Argument....86
Calculating the Zero-constant Regression....88
Partial and Semipartial Correlations....90
Partial Correlation....91
Understanding Semipartial Correlations....95

4 Using the LINEST( ) Function....103
Array-Entering LINEST( )....103
Understanding the Mechanics of Array Formulas....104
Inventorying the Mistakes....105
Comparing LINEST( ) to SLOPE( ) and INTERCEPT( )....107
The Standard Error of a Regression Coefficient....109
The Meaning of the Standard Error of a Regression Coefficient....109
A Regression Coefficient of Zero....110
Measuring the Probability That the Coefficient is Zero in the Population....112
Statistical Inference as a Subjective Decision....113
The t-ratio and the F-ratio....116
Interval Scales and Nominal Scales....116
The Squared Correlation, R^2....117
The Standard Error of Estimate....120
The t Distribution and Standard Errors....121
Standard Error as a Standard Deviation of Residuals....125
Homoscedasticity: Equal Spread....128
Understanding LINEST( )'s F-ratio....129
The Analysis of Variance and the F-ratio in Traditional Usage....129
The Analysis of Variance and the F-ratio in Regression....131
Partitioning the Sums of Squares in Regression....133
The F-ratio in the Analysis of Variance....136
The F-ratio in Regression Analysis....140
The F-ratio Compared to R^2....146
The General Linear Model, ANOVA, and Regression Analysis....146
Other Ancillary Statistics from LINEST( )....149

5 Multiple Regression....151
A Composite Predictor Variable....152
Generalizing from the Single to the Multiple Predictor....153
Minimizing the Sum of the Squared Errors....156
Understanding the Trendline....160
Mapping LINEST( )'s Results to the Worksheet....163
Building a Multiple Regression Analysis from the Ground Up....166
Holding Variables Constant....166
Semipartial Correlation in a Two-Predictor Regression....167

Finding the Sums of Squares 169
R^2 and Standard Error of Estimate 170
F-Ratio and Residual Degrees of Freedom 172
Calculating the Standard Errors of the Regression Coefficients 173
Some Further Examples 176

Using the Standard Error of the Regression Coefficient 181
Arranging a Two-Tailed Test 186
Arranging a One-Tailed Test 189

Using the Models Comparison Approach to Evaluating Predictors 192
Obtaining the Models' Statistics 192
Using Sums of Squares Instead of R^2 196

Estimating Shrinkage in R^2 197

6 Assumptions and Cautions Regarding Regression Analysis 199

About Assumptions 199
Robustness: It Might Not Matter 202
Assumptions and Statistical Inference 204

The Straw Man 204

Coping with Nonlinear and Other Problem Distributions 211

The Assumption of Equal Spread 213
Using Dummy Coding 215
Comparing the Regression Approach to the t-test Approach 217
Two Routes to the Same Destination 218

Unequal Variances and Sample Sizes 220
Unequal Spread: Conservative Tests 220
Unequal Spread: Liberal Tests 225
Unequal Spreads and Equal Sample Sizes 226
Using LINEST( ) Instead of the Data Analysis Tool 230
Understanding the Differences Between the T.DIST( ) Functions 231
Using Welch's Correction 237
The TTEST( ) Function 243

7 Using Regression to Test Differences Between Group Means 245

Dummy Coding 246
An Example with Dummy Coding 246
Populating the Vectors Automatically 250
The Dunnett Multiple Comparison Procedure 253

Effect Coding 259
Coding with −1 Instead of 0 260
Relationship to the General Linear Model 261
Multiple Comparisons with Effect Coding 264

Orthogonal Coding 267
Establishing the Contrasts 267
Planned Orthogonal Contrasts Via ANOVA 268
Planned Orthogonal Contrasts Using LINEST( ) 269

Factorial Analysis 272
Factorial Analysis with Orthogonal Coding 274
Factorial Analysis with Effect Coding 279
Statistical Power, Type I and Type II Errors 283
Calculating Statistical Power 285
Increasing Statistical Power 286
Coping with Unequal Cell Sizes 288
Using the Regression Approach 289
Sequential Variance Assignment 291

8 The Analysis of Covariance 295
Contrasting the Results 297
ANCOVA Charted 305
Structuring a Conventional ANCOVA 308
Analysis Without the Covariate 308
Analysis with the Covariate 310
Structuring an ANCOVA Using Regression 315
Checking for a Common Regression Line 316
Summarizing the Analysis 320
Testing the Adjusted Means: Planned Orthogonal Coding in ANCOVA 321
ANCOVA and Multiple Comparisons Using the Regression Approach 328
Multiple Comparisons via Planned Nonorthogonal Contrasts 330
Multiple Comparisons with Post Hoc Nonorthogonal Contrasts 332
Index 337

About the Author

Conrad Carlberg (www.conradcarlberg.com) is a nationally recognized expert on quantitative analysis and on data analysis and management applications such as Microsoft Excel, SAS, and Oracle. He holds a Ph.D. in statistics from the University of Colorado and is a many-time recipient of Microsoft's Excel MVP designation.

Carlberg is a Southern California native. After college he moved to Colorado, where he worked for a succession of startups and attended graduate school. He spent two years in the Middle East, teaching computer science and dodging surly camels. After finishing graduate school, Carlberg worked at US West (a Baby Bell) in product management and at Motorola.

In 1995 he started a small consulting business that provides design and analysis services to companies that want to guide their business decisions by means of quantitative analysis—approaches that today we group under the term "analytics." He enjoys writing about those techniques and, in particular, how to carry them out using the world's most popular numeric analysis application, Microsoft Excel.

Acknowledgments

I thank Joan Murray for the skill and tact she employed in seeing this book through from a proposed table of contents to a bound volume (or, if your preferences in media run that way, an electronic gadget). Bill Pate provided a careful technical edit, not a straightforward task when the platform gets as idiosyncratic as Excel does from time to time. Paula Lowell's copy edit unmixed my metaphors, got my subjects to reach agreement with my verbs, and gently pointed out those occasions when I accidentally duplicated figure captions. Seth Kerney, Mandie Frank and Mark Renfrow kept the whole thing from careening out of control, and I'm the sole owner of any errors remaining in this book.

We Want to Hear from You!

As the reader of this book, *you* are our most important critic and commentator. We value your opinion and want to know what we're doing right, what we could do better, what areas you'd like to see us publish in, and any other words of wisdom you're willing to pass our way.

We welcome your comments. You can email or write to let us know what you did or didn't like about this book—as well as what we can do to make our books better.

Please note that we cannot help you with technical problems related to the topic of this book.

When you write, please be sure to include this book's title and author as well as your name and email address. We will carefully review your comments and share them with the author and editors who worked on the book.

Email: feedback@quepublishing.com

Mail: Que Publishing
ATTN: Reader Feedback
800 East 96th Street
Indianapolis, IN 46240 USA

Reader Services

Register your copy of Regression Analysis Microsoft Excel at quepublishing.com for convenient access to downloads, updates, and corrections as they become available. To start the registration process, go to quepublishing.com/register and log in or create an account*. Enter the product ISBN, 9780789756558, and click Submit. Once the process is complete, you will find any available bonus content under Registered Products.

*Be sure to check the box that you would like to hear from us in order to receive exclusive discounts on future editions of this product.

INTRODUCTION

Like a lot of people, I slogged through my first undergraduate classes in inferential statistics. I'm not talking here about the truly basic, everyday stats like averages, medians, and ranges. I'm talking about things you don't commonly run into outside the classroom, like randomized block designs and the analysis of variance.

I hated it. I didn't understand it. Assistant professors and textbooks inflicted formulas on us, formulas that made little sense. We were supposed to pump data through the formulas, but the results had mysterious names like "mean square within." All too often, the formulas appeared to bear no relationship to the concept they were supposed to quantify. Quite a bit later I came to understand that those formulas were "calculation formulas," meant to be quicker to apply, and less error-prone, than the more intuitively useful definitional formulas.

Eventually I came to understand why the analysis of variance, or ANOVA, is used to evaluate the differences between means—as counterintuitive as that sounded—but all those sums of squares between and sums of squares within and degrees of freedom just did not make sense. I knew that I had to calculate them to satisfy a requirement, and I knew how to do so, but I did not understand why.

Eventually I came across a book on regression analysis. Another student recommended it to me—it had clarified many of the issues that had confused him and that were still confusing me. The book, now long out of print, discussed the analysis of variance and covariance in terms of regression analysis. It resorted to computer analysis where that made sense. It stressed correlations and proportions of shared variance in its explanations. Although it also discussed sums of squares and mean squares, the book talked about them principally to help show the relationship between conventional Fisherian analysis and the regression approach.

The concepts began to clarify for me and I realized that they had been there all the time, but they were hidden behind the arcane calculations of ANOVA. Those calculations were used, and taught, because they were developed during the early twentieth century, when twenty-first century computing power wasn't merely hard to find, it just didn't exist. It was much easier to compute sums of squares (particularly using calculation formulas) than it was to calculate the staples of regression analysis, such as multiple correlations and squared semipartials. You didn't have to find the inverse of a matrix in traditional ANOVA, as was once necessary in regression. (Calculating by hand the inverse of any matrix larger than 3×3 is a maddening experience.)

Today, all those capabilities exist in Excel worksheets, and they make the concepts behind the analysis of variance much more straightforward. Furthermore, the Excel worksheet application makes things much easier than was hinted at in that book I read. The book was written long before Excel first emerged from its early shrink-wrap, and I shake my head that once upon a time it was necessary to pick individual pieces from the inverse of a matrix and fool around with them to get a result. Today, you can get the same result in an Excel worksheet just by combining fixed and relative addressing properly.

We still rely heavily on the analysis of variance and covariance in various fields of research, from medical and pharmaceutical studies to financial analysis and econometrics, from agricultural experiments to operations research. Understanding the concepts is important in those fields—and I maintain that understanding comes a lot easier from viewing the problems through the prism of regression than through that of conventional ANOVA.

More important, I think, is that understanding the concepts that you routinely use in regression makes it much easier to understand even more advanced methods such as logistic regression and factor analysis. Those techniques expand your horizons beyond the analysis of one-variable-at-a-time methods. They help you move into areas that involve latent, unobserved factors and multinomial dependent variables. The learning curve is much steeper in principal components analysis if you don't already have the concept of shared variance in your hip pocket.

And that's why I've written this book. I've had enough experience, first as a suit and then in my own consulting practice, with inferential statistics to know how powerful a tool it can be, if used correctly. I've also been using Excel to that end for more than 20 years. Some deride Excel as a numeric analysis application. I think they're wrong. On the other hand, Microsoft's history as Excel's publisher is, well, checkered. Not long ago a colleague forwarded to me an email in which his correspondent wondered, a little plaintively, whether it was "safe" to use Excel's statistical functions. At the time I was finishing this book up, and much of the book has to do with the use of Excel's LINEST() worksheet function. Here's what I wrote back:

> The question of whether it's "safe" to use Excel for statistical analysis is a messy one. Microsoft is at fault to some degree, and those who rant that it's dangerous to use Excel for statistical analysis share that fault.
>
> Since 1995, MS has done nothing to improve the Data Analysis add-in (aka the Analysis Toolpak) other than to convert it from the old V4 macro language to VBA.

That's a shame, because the add-in has plenty of problems that could easily be corrected. But the add-in is not Excel any more than the old Business Planner add-in is Excel. Nevertheless, I've seen plenty of papers published both privately and in refereed journals that rightly complain about the add-in's statistical tools, and then lay the blame on the actual application.

There were, through either 2003 or 2007—I can't now recall which two principal problems with LINEST(). One had to do with the way that the regression and residual sums of squares were calculated when LINEST()'s third, *const* argument is set to FALSE. This was known as early as 1995, but MS didn't fix it until much later.

Another was that LINEST() solved what are termed the "normal equations" using matrix algebra—long the preferred method in statistical apps. But on rare occasions it's possible for multicollinearity (the presence of strong correlations among the predictor variables) to result in a matrix with a zero determinant. Such a matrix cannot be inverted, and that makes it impossible to return LINEST()'s usual results. In 2003 or 2007, MS fixed that by replacing it with something called *QR decomposition*.

But the multicollinearity problem caused LINEST() to return the #NUM! error value. No one could be led down an unsafe, dangerous path by that. And the problem with the third, *const* argument resulted in such things as a negative R-squared. Only someone utterly untutored in regression analysis could be misled by a negative R-squared. It cannot come about legitimately, so something must be wrong somewhere.

Finally, various bread-and-butter statistical functions in Excel have been improved to enhance their accuracy when they're pointed at really extreme values. This is useful—more accuracy is always better than less accuracy. But it's an instance of what Freud called the "narcissism of small differences." If I'm a biostatistician and I'm called upon to make a decision based on a difference between 10^–16 and 10^–17, I'm going to replicate the experiment. The difference is too small, both substantively and technically, to use as the basis for an important decision—regardless of whether I'm using SAS, R, or Excel.

Which brings me to the Chicken Little alarmists who scare people with lengthy screeds regarding this stuff. Badmouthing sound statistical applications has a long, dishonorable history. When I was still in school, other students who had sweated blood to learn an application named BMD said that it was a bad idea to use a different application. They said the competing app wasn't accurate, but their real motive was to prevent the erosion of their own hard-acquired expertise—more precisely, the perception of that expertise. (The new, competing application was SPSS.)

I spend some ink in the introduction to my book *Statistical Analysis Excel 2013*, and in its sixth chapter, on these and closely related matters. If it's unsafe to use Excel for statistical analysis, the danger lies in the use of an accurate tool by someone who hasn't a clue what he's doing, either with inferential statistics or with Excel.

That's my screed for 2016. I hope you enjoy this book as much as I enjoyed revisiting old friends.

Measuring Variation: How Values Differ

1

IN THIS CHAPTER

How Variation Is Measured 5
The Standard Deviation 14
The Standard Error of the Mean 15

If you bear with me through this entire book, or even just some of it, you'll read a considerable amount of material about variability: how numbers that go together in some way also differ from one another. That's because variation—actually, *shared variation*—is at the heart of regression analysis. When two or more variables share variance, that means they're quantitatively related. For example, if the correlation coefficient between two variables is 0.5, they share 25% of their variance: The proportion of shared variance is the square of the correlation coefficient.

So it's hardly exaggerating to say that variation and covariation are so fundamental to regression analysis that understanding them is absolutely essential to understanding regression analysis. Even if this chapter, on variation, and the next chapter, on correlation, are only a review for you, the information on Excel's variance and correlation functions might prove new and useful. Let's start with a brief overview of the ways to measure variation, and why you might choose to use a particular method.

How Variation Is Measured

Statisticians have several approaches that they can use to measure the amount of variability in a given set of numbers. Probably the most familiar is the range, which has lesser-known subdivisions such as the semi-interquartile range and the midhinge. These measures can provide insight into how a set of numbers are distributed between their minimum and maximum values. Furthermore, you can calculate them quickly and accurately (because the math isn't at all cumbersome). You often want to know the first quartile, the median, and the third quartile (25% of the values lie below the first quartile, and 25% of the values lie above the third quartile).

As useful as these approaches can be in many situations—particularly when your purpose is to describe the attributes of a data sample—they aren't so helpful when you're after more information. For example:

- These methods tend to ignore all but a few observations in a sample. The important values are the minimum, the maximum, the median, and the first and third quartiles. But all the other values in the data set are ignored, and that's a lot of information to ignore.
- For the past 20 years, most of us have been able to calculate a standard deviation by typing a few letters and dragging through some cells on a worksheet. The more sensitive statistics no longer require a tolerance for more tedium. Ease of calculation is no longer a reason to prefer a range to a standard deviation.
- Measuring variation using, say, the variance instead of the range makes it possible to draw quantitative comparisons between a sample and the population from which you draw the sample. Those comparisons put you in a position to make inferences about an otherwise inaccessible population.

Methods of measuring variation that do not depend on ranges and quartiles depend instead on the differences between all the individual observations and their mean. Then those differences, or *deviations*, are aggregated into a single number that expresses the total amount of variation in the data, not just the difference between the minimum and maximum values.

Sum of Deviations

It's conventional to represent each of the original observations in a data set by the capital letter "X." The deviation of the observation from the mean of the data set is represented by the lowercase letter "x." I make use of that convention in the following proof and in various similar situations throughout this book.

The way to summarize the deviations, the differences between the individual observations and their mean, that occurs to most of us when we start to study statistics is to total them up. Unfortunately, the sum of the simple deviations of each value in a data set from their mean is always 0.0. It's equally simple to prove that's so:

$$\sum_{i=1}^{n} x_i = \sum_{i=1}^{n} (X_i - \overline{X})$$

$$= \sum_{i=1}^{n} X_i - \sum_{i=1}^{n} \overline{X}$$

$$= \sum_{i=1}^{n} X_i - n\overline{X}$$

$$= \sum_{i=1}^{n} X_i - n\left(\sum_{i=1}^{n} X_i\right)/n$$

$$= \sum_{i=1}^{n} X_i - \sum_{i=1}^{n} X_i = 0.0$$

The sum of the simple deviations from the mean is 0.0 whether the set of values is {1, 3, 5} or {1, 3, 5000} so it's useless as a measure of the variability in a set of scores.

Summing Squared Deviations

The squares of the deviations are another matter. If you square all the deviations and sum them, you inevitably wind up with a positive number. And that's the fundamental calculation used in the statistical analysis of variability, from simple single-variable analyses such as the standard deviation and variance to multivariate analyses such as factor analysis.

Another way to deal with the problem of deviations that sum to 0.0 is to convert them to absolute values instead of squaring them. (The absolute value of a negative number is its positive counterpart: The absolute value of –5.5 is +5.5. The absolute value of a positive number is the number itself. The absolute value of 3.1416 is 3.1416.) Taking the average of a set of absolute values results in a measure termed the *mean absolute deviation* or *MAD*. The MAD is conceptually simple and easy to calculate—you can use Excel's ABS() function to convert a value to its absolute value. You see the MAD infrequently, but it does show up in some forecasting work. Even there, though, the MAD tends to be used principally to confirm the results of other measures of variability. The MAD lacks some characteristics of other measures of variability that prove useful in more advanced statistical analysis.

So we start to solve the problem presented by the simple deviations by squaring them. Then we can describe the variability among the individual values in a set of numbers using statistics that are based on the sum of the squared deviations in that set. Figure 1.1 has an example.

Cells B2:B11 of Figure 1.1 show the ages of ten people, ranging roughly from 20 to 70. The figure also shows the slow, tedious way to go about calculating the sum of the squared deviations, partly so that you can see how to accumulate them. Cell B13 calculates the mean of the ten age values, using Excel's AVERAGE() function:

```
=AVERAGE(B2:B11)
```

Then, cells C2:C11 are linked to cell B13 with this simple assignment, which is used in each cell in the range C2:C11 to pick up the mean of the original values:

```
=$B$13
```

The dollar signs in the formula anchor its link to row 13 in column B. Their effect is to retain the reference to cell B13 no matter which worksheet cell the formula is copied into.

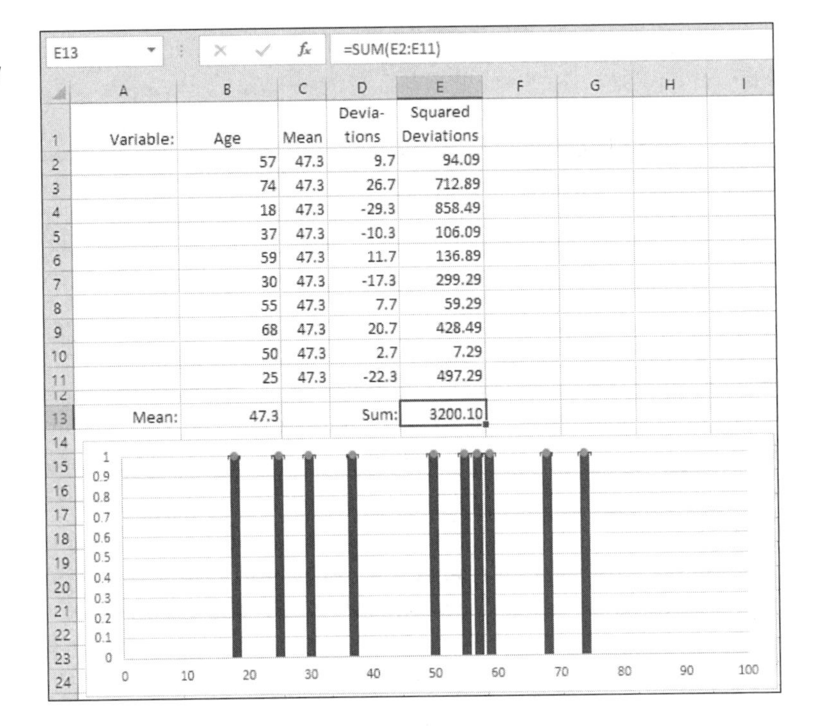

E13 =SUM(E2:E11)

	A	B	C	D	E
1	Variable:	Age	Mean	Devia-tions	Squared Deviations
2		57	47.3	9.7	94.09
3		74	47.3	26.7	712.89
4		18	47.3	-29.3	858.49
5		37	47.3	-10.3	106.09
6		59	47.3	11.7	136.89
7		30	47.3	-17.3	299.29
8		55	47.3	7.7	59.29
9		68	47.3	20.7	428.49
10		50	47.3	2.7	7.29
11		25	47.3	-22.3	497.29
12					
13	Mean:	47.3		Sum:	3200.10

Figure 1.1
You very seldom actually go through these steps: They're usually done for you.

Cells D2:D11 calculate the deviation of each value in B2:B11 from the mean of the values in C2:C11. For example, the formula in cell D2 is:

=B2 – C2

NOTE It doesn't make any difference here, but there are many situations in which it matters whether you subtract the mean from the individual values, or vice versa. You won't go wrong if you subtract the mean from the individual values. For example, the calculation of standard scores (or *z-scores*) assumes that the mean has been subtracted from the observed value, so that the calculated z-score is positive if the original observation is greater than the mean.

The deviations are squared in column E, using formulas such as this one in cell E2:

=D2^2

(The caret is Excel's exponentiation operator. You can get, for example, the cube of 2 by entering =2^3.)

Finally, the squared deviations are summed in cell E13 using this formula:

=SUM(E2:E11)

which returns the value 3200.10. That's the sum of the squared deviations of each person's age from the mean of the ten ages in column B.

So, we've taken these steps:

1. Entered ten age values on the worksheet.
2. Calculated their mean.
3. Subtracted the mean from each individual age value.
4. Squared each deviation in step 3.
5. Totaled the squared deviations calculated in step 4.

Because these steps are tedious to carry out, at some point you'll be delighted to learn that Excel has a function that does it all except for step 1. Given that you have actual observations in the range B2:B11, this formula calculates their sum of squares for you:

```
=DEVSQ(B2:B11)
```

Everything's great about the DEVSQ() function except its name. It's short, I guess, in some sort of perverse way, for "squared deviations," and that's a shame, but anyone who's done much regression work in Excel knows that the function itself can be a real time- and error-saver.

Figure 1.1 also shows how the individual values in the range B2:B11 disperse along the horizontal axis in the chart. Bear in mind what that pattern looks like, and that the sum of squares is 3200.10, and have a look at Figure 1.2.

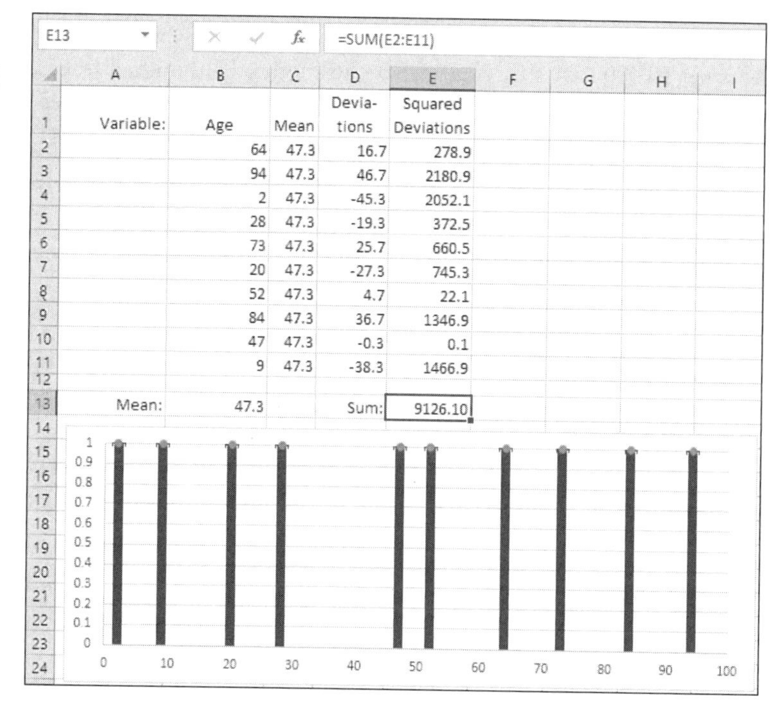

	A	B	C	D	E
1	Variable:	Age	Mean	Devia-tions	Squared Deviations
2		64	47.3	16.7	278.9
3		94	47.3	46.7	2180.9
4		2	47.3	-45.3	2052.1
5		28	47.3	-19.3	372.5
6		73	47.3	25.7	660.5
7		20	47.3	-27.3	745.3
8		52	47.3	4.7	22.1
9		84	47.3	36.7	1346.9
10		47	47.3	-0.3	0.1
11		9	47.3	-38.3	1466.9
12					
13	Mean:	47.3		Sum:	9126.10

Figure 1.2
Both the degree of dispersion in the chart and the sum of squares are larger than in Figure 1.1.

Compare what's shown in Figure 1.1 with that in Figure 1.2. The range of the observed values has increased from about 55 years (from 18 to 74) to over 90 years (from about 2 to 94). The number of observations is the same, and so is the mean age. The range displayed on the chart's horizontal axis is the same in both charts. But just from looking at the placement of the individual ages on the two charts, you can see that they are a good bit further apart in Figure 1.2 than in Figure 1.1.

That difference is echoed by the size of the sum of squares. It's 3200.10 in Figure 1.1 and 9126.10 in Figure 1.2: nearly a threefold increase. Of course, the sum of squares is the important bit of information here: It's an objective summary of the amount of variability in a set of values, whereas the visual appearance of the placement of data markers on a chart is highly subjective.

From the Sum of Squares to the Variance

As you'll see in later chapters of this book, sometimes the sum of squares is most important, and sometimes a closely related statistic called the *variance* is the one to attend to. Much depends on the context of the analysis. But to appreciate how closely the variance is related to the sum of squares, consider that the variance is the *mean* of the squares. It's the average squared deviation.

One reason that the variance is such a useful measure of the amount of dispersion in a set of scores is that it is an average, and therefore is not sensitive to the number of values in the set. If you were told that Set A has a sum of squares of 10, and Set B has a sum of squares of 98, you would not know which set has the greater variability. It's entirely possible that Set A has 10,000 values that vary from one another by thousandths of a point, and Set B has only two values, 2 and 16. Clearly, Set B has greater dispersal than Set A, but you can't tell that from the sum of squares alone.

On the other hand, if you know that two sets of values have the same variance—5, say—then you know that the degrees of dispersal are identical, regardless of the number of values in each set: The variance is an average, which removes the effect of the sample size. See, for example, Figure 1.3.

Figure 1.3 shows two data sets, one with ten values in the range B2:B11 and one with 15 values, in the range E2:E16. The two sets have the same degree of dispersion of the values around their means (which are identical). The variance of Set A is almost identical to that of Set B. But their sums of squares are far apart: The sum of squares for Set B is half again as large as the sum of squares for Set A. The difference in the two sums of squares is due entirely to the difference in the number of observations in each data set.

For several reasons which will appear in later chapters, it's useful to have a measure of the variability in a set of values that is not sensitive to the number of values in the set. And as will also appear, it's often useful to work with the sum of squares instead of the variance. Both statistics are useful in different situations.

You've already met the DEVSQ() worksheet function, which appears in cells B14 and E19 in Figure 1.3, and the AVERAGE() function in cells B13 and E18. Figure 1.3 introduces the VAR.P() function, which returns the variance of the observations, in cells B15 and E20. I discuss the VAR.P() function in the next section.

Figure 1.3
Excel's worksheet functions, such as VAR.P( ), do much of the heavy lifting for you.

E20 | =VAR.P(E2:E16)

	A	B	C	D	E	F
1	Variable:	Age, Set A			Age, Set B	
2		57			58	
3		74			72	
4		18			18	
5		37			37	
6		59			60	
7		30			30	
8		55			56	
9		68			71	
10		50			51	
11		25			25	
12					28	
13	Mean:	47.3	=AVERAGE(B2:B11)		56	
14	Sum of Squares:	3200.1	=DEVSQ(B2:B11)		71	
15	Variance:	320.01	=VAR.P(B2:B11)		51	
16					25	
17						
18				Mean:	47.3	=AVERAGE(E2:E16)
19				Sum of Squares:	4818.9	=DEVSQ(E2:E16)
20				Variance:	321.26	=VAR.P(E2:E16)

One matter of terminology to bear in mind: When statisticians talk about a "sum of squares," they are almost certainly referring to the sum of the squared *deviations from the mean*, not the sum of the squares of the original values. The actual term doesn't make that distinction clear. What's worse, in years past, confusion arose because several formulas for the variance were recommended for use with pencil-and-paper and hand calculators, and those formulas *did* employ the sums of the squares of the original values. Those days are behind us, thanks be, and you should quickly become comfortable with the interpretation of "sum of squares" as referring to the sum of the squared deviations.

Using the VAR.P( ) and VAR.S( ) Functions

Excel has two useful functions that return the variance: VAR.P() and VAR.S(). The distinction between them is based on the nature of the values whose variance you want to calculate:

- Use VAR.P() when the values you point the function at constitute the full population of interest. Of course, the "P" in the function name stands for "population." For example, VAR.P(A2:A21) implies that the values in the range A2:A21 are the only values you're interested in. You don't intend to generalize the finding to another or larger set of data.
- Use VAR.S() when the values you point the function at constitute a sample from a larger population. The "S" in the function name stands for "sample." For example, VAR.S(A2:A21) implies that there are more values unaccounted for whose variance you would like to estimate.

Of course, you could also think of VAR.P() as returning a parameter and VAR.S() as returning a statistic.

> **NOTE** If you calculate a value such as a mean or correlation coefficient on a population, that value is termed a *parameter*. If you calculate it on a sample from a population, it's termed a *statistic*.

There are two other Excel worksheet functions that return the variance: VARA() and VARPA(). The former assumes that you provide a sample of values as its argument, and the latter assumes a population of values. In both cases, the "A" at the end of the function name indicates that you can include both text values and logical values along with numeric values in the function's argument. Text values and the logical FALSE are treated as zeroes, and the logical value TRUE is treated as a 1. The functions VAR.P() and VAR.S() merely ignore text values. I've never grasped the rationale for including VARA() and VARPA() in Excel's function library, other than as an ineffective way of compensating for sloppy worksheet design or badly planned mapping of values to observations.

VAR.P() and VAR.S() are a different matter. They're valuable functions and this book's examples make extensive use of both. It's important to understand both the difference between the two functions and the reason for that difference. I explore those questions in some depth in Chapters 2 and 3 of *Statistical Analysis: Microsoft Excel 2013*, published by Que. But briefly:

The arithmetic mean (which is calculated by Excel's AVERAGE() function) and the variance are closely related. That's in large measure due to the fact that the variance is the average of the squared deviations from the mean. Another property of the relationship between the two statistics, the mean and the variance, is this: Taking the deviations of the values in a set from their mean, and squaring them, results in a smaller sum of squared deviations than you get by deviating the values from *any other number*. The same is true of the variance of those values: The variance is just the average of the squared deviations rather than their sum.

So, the variance of the values 1, 4, and 7 is 6. Try calculating the deviations of those values from any number other than their mean, squaring the results and averaging them. You cannot find a number other than their mean value, 4, that results in an average squared deviation smaller than 6. You can demonstrate this property for yourself by setting up a worksheet to calculate the sum of squares deviated from some number other than the mean of a set of values. Then use Solver to minimize that sum of squares by changing the value you use in place of the mean. Solver will minimize the sum of squares—and therefore their average, the variance—by deviating the values from their mean. Guaranteed.

Why is this important? Suppose you measure the height of 100 randomly selected adult males and calculate their mean height. The mean of this random sample is likely to be fairly close to the mean of the entire population, but the mean of a sample is wildly unlikely to be exactly the same as the mean of the population. The larger the sample, the more accurate the statistic. But there's no sample size, other than the size of the population itself, beyond which you can be certain that the statistic equals the parameter.

So you have a sample of 100 height values, whose mean value is different from—however slightly—the population mean. When you calculate the variance of the sample, you get a value of, say, 4. As just discussed, that result is smaller than you would get if you deviated the 100 individual values from any other number, *including the actual mean of the full population*. Therefore, the variance you calculate is different from, and smaller than, the average squared deviation you would get if you somehow knew and used the mean of the population instead of the mean of the sample.

The average sum of squares from the sample underestimates the population variance: The variance as calculated in that fashion is said to be a *biased* statistic.

It turns out that if you divide the sum of the squared deviations by (n – 1), the sample size minus 1, instead of by n, you remove the bias from the statistic.

NOTE

The quantity (n – 1) is termed the *degrees of freedom*, or *df*. The term is construed as singular, so "The degrees of freedom in this analysis is 99." As we get further into the topic of regression analysis, you'll see that there are other ways to calculate the df, although they all involve either subtracting a number from a sample size, or counting the number of categories into which the observations fall.

So in terms of the arithmetic involved, VAR.P() and VAR.S() differ only in this way:

- VAR.P() divides the sum of squares by the number of observations, and therefore is a "true" variance: the average squared deviation.
- VAR.S() divides the sum of squares by the number of observations minus 1—that is, by the degrees of freedom, and in so doing provides you a more accurate estimate of the variance of the population from which the sample was taken.

Figure 1.4 illustrates the distinction.

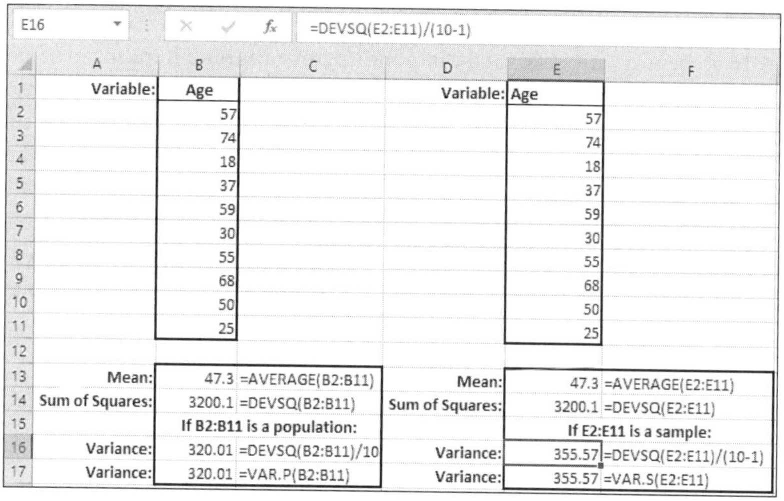

Figure 1.4
The values in B2:B11 are identical to those in E2:E11, but the actual population variance is smaller than the estimated population variance.

E16 =DEVSQ(E2:E11)/(10-1)

	A	B	C	D	E	F
1	Variable:	Age		Variable:	Age	
2		57			57	
3		74			74	
4		18			18	
5		37			37	
6		59			59	
7		30			30	
8		55			55	
9		68			68	
10		50			50	
11		25			25	
12						
13	Mean:	47.3	=AVERAGE(B2:B11)	Mean:	47.3	=AVERAGE(E2:E11)
14	Sum of Squares:	3200.1	=DEVSQ(B2:B11)	Sum of Squares:	3200.1	=DEVSQ(E2:E11)
15		If B2:B11 is a population:			If E2:E11 is a sample:	
16	Variance:	320.01	=DEVSQ(B2:B11)/10	Variance:	355.57	=DEVSQ(E2:E11)/(10-1)
17	Variance:	320.01	=VAR.P(B2:B11)	Variance:	355.57	=VAR.S(E2:E11)

Figure 1.4 shows the result of using VAR.P() in cell B17. That result, 320.01, is identical to the result of dividing the sum of squares by the number of observations, shown in cell B16.

The figure also shows the result of using VAR.S() in cell E17. The result, 355.57, is identical to dividing the sum of squares by the degrees of freedom, or (10 – 1).

Clearly, the difference between the values of the parameter (cell B17) and the statistic (cell E17) is due exclusively to the denominator. In Figure 1.4's example, *n* is 10 and the df is 9, so there's an 11% difference in the denominators and therefore in the functions' results.

Suppose that *n* were 100 instead of 10. Then the df would be 99 and we would have a 1% instead of an 11% difference in the denominators and in the functions' results. So, the larger the sample size, the less bias exists in estimating the population parameter from the sample statistic, and the smaller the required correction for the bias.

I'll anticipate this chapter's next section, on standard deviations, by noting that even though VAR.S() returns an unbiased estimator of the population variance, its square root is not an unbiased estimator of the population standard deviation. The standard deviation is simply the square root of the variance, but it's not necessarily true that the square root of an unbiased estimator is also unbiased. The standard deviation is a biased statistic, but the amount of bias that remains after using the degrees of freedom to calculate the variance is vanishingly small and is usually regarded as insignificant.

The Standard Deviation

In one sense, the standard deviation is simply the square root of the variance:

$$SD = \sqrt{Variance}$$

In fact, the usual symbol for the variance is s^2, where "s" stands for "standard deviation," so—in symbols anyway—the variance is defined by the standard deviation.

In some ways it's easier and more informative to think in terms of standard deviations than in terms of variances. (You'll see how the opposite is true when we start digging into what makes regression analysis work.) For example, both the variance and the standard deviation use the sum of squared deviations as their building blocks. Squaring the deviations takes the scale of measurement to a different metric: the original metric, squared. Square feet instead of feet, square dollars instead of dollars, square miles per gallon instead of miles per gallon. As it happens, it's easy enough to think in terms of square feet, but neither square dollars nor square miles per gallon qualify as standard, everyday measurements.

But it's easy enough to think in terms of standard deviations. Because the standard deviation is the square root of the variance, it returns to the original unit of measurement. If the variance is 100 squared dollars, the standard deviation is 10 dollars. The standard deviation still expresses the amount of variability in a set of numbers, though, so we can use it to express the difference between two values.

For example, the standard deviation of the miles per gallon achieved by a broad range of cars might be 10 mpg. In that case, if a tiny, vulnerable subcompact gets 28 mpg and a thirsty SUV gets 18 mpg, the difference between the cars in fuel efficiency is 1 standard deviation: (28 – 18)/10. Is 1 standard deviation a lot or a little? That's a subjective judgment. Judge for yourself after you glance at Figure 1.5.

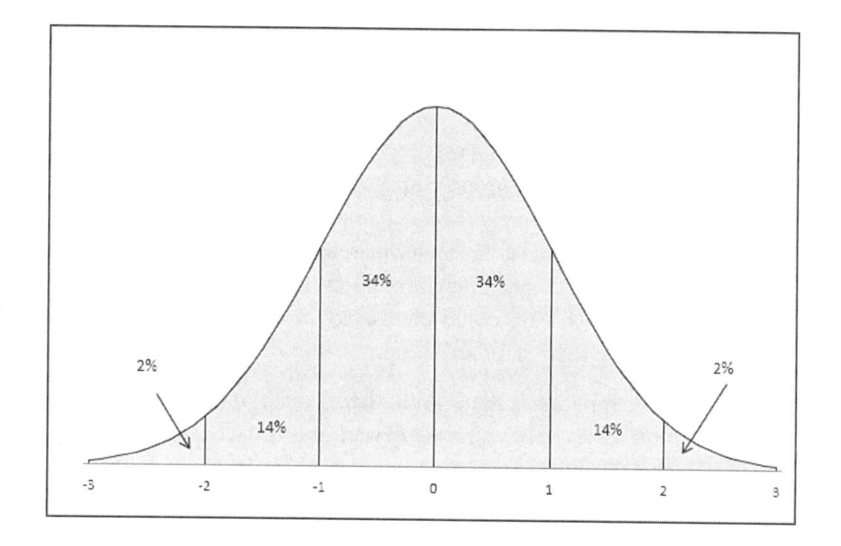

Figure 1.5
These probabilities are typical of any normally distributed variable.

Figure 1.5 shows how the width of a standard deviation on the horizontal axis can partition a distribution of values into six areas. Two areas each contain 34% of the area under the curve, two each contain 14% of the area, and two each contain 2% of the area. So while the answer to the question of the distance between the subcompact and the SUV is 1 standard deviation, the meaning of the answer depends on which portion of the curve divides the two vehicles. A difference of one standard deviation could mean that two observations are separated by only 2% of the population or by as much as 34% of the population.

The Standard Error of the Mean

Standard deviations come in more than just one flavor. It often happens that you want to calculate the standard deviation of data after it's been manipulated in some way. Although the results are definitely standard deviations, it's customary to refer to them as *standard errors*.

One good example, which I discuss much more in various later chapters, is the standard error of estimate. That statistic has to do with the accuracy of a regression equation. One of the purposes of regression is to predict the value of one variable, given knowledge of another variable. If you're interested in commercial real estate, you know that there's a quantitative relationship, a formula, between the number of square feet in an office and the monthly cost of leasing it: dollars per square foot. But you also know that the relationship isn't exact. If you apply the formula to one hundred different offices, you'll estimate their monthly lease costs closely in some cases, not so closely in a few more, and you'll be way off in others.

You can quantify the errors of estimate by subtracting the figure that the formula gives you from the offices' actual lease costs. If you take all those errors of estimate and calculate their standard deviation, you have the standard error of estimate. The smaller it is, the smaller the individual errors, and the more accurate the formula. It's a standard deviation, but because you have manipulated the observations to arrive at the errors of estimate, it's termed a standard error.

There are several sorts of standard error, including the standard error of measurement and the standard error of a proportion. Another example, pertinent to this discussion of variability, is the *standard error of the mean*. Suppose that you had the resources to collect data regarding the height of 25 randomly selected adult males in each of 50 states. You could then calculate the mean height of adult males in Alabama, Alaska, Arizona, . . . , and Wyoming. Those 50 means could themselves be considered to be observations. In that case, you could calculate their standard deviation, and that would be the *standard error of the mean*.

Why would one care about that? Well, there's a fair amount of variation in the heights of individual adult males: The standard deviation is in fact about 2.5 inches and the mean is about 70 inches. Going down 2 standard deviations from the mean gets you to 70 – 5 or 65 inches, and going up 2 standard deviations from the mean gets you to 70 + 5 or 75 inches. Take any given adult male who is a resident of the United States. Because of the relationship between standard deviations in a normal curve and areas under the curve, we know that the probability is close to 96% that the randomly selected male stands between 5'5" and 6'5".

But all that variability is swallowed up in the process of converting the individual observations to means. The 25 males you sample from South Dakota might have a standard deviation of 2.5 inches, but the standard deviation of the mean of those 25 values is 0: A single value can have no variability. If you take the standard error of the 50 state means, the variability due to differences among individuals is absent, and all you're left with is differences associated with individual states of residence. There might be some such state-to-state variation, but it's just about certain to be a good bit less than the variation among individual people.

Figure 1.6 compares the distribution of the 1,250 (that's 50 states times 25 men from each state) original, individual observations to the 50 state mean values.

Figure 1.6 shows two curves. The broader curve, which extends from 62.5 to 77.4 inches on the horizontal axis, illustrates the relative frequency of different heights of the individual males in the study. The narrower curve, which extends from 68.2 to 71.8 inches on the horizontal axis, illustrates the relative frequency of different mean heights associated with the states in the study.

The most conspicuous difference between the two curves is that the curve showing individual frequencies for different heights is much wider than the curve showing state frequencies for different means. This is consistent with the argument I advanced earlier in this section, that the individual variation disappears when you take the 50 means of the original 1,250 observations.

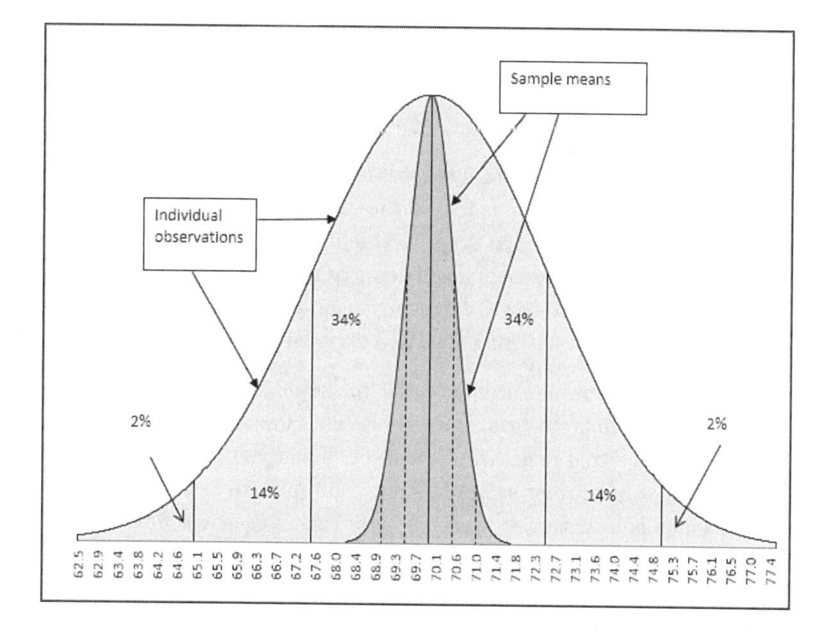

Figure 1.6
The state means vary from one another much less than do the individual observations.

In fact, there's a formula that tells you how much the variability will shrink when you go from differences among individual observations to differences among the means of samples. The formula to estimate the standard error of the mean—the standard deviation of the means rather than of the individual observations—is:

$$s_{\overline{X}} = s/\sqrt{n}$$

where:

$s_{\overline{X}}$ is the standard error of the mean.

s is the standard deviation of the original observations.

n is the number of observations in a sample.

> **NOTE**
>
> Another convention used in statistics concerns the distinction between Greek and Roman letters. Greek letters are used to represent population parameters, and Roman letters to represent sample statistics. So if we're considering the standard deviation of a population, we would write σ; if we are considering the standard deviation of a sample, we would write s.
>
> The symbols used for means aren't quite so neat. We use the Greek letter μ, pronounced "mew," to represent a population mean. But tradition has us use the symbol $\overline{X}$, pronounced "X-bar," to represent a sample mean.

The prior formula uses Roman letters for the standard error and standard deviation, so it assumes that you're estimating the population parameters by means of statistics calculated on a sample. If you're fortunate enough to know the parameter σ, you would write the formula with Greek letters, to indicate that parameters are in use:

$$\sigma_{\overline{X}} = \sigma\sqrt{n}$$

In the present example, *s*, the standard deviation of the original observations, is 2.5 inches. The number per sample, *n*, is 25. So the standard error of the mean in this case is 2.5 divided by the square root of *n*, or 5, which equals 0.5 inches. If you had sampled 16 men per state instead of 25, the standard error of the mean would have been 2.5 inches divided by the square root of 16, or 4, resulting in a standard error of about 0.6. And a sample size of 100 per state brings about a standard error of 2.5/10, or 0.25.

In general, as the sample size increases, the standard error of the mean decreases as a function of the square root of the sample size. Suppose that you obtained the heights of 10 randomly selected males in one sample. Then you obtain the heights of 100 randomly selected males in another sample. Then you repeat the two samples, using the same sample sizes. Which pair of sample means do you expect to be closer to one another? The two 10-member samples or the two 100-member samples?

I think that you would intuitively expect the two 100-member samples to have closer means than the two 10-member samples. The larger the sample size, the more accurate the estimate of the population mean, and therefore the closer that the two mean values will be to one another. This is just another way of saying that the larger the sample size, the smaller the standard error of the mean.

About z-Scores and z-Values

One of the ways that we make use of standard deviations and standard errors is to define certain "standard scores." These scores tell you immediately the position of an observation in a normal curve. A *z-score* expresses an observation's position in the distribution as a number of standard deviations above or below the mean. For example, a z-score of 1.5 indicates that the observation is 1.5 standard deviations above the mean, whereas a z-score of –2.0 indicates that the observation is 2 standard deviations below the mean.

You calculate a z-score by subtracting the mean of a set of values from the value of an observation, and dividing the difference by the standard deviation. So, if a subcompact car gets 28 mpg, if the mean for subcompacts is 22 mpg, and if the standard deviation is 10 mpg, the subcompact's z–score is (28 – 22)/10, or 0.6.

It's usual to reserve the term *z-score* for individual scores, measurements assigned to individual members of a sample. You can use much the same calculation with statistics such as means in place of individual scores, and in that case you'll generally see the term *z-value*. I'll follow that fairly informal distinction in this book.

We'll be working with standard deviations somewhat less than with sums of squares and variances, but standard deviations are nevertheless crucial when we look at the standard error of estimate and the standard errors of regression coefficients. Solving problems about those standard errors are really just variations on a theme.

Suppose you're the club pro at a golf course. You've occupied that position for several years and one of your responsibilities is to oversee the maintenance of the driving range, including the retrieval of driven balls from the range several times per day. You've had the opportunity to gauge the average length of drives made at the range, and you're confident that the average drive is 205 yards and the standard deviation is 36 yards.

One day a sales rep for a sports equipment manufacturer comes calling and tells you that his company has a new, and legal, driver that is adding 10 yards to their customers' drives. He leaves one of the new drivers with you and out of curiosity you arrange to have the next 81 players who use the range take one practice swing with the new club, and you make a note of the distance of each of those drives.

It turns out that the average length of the drives made with the new club is 215 yards. What's the probability that the difference of 10 yards (215 versus 205) is just due to sampling error? Asked another way: Over a large test of thousands of sample drives, how likely is it that the new club turns out to add no distance to the existing long-term average of 205 yards?

We can test that by forming a z-value. First we need the standard error of the mean, because means are what we're comparing: the mean of the 81-shot sample and the long-term mean of driving distances on the golf range. Using the formula for the standard error of the mean, we get:

$$\sigma_{\overline{X}} = \sigma / \sqrt{n}$$

or:

$$4 = 36 / \sqrt{81}$$

And the z-value, the difference between the sample mean and the population mean, divided by the standard error of the mean, is:

(215 – 205) / 4

or 2.5.

Is the apparent additional 10 yards per drive a reliable outcome, one that you can expect other members of the golf club more often than not to replicate? Or did you just happen to sample 81 members who drive golf balls farther than other members? Is the outcome real or just sampling error?

Because of what we know about how the normal curve behaves, an Excel worksheet function can tell us the probability of obtaining a z-value of 2.5 when the mean distance for the new club—the parameter, which we have not directly observed, rather than the statistic, which we have—in fact *equals* the long-established mean of 205 yards.

Put another way: Suppose that the average drive with the new club, measured for all golfers at your driving range, were to travel 205 yards. That's the same as the long-term average you have observed with older golf clubs. If the two averages are in fact the same in the full population, what's the probability of observing a z-value of 2.5 calculated on a sample of 81 golfers?

In Excel, one way to calculate that probability is by means of the NORM.S.DIST() function. The function's name, of course, refers in part to the *norm*al *dist*ribution. The *S* in the middle of the function name refers to *s*tandard, a distinction I'll explain shortly.

If you submit the obtained z-value to the NORM.S.DIST() function in this way:

=NORM.S.DIST(2.5,TRUE)

the result is 99.38%. That's actually the complement of the probability we're after. It's easier to understand the meaning of the result if you have a look at Figure 1.7.

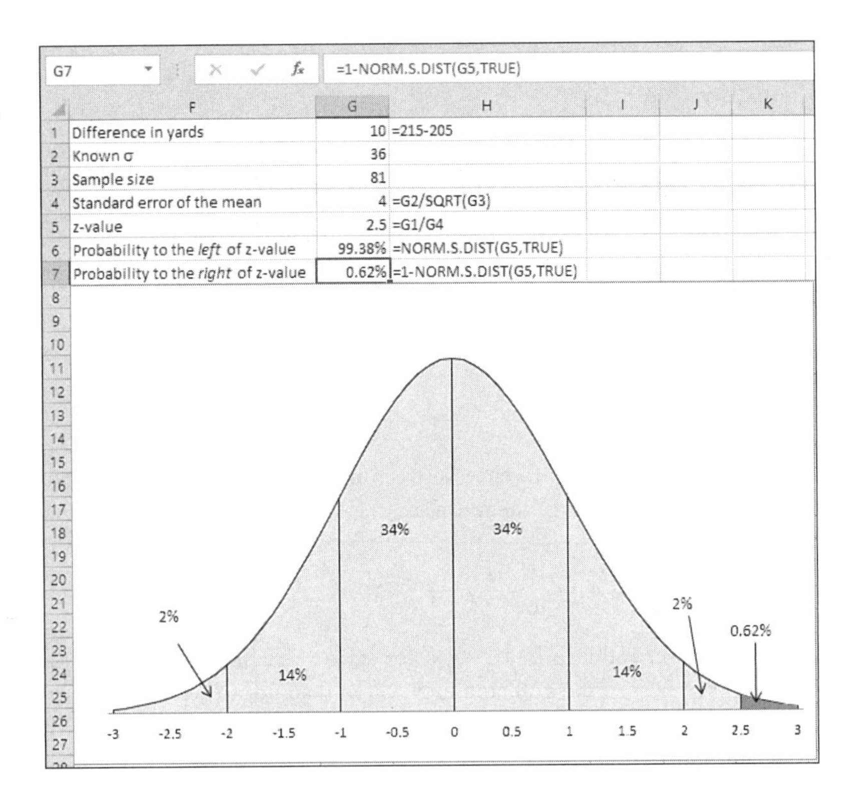

Figure 1.7
The NORM.S.DIST() function returns the area under the curve to the *left* of the z-value.

The information on the worksheet and chart shown in Figure 1.7 summarizes the process described in the prior few paragraphs. Let's walk through it:

1. Cell G1 returns the difference between the mean driving distance you obtained from your 81-person sample, and the long-term average that you established from years of

watching members use the driving range. (The formula used in cell G1 is repeated as text in cell H1.)

2. Cell G2 contains the standard deviation of the length of the drives that you've observed over the years. This is the standard deviation of the lengths of the thousands of individual drives. Like the long-term mean length, it's a parameter, a known quantity.
3. Cell G3 contains the size of the sample of club members you asked to try a shot with the new driver.
4. Cell G4 uses the information in cells G2 and G3 to calculate the standard error of the mean. Again, the formula is shown in cell H4 as text. The expectation is that if you repeated your 81-person sample over and over again, the standard deviation of the sample means would equal cell G4's standard error of the mean.
5. Cell G5 contains the z-value, the ratio of the difference between the sample mean of 215 and the long-term mean of 205, divided by the standard error of the mean.

Shift your focus now from the steps in the analysis to the chart. The curve in the chart is very similar to the one shown in Figure 1.5. The main difference is the new, shaded wedge in the right tail of the distribution. Its left edge, which separates the wedge from the remainder of the area under the curve, is at 2.5 on the horizontal axis. Recall from the discussion of Figure 1.5 that the labels on the horizontal axis represent z-values, or the number of standard deviations—in this case, standard errors—between a given point and the distribution's mean value of 0.0.

The z-value we calculated is 2.5, so we know that your sample outcome, a driving distance of 215 yards, is 2.5 standard errors above the mean. The distribution's mean value of 0.0 represents a reality in which the difference between the long term mean of 205 and the actual mean distance—the population parameter—for the new driver is 0.0 yards. Under that assumption, the sample you took returned a result that's 2.5 standard errors above the hypothetical mean difference of 0.0 yards.

6. Back to the worksheet. Cell G6 shows the percent of the area under the curve that's to the *left* of 2.5 standard errors above the mean. The value is 99.38%. That's useful information, but it's not quite what we're looking for. It's the probability of mean distances less than 10 yards longer than the long-term mean. We want the probability of mean distances *more* than 10 yards longer than the long-term mean.
7. We get that in cell G7 by subtracting the result of the NORM.S.DIST() function from 1.0. The result is 0.62%, which is the size of the shaded wedge in the right tail of the charted distribution.

NOTE

The second argument to the NORM.S.DIST() function should be given as either TRUE or FALSE. A TRUE argument tells Excel that you want it to return the area under the curve that's to the left of the value of the first argument—in this case, 2.5. A FALSE argument tells Excel to return the height of the curve at the point specified by the first argument.

NOTE

The ".S." in NORM.S.DIST( ) stands for "standard." There's an uncountable number of normal curves, with different means and different standard deviations. What they all have in common is their *shape*. One normal curve is particularly important, in large measure because it is the reference distribution to interpret z-scores and z-values. That normal curve has a mean of 0.0 and a standard deviation of 1.0, and it's called the *standard normal distribution* (or, often, the *unit normal distribution*). Hence the ".S." for "standard" in the function name. You'll also make use of Excel's NORM.DIST( ) function, which has no ".S." in its name. You supply the mean and standard deviation in the arguments to NORM.DIST( ), so that you can get the same information as you do with NORM.S.DIST( ) with distributions other than the standard normal distribution.

The design of the experiment with golf clubs that I've just described is shockingly weak. But it allows me to focus on the statistical test. (I feel somewhat guilty nevertheless, because statistical error is just one of a variety of reasons that experimental results can mislead. You rely on a strong experimental design to protect you from those threats to the validity of an experiment.)

Still, I ask you to notice that the probability of getting a z-value of 2.5 or greater, if the population parameter is 0.0, is very small: 0.62% is just a little larger than half a percent. The shaded wedge in the right tail of the curve, 0.62% of the area under the curve, in Figure 1.7 reinforces that finding visually. It shows that an average difference of ten yards is very unlikely when the population means for the two clubs are identical.

What if a different reality is in place? What if, in the full population from which you took your sample of 81, the mean driving distance for the new golf club is *not* the same as the long-term mean distance? If the two mean values are the same, the difference in the population means, the parameters, is 0.0 and the difference you observed is very unlikely to transpire. The analysis says that it will happen by chance just 6.2 times in a thousand opportunities.

Is 6.2 times in 1,000 so small a likelihood that you should abandon the hypothesis that the new driver's mean driving distance is the same as you've observed with older clubs for the past several years? That's a personal, subjective assessment, but it's also entirely legitimate. If the probability of 0.62% seems sufficiently small to you, then you would probably reject the hypothesis that the new club delivers the same distance as existing models. You don't yet know for sure what the difference in the population means is, but you can conclude with reasonable confidence that the difference isn't 0.0.

Now, what's all this discussion of the probability of z-values doing in a book about regression analysis? It turns out that t-tests, which are very similar to tests of z-values as discussed in this section, are useful in helping you determine whether you should retain a variable in a regression analysis. And when we get to the point of using regression analysis in place of traditional analysis of variance and covariance, in later chapters, similar techniques are often used to derive multiple comparisons.

Using t-values instead of z-values involves a complication, although it's a minor one. The complication has to do with whether you know for sure, or have to estimate, the standard deviation of the population from which you got your observations. Untangling the complication involves the calculation and assessment of t-values, coming up next.

About t-Values

When you use a z-value, you depend on the normal distribution. Standard deviations, and therefore z-values, divide a true, normal distribution into areas that account for 2%, 14%, 34%, 34%, 14%, and 2% of the curve's total area.

That's all well and good if you know the actual value of the standard deviation in the population: that is, the parameter. If you don't know it, you need to estimate the standard deviation, and aside from pulling it out of your hat the best source is a sample from the appropriate population. But all samples, even very large ones, share the same defect: They don't tell you the exact value of the parameter in the population. The larger the sample, of course, the better the estimate, but some sampling error always enters the picture. (Sampling error, please note, not the bias discussed earlier in this chapter. That's a different animal entirely.)

The z-value, discussed in the prior section, assumes that both the population mean μ and the population standard deviation σ are known. So if we want to calculate a z-score or z-value, we need to know the values of both parameters. This knowledge normally comes from long experience with or extensive research into a phenomenon. One example is the annual nationwide average number of traffic fatalities, calculated from records maintained by local agencies. Another is the standard deviation of systolic blood pressure measurements, which can be determined from aggregate data maintained by both private and public hospitals and health organizations. The IRS routinely maintains data on the average number of arithmetic errors that filers make in filling out Form 1040.

But all too often we don't know the values of those population parameters. In that case we have to estimate them, usually on the basis of the same sample we use to calculate a mean that we want to test. In the golf club example from the prior section, I posited that you knew the population standard deviation from long observation of tee shots made on a driving range. Without that *deus ex machina*, you could have estimated the standard deviation of the length of 81 tee shots by using the same data used to calculate their mean of 215 yards.

Now, here's the complication I mentioned at the end of the prior section. The sampling error that enters the picture when you estimate a population standard deviation using sample data results in a distribution that's not quite normal. See Figure 1.8.

Suppose you suspect that a make of diesel-powered cars put out more nitrogen oxide (N_{ox}) than its advertising claims. You know that the standard imposed by the federal government is 0.4 grams of N_{ox} per mile driven. You would like to compare empirically measured results to that standard, rather than to a population mean, so in this case you know the value of the comparison: 0.4 grams per mile. That's a target value rather than a known population parameter as was used in the golf club example.

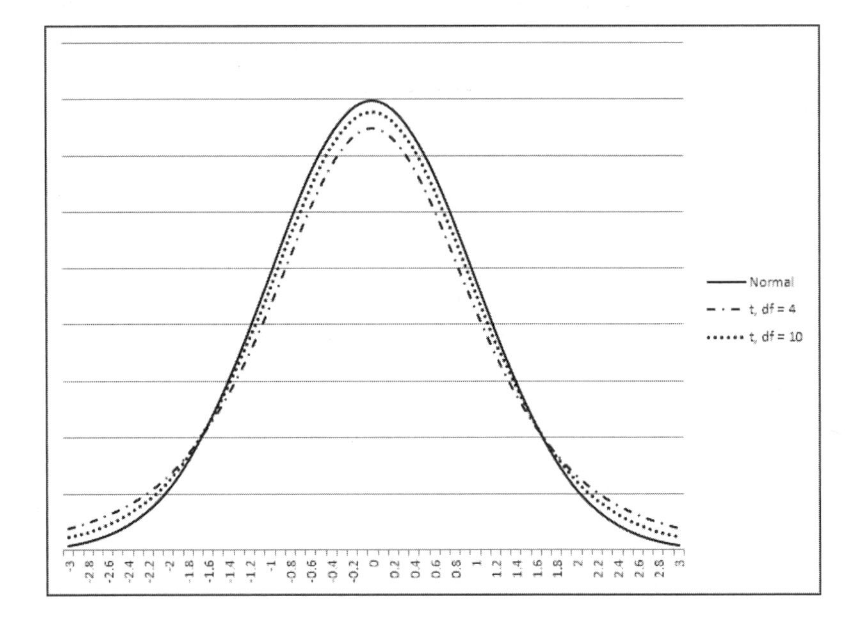

Figure 1.8
The t distributions that result when σ is not known are shaped somewhat differently from the normal distribution.

You test-drive five cars of the make in question and measure their N_{ox} levels on the road rather than in a test lab. You calculate the mean N_{ox} measurement for the five cars and find the standard deviation of the five N_{ox} measures. Now you calculate a ratio of this form:

$$(\overline{X} - 0.4)/s$$

where:

- $\overline{X}$ is the average N_{ox} level for the five cars.
- 0.4 is the mandated standard N_{ox} level.
- s is the standard deviation of the N_{ox} level for the five cars.

That ratio looks very similar to the formula for a z-value but in fact it's a *t-value*. The difference between the structure of a z-value and that of a t-value is the t-value's use of a standard deviation that's derived from a sample, rather than a known population value. The use of the Roman letter s for the standard deviation, rather than the Greek σ, tends to confirm that the standard deviation is an estimate from the sample rather than a known parameter.

If you repeated this procedure thousands of times, and charted the frequency of each of the t-values you obtain, you would get a distribution very similar to the one shown in Figure 1.8 and labeled, in the chart legend, as "t, df = 4." The chart shows three distributions, and you can see that the distribution in question is lower than the other two in the center of the curves and higher than the other two in the tails of the curve.

As I mentioned earlier, the normal curve has only one shape, although it takes many different values for its mean and its standard deviation. But the distribution of t-values has slightly different shapes, depending on the number of degrees of freedom (df) in the samples that it's made from. Recall from this chapter's section titled "Using the VAR.P() and VAR.S() Functions" that for the purpose of testing its mean, a sample's df equals $n - 1$.

You can also see from Figure 1.8 that, if you had included not 5 but 11 cars in each of thousands of samples, the curve that would result has a different shape from the curve based on 5-member samples. The curve based on samples with 11 observations each, labeled "t, df = 10" on the chart, is a little taller at the center and a little lower in the tails, compared to "t, df = 4."

The third curve displayed in Figure 1.8 is labeled "Normal" and is shown by the solid black line. It has the unique shape of the normal curve, and is the tallest of the three curves in the center and the lowest in the tails.

The differences between the curves, evidenced by the heights of the curves at different points along the horizontal axis, aren't just an interesting visual phenomenon. They have consequences for the probability of observing a given value.

We saw in the prior section that the z-value of 2.5 or larger has a probability of 0.62% of occurring in a normal curve. What if we were to obtain the same figure of 2.5 as a t-value, using the sample standard deviation, rather than as a z-value, using a known population standard deviation? Because it's a t-value, we would not use the NORM.S.DIST() function, but the T.DIST() function instead:

=1 – T.DIST(2.5,4,TRUE)

As I noted earlier, the NORM.S.DIST( ) function returns the area under a normal curve that lies to the left of a given z-value. To get the area to the *right* of that value, we need to subtract the result of the NORM.S.DIST( ) function from 1.0. The same is true of the T.DIST( ) function, but Excel supplies a version of the T.DIST( ) function that makes things a little easier. You can use T.DIST.RT( ) to get the area under the curve and to the right of the supplied t-value, and avoid having to subtract the result of T.DIST( ) from 1.0.

In the various t distribution functions supported by Excel, the second argument is the degrees of freedom. Excel needs to know the df in order to evaluate that shape of the t distribution properly, and thus to return the correct probability given the t-value in the first argument. That formula returns 3.34%, whereas this formula:

=1 – NORM.S.DIST(2.5,TRUE)

returns 0.62%. Referencing the same value, 2.5, to the normal distribution, versus referencing it to the t distribution with 4 df, results in a probability under the t distribution that's more than 5 times greater than under the normal distribution. Why is that?

Let's take a closer look at the right tails of the three distributions in Figure 1.8. See Figure 1.9.

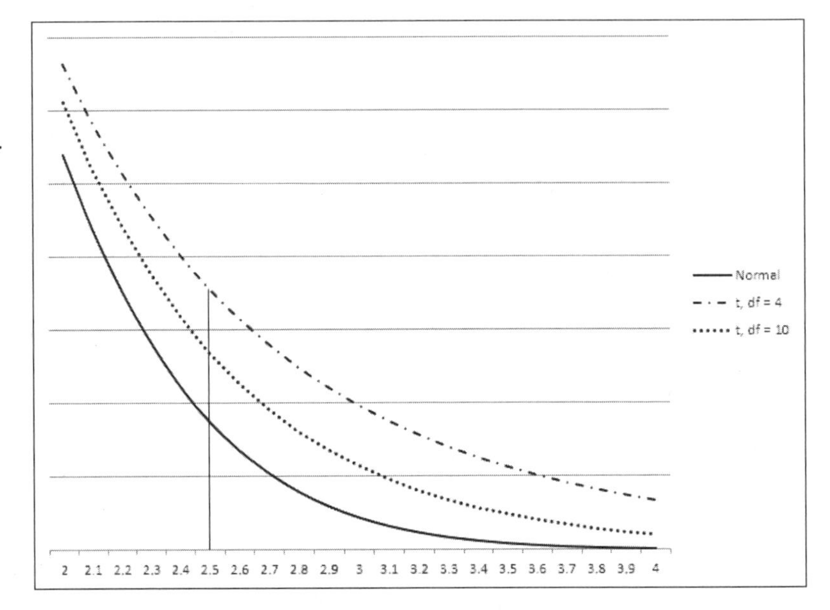

Figure 1.9
The probability of a given z-value or t-value depends on which distribution you reference.

Figure 1.9 shows the portions of the normal distribution, the t distribution with 4 df, and the t distribution with 10 df, to the right of the distribution's t- or z-value. More of the area under the t distribution with 4 df is found to the right of the t-ratio of 2.5 than is found to the right of the normal distribution with a z-ratio of 2.5. If you glance back at Figure 1.8 you can see where that additional area under the t distribution comes from: The center of the t distribution is pushed down, and the tails get taller to make room for the area lost in the center.

The NORM.S.DIST() and the T.DIST() functions therefore return different results for the probability of a given z-value or t-value. Generally, the larger the sample size, the larger the df, and the closer the t distribution approximates the normal distribution. With 4 df, 3.34% of the area under the t distribution lies to the right of a t-value of 2.5. Increase the sample size to 11 so that the df is 10, and the area drops to 1.57% of the total area. Increase the df to 999 and the area is 0.629%, as compared to 0.621% for a z-value of 2.5 on the normal curve.

Bear in mind that the necessity of using the t distribution rather than the normal curve is due to estimating the standard deviation because its population parameter is unknown. The effect of using the t distribution with a given df, rather than the normal curve, is that the probability of observing a t-value differs from the probability of observing the identical z-value.

This outcome has major implications for design choices, such as how large a sample to take, whether to adopt a directional or non-directional research hypothesis, and the statistical power of the test you're running. You're generally in a position to exert control over these matters, whereas all you can do about sampling error is to gauge how badly it might mislead you. I will touch on the factors that influence statistical power in later chapters. In the meantime, I believe that it's important that you grasp how it is that standard errors affect your decision making. It's the basis for understanding how regression coefficients and their standard errors are evaluated and used, a matter that I'll discuss in Chapter 5, "Multiple Regression."

First, though, it's important to deconstruct the other major building block of regression analysis, correlation. Chapter 2, "Correlation," connects the dots between the variances that this chapter has discussed and the correlations that express how variance is shared.

Correlation

2

IN THIS CHAPTER

Measuring Correlation 29
Calculating Correlation 34
Correlation and Causation 53
Restriction of Range 55

This book is about regression analysis. Nevertheless, I'm going to discuss correlation before digging into regression. Correlation is an essential building block in any regression analysis. Furthermore, it's impossible to understand regression analysis without an understanding of correlation.

So, even though I suspect that you already have a good sense of the meaning of correlation, I want to spend some pages reviewing it here. You might even find something that's new to you.

Measuring Correlation

It's often said that a correlation measures the strength of the relationship between the members of a set of ordered pairs. I don't care much for that definition (although it's pithy and I like that). You need pairs, yes, but there's no need for them to be "ordered." The observations come in pairs, and as long as you have some way to match one member of a pair with the other member, you're in good shape.

It's important to bear in mind that the field of statistical analysis uses correlation to measure the strength of the relationship between variables measured on an *interval* or *ratio* scale. An interval scale is a numeric scale of values that expresses the size of the differences between beings or things. The classic example is temperature: The difference between 30 degrees and 40 degrees centigrade is the same as the difference between 70 degrees and 80 degrees. The same interval represents the same difference.

A ratio scale is the same as an interval scale, but it has a true zero point. Because the centigrade scale has no point identified as zero degrees that represents the complete absence of molecular motion, it is not a ratio scale. The Kelvin scale, which does have a zero point where there is no heat at all, is a ratio scale, on which 100 degrees Kelvin is twice as hot as 50 degrees Kelvin.

In informal conversation you occasionally hear people speak of any relationship as a "correlation." For example, a TV news reporter might speak of a correlation between political party and preference for a bill in Congress. Just keep in mind that when you think of a correlation in the context of regression, or any statistical analysis, you're very likely thinking of the relationship between two interval or ratio variables. (There are correlations intended for use with nominal or ordinal scales, but when we speak formally of them we use terms such as *rank correlation* or *biserial correlation*.) The term *correlation* used by itself in a formal context means the Pearson correlation discussed in this book generally and this chapter in particular.

Expressing the Strength of a Correlation

You express the strength of the relationship by means of what's termed the *correlation coefficient*, or r. The range of values that r can assume runs from –1.0 to +1.0. The closer that r is to 0.0, the weaker the relationship. The closer r is to either –1.0 or +1.0, the stronger the relationship. I'll return to the issue of a positive or a negative r shortly, but first let's take a look at a moderately strong correlation coefficient. See Figure 2.1.

Figure 2.1 shows the 2014 average home sales price and the 2014 median family income for each of the 50 states plus Washington D.C. It also shows the correlation between price and income in cell G2. And the chart shows how income and sales prices move together: the higher a state's median family income, the higher the average sales price for a house.

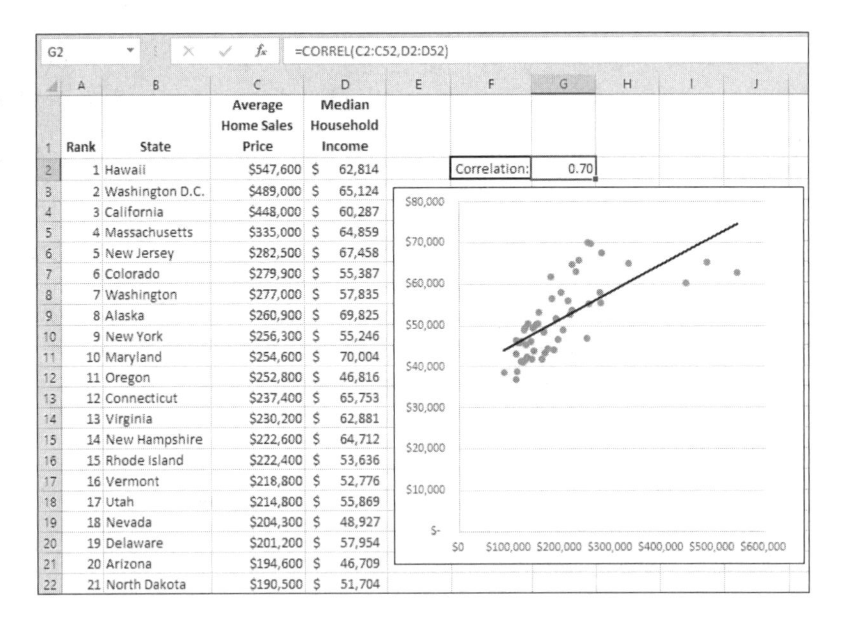

G2 =CORREL(C2:C52,D2:D52)

	A	B	C	D	E	F	G
1	Rank	State	Average Home Sales Price	Median Household Income			
2	1	Hawaii	$547,600	$ 62,814		Correlation:	0.70
3	2	Washington D.C.	$489,000	$ 65,124			
4	3	California	$448,000	$ 60,287			
5	4	Massachusetts	$335,000	$ 64,859			
6	5	New Jersey	$282,500	$ 67,458			
7	6	Colorado	$279,900	$ 55,387			
8	7	Washington	$277,000	$ 57,835			
9	8	Alaska	$260,900	$ 69,825			
10	9	New York	$256,300	$ 55,246			
11	10	Maryland	$254,600	$ 70,004			
12	11	Oregon	$252,800	$ 46,816			
13	12	Connecticut	$237,400	$ 65,753			
14	13	Virginia	$230,200	$ 62,881			
15	14	New Hampshire	$222,600	$ 64,712			
16	15	Rhode Island	$222,400	$ 53,636			
17	16	Vermont	$218,800	$ 52,776			
18	17	Utah	$214,800	$ 55,869			
19	18	Nevada	$204,300	$ 48,927			
20	19	Delaware	$201,200	$ 57,954			
21	20	Arizona	$194,600	$ 46,709			
22	21	North Dakota	$190,500	$ 51,704			

Figure 2.1
The stronger the correlation, the closer the individual observations are to the regression line.

The straight diagonal line in Figure 2.1 is called the *regression line*. (Excel terms it a *trendline*, and that's a perfectly acceptable alternative.) We'll deal with regression lines and their properties repeatedly in this book. For now it's enough to know that linear regression results in a straight regression line, that its direction tells you something of the nature of the relationship between the two variables, and that its position in the chart minimizes the sum of the squared differences between the line and the charted data points.

The result of analyzing the relationship between family incomes and housing prices is about what you would expect. States where the family income is not high enough to drive prices up will have less expensive housing. Contrast the findings in Figure 2.1 with those in Figure 2.2.

Figure 2.2
The relationship between state median income and state area is a random one.

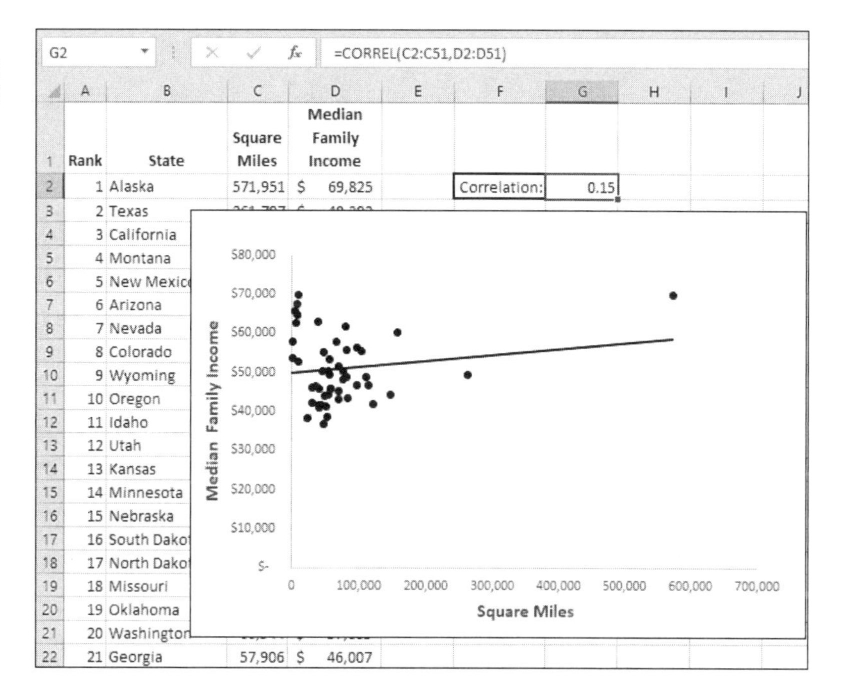

Figure 2.2 analyzes the relationship between two variables that we expect to be unrelated. There's no special reason to expect that the number of square miles in a state is related to the state's median family income. That's borne out by the correlation between the two variables, shown at 0.15 in cell G2. A correlation of 0.15 is a weak correlation. You can see the weakness of the relationship in Figure 2.2's chart. The individual data points in the chart form a random scatter around the regression line. One outlier on the chart, Alaska, has more than twice the area as the next largest state, Texas, and that outlier pulls the right end of the regression line up—and pulls the correlation coefficient away from a completely random 0.0.

At the other side of the continuum is such a strong relationship that it just doesn't come about in nature. See Figure 2.3.

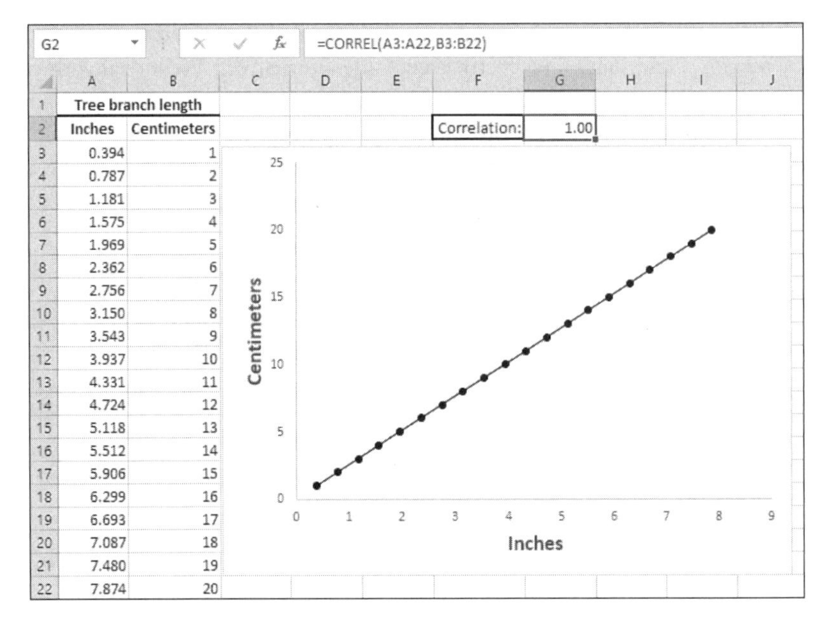

Tree branch length	
Inches	Centimeters
0.394	1
0.787	2
1.181	3
1.575	4
1.969	5
2.362	6
2.756	7
3.150	8
3.543	9
3.937	10
4.331	11
4.724	12
5.118	13
5.512	14
5.906	15
6.299	16
6.693	17
7.087	18
7.480	19
7.874	20

Figure 2.3
This sort of purely mathematical relationship tends to be completely uninformative, because it's so strong that it's trivial.

The relationship between the length of tree branches measured in inches and the length of the same branches measured in centimeters results in a correlation that's as strong as one can get: 1.00 (see cell G2). Notice in the chart that the individual data points don't just cluster close to the regression line, they lie directly on it.

But as strong as the relationship is, that strength derives from the way that the measures are defined, not to any intrinsic attribute shared by the two variables (such as the heights of siblings, presumably influenced by both similar genetic heritage and early nutrition). So these relationships tend to be of little substantive interest. But they serve to show the characteristics of strong correlations, just as relationships that are close to random are useful for understanding the nature of weak relationships.

Determining a Correlation's Direction

The correlation coefficient's range of values runs from –1.0 to +1.0, and it's not just possible but common for correlation coefficients to be negative. The issue of whether the correlation coefficient is negative or positive has nothing at all to do with the strength of the relationship. It has everything to do with how the variables are measured.

Consider the runners in a 10K race. Suppose you collect information on the number of minutes it took each person to run 10,000 meters and that person's age. You would find that the fewer minutes it took for a runner to finish, the lower the runner's age. Of course, you would find plenty of counterexamples. A typical 10-year-old is not going to run 10,000 meters as fast as a typical 18-year-old. But you would very likely find plenty of 60-year-olds who simply don't run as fast as 20-year-olds.

That implies a *positive* or *direct* correlation. The higher the score on one of the variables (here, the runner's age), the higher the score on the other variable (the number of minutes it took him or her to finish).

What if you also had access to the number of hours each runner trained weekly during the months leading up to the race? If you examined the relationship between running time in the 10K and the number of weekly training hours, you would very likely find that the more hours spent in training, the less time it took to finish the actual race. That implies a *negative* or *inverse* relationship between the variables.

Notice that both analyses race time with age, and race time with training hours—examine the relationship between measures of elapsed time. But two of the measures of elapsed time have different meanings. Beyond a certain point increasing age ceases to help a person run fast. But a relatively high number of training hours is usually associated with faster running times. The resulting difference between a positive (age with number of minutes to run 10K meters) and a negative (training hours with number of minutes in the 10K) correlation is caused not by the strength of the relationship between the variables, but by the meaning and directionality of the scale on which they're measured.

Figure 2.4 shows what those relationships look like when they're charted.

In Figure 2.4, the runner's age is positively correlated with the number of minutes it took to finish the race: generally, the older the runner, the more minutes to finish the race. The correlation coefficient is in cell J2. Note that the regression line slopes from the lower left to the upper right.

Also in Figure 2.4, the length of time to finish is negatively correlated with the training hours per week. Generally, the more time spent in training, the shorter the time to finish. The correlation coefficient is in cell J14. Notice that the regression line slopes from upper left to lower right.

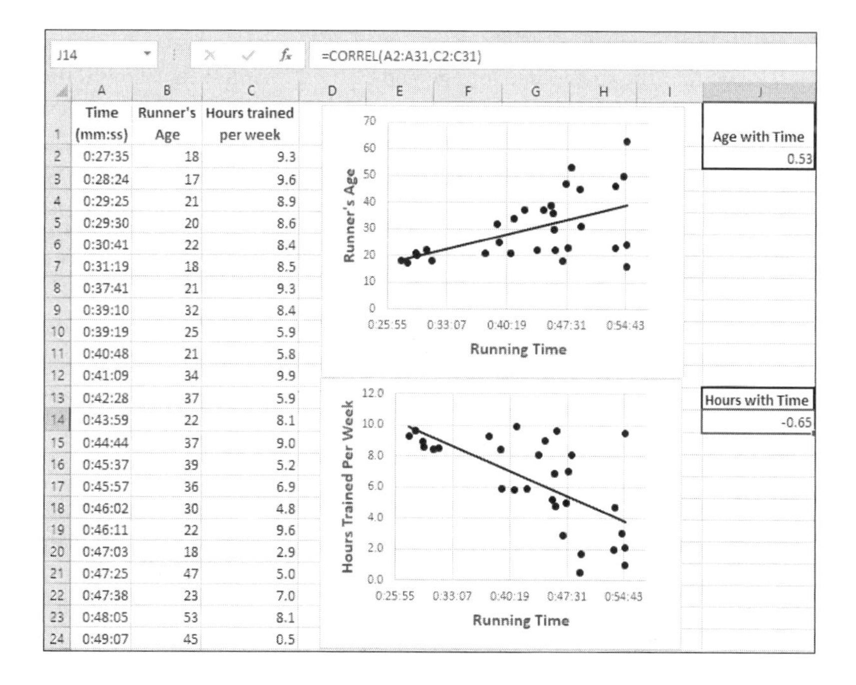

J14 =CORREL(A2:A31,C2:C31)

	A	B	C
1	Time (mm:ss)	Runner's Age	Hours trained per week
2	0:27:35	18	9.3
3	0:28:24	17	9.6
4	0:29:25	21	8.9
5	0:29:30	20	8.6
6	0:30:41	22	8.4
7	0:31:19	18	8.5
8	0:37:41	21	9.3
9	0:39:10	32	8.4
10	0:39:19	25	5.9
11	0:40:48	21	5.8
12	0:41:09	34	9.9
13	0:42:28	37	5.9
14	0:43:59	22	8.1
15	0:44:44	37	9.0
16	0:45:37	39	5.2
17	0:45:57	36	6.9
18	0:46:02	30	4.8
19	0:46:11	22	9.6
20	0:47:03	18	2.9
21	0:47:25	47	5.0
22	0:47:38	23	7.0
23	0:48:05	53	8.1
24	0:49:07	45	0.5

J	
Age with Time	0.53
Hours with Time	-0.65

Figure 2.4
The direction of the correlation is reflected in the direction of the regression line.

To summarize, variables that are correlated have either a positive or a negative correlation coefficient:

- If the correlation coefficient is positive, the slope of the regression line on the chart runs lower left to upper right. High values on one variable go with high values on the other variable, and low values go with low values.
- If the correlation coefficient is negative, the slope of the regression line on the chart runs upper left to lower right. High values on one variable are paired with low values on the other variable.

By the way, nothing is implied by the order in which the variables are named. The correlation between height and weight, say, is the same as the correlation between weight and height, as to both the strength and the direction of the correlation.

Calculating Correlation

Chapter 1, "Measuring Variation: How Values Differ," discussed how useful it can be to have standard scores, such as z-scores, available to describe the location of a value in a distribution according to a particular standard. If a person living in Warsaw told you that he makes 4,000 zlotys per year, it's not likely you would get any useful information from that unless you were familiar with the Polish economy. But if he told you that the z-score for his salary in Poland was +0.10, you would immediately know that his salary was a bit above average. (Z-scores, recall, have a mean of 0.0 and a standard deviation of 1.0.)

The same is true of the correlation coefficient. One way to describe the relationship, discussed in the prior section, between number of minutes to run a 10K and the runner's age, is by way of what's called the *covariance*. A person who studies these matters might tell you that the covariance for age and elapsed time in this 10K is 0.04. The problem is that the value of the covariance is in part a function of the standard deviation of both the variables that it comprises. So the covariance's value of 0.04 might mean something to a professional footrace analyst, in the same way that 4,000 zlotys means something to a European economist. But it has no meaning for me and, probably, not for you either.

The correlation coefficient is a different matter. It's standard, and it's not affected by differences in measurement scales. A correlation coefficient of 0.70 describes a moderately strong positive relationship whether you're talking about the relationship between housing prices and family income, or between the tensile strength of paper and its percentage of hardwood concentration. It's a standard measure of the strength and direction of a numeric relationship.

Step One: The Covariance

Even though the covariance isn't as intuitively friendly as the correlation, it's a useful place to start exploring how a correlation gets that way. Here's one formula for the covariance:

$$s_{xy} = \Sigma_{i=1}^{N}(X_i - \overline{X})(Y_i - \overline{Y})/N$$

That formula might remind you of a formula for the variance in Chapter 1:

$$S^2 = \Sigma_{i=1}^{N}(X_i - \overline{X})^2/N$$

The variance multiplies each deviation from the mean by itself—it squares them. In contrast, the covariance multiplies a deviation from the mean on one variable by the deviation from the mean on the other variable. The covariance of variable X with itself is its variance.

Just as the variance of variable X is the average of its squared deviations from the mean of X, so the covariance is the average of the product of the deviations on X and the deviations on Y.

Let's see how this works out with some actual data. Suppose that you're running a medical experiment to see whether a medication is able to lower the LDL (low density lipoprotein, the so-called "bad cholesterol") levels in patients with coronary heart disease. Reviewing some baseline data, you see that two patients, Jim and Virginia, have weight and LDL data as shown in Figure 2.5.

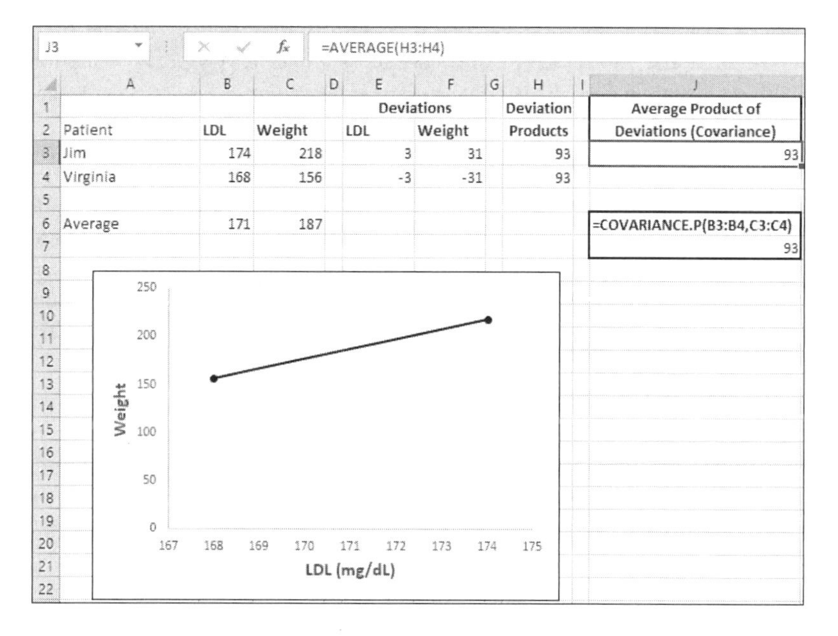
J3 =AVERAGE(H3:H4)

	A	B	C	D	E	F	G	H	I	J
1					Deviations			Deviation		Average Product of
2	Patient	LDL	Weight		LDL	Weight		Products		Deviations (Covariance)
3	Jim	174	218		3	31		93		93
4	Virginia	168	156		-3	-31		93		
5										
6	Average	171	187							=COVARIANCE.P(B3:B4,C3:C4)
7										93

Figure 2.5
It takes at *least* two records and two variables to calculate a covariance and, therefore, a correlation.

Jim's LDL measure and weight are in row 3, Virginia's are in row 4.

> **TIP**
> When you're setting up a correlation analysis in Excel, arrange your data so that you have the values for different variables on one record in the same row. So, in Figure 2.5, Jim's LDL is in row 3 and his Weight is also in row 3. It's best to avoid empty rows. If you anticipate using the Correlation or Covariance tool to create a matrix of correlations or covariances, avoid empty columns in your data set. Otherwise, non-adjacent columns are acceptable (imagine Figure 2.5 with Weight in column D and an empty column C), but they can create problems later.

Figure 2.5 shows the basic arithmetic to calculate a covariance. The underlying raw data is in the range B3:C4. To calculate the deviations, we need the average of each variable, which appear in cells B6 and C6, using Excel's AVERAGE() function.

The deviations themselves are in the range E3:F4, and each cell subtracts the variable's mean from its observed value. So, in cell E3, Jim's LDL deviation is B3 – B6, or 174 – 171, or 3. Virginia's Weight deviation in cell F4 is C4 – C6, or 156 – 187, or –31.

The products of the deviations are calculated in the range H3:H4, using these formulas:

H3: =E3*F3

H4: =E4*F4

Finally, the covariance is calculated by totaling the products of the deviations and dividing the result by the number of observations. Again, the covariance is the average of the products of the deviations for two variables, just as the variance is the average of the squared deviation for one variable.

Just to corroborate these calculations (and to point out to you that there's a quicker way to calculate the covariance), I've used Excel's COVARIANCE.P() function in cell J7 to get the covariance directly. Notice that it returns the same value as you see in cell J3, where it's calculated the slow way.

Just as Excel has an S version and a P version of its variance and standard deviation functions—VAR.S(), VAR.P(), STDEV.S(), and STDEV.P()—Excel also uses the P and S tags to distinguish between the sample and the population forms of the COVARIANCE() function. Their effect is the same as with the variance and the standard deviation. If you're working with a sample of data and you want the function to estimate the population parameter, use the S form. For example:

=COVARIANCE.S(M1:M20,N1:N20)

In that case, Excel uses N – 1 in the denominator rather than N, so it doesn't return strictly the average of the products of the deviations. If you're working with a population, or if you're not interested in estimating a population's covariance (as in the present example), use the P form:

=COVARIANCE.P(M1:M20,N1:N20).

The COVARIANCE.P() function uses N in the denominator and returns a true average of the deviation products.

Watching for Signs

Still working with Figure 2.5, we can get a sense for the reason that a covariance (and therefore a correlation) is positive or negative. Have another look at the four deviation scores in the range E3:F4. The deviations are always calculated by subtracting the mean of one variable from that variable's observed values. So we get a positive deviation of 31 in cell

F3 by subtracting the mean Weight, 187, from Jim's observed Weight, 218. Similarly, we get a negative deviation of –3 in cell E4 by subtracting the mean LDL, 171, from Virginia's measured LDL of 168.

In Jim's case, his observed values on both LDL and Weight are above their respective means, so his deviation scores are both positive—therefore, the product of his deviation scores is positive. In Virginia's case, her two observed values are both below the mean of their variables, and so her deviation scores are both negative. Their product is positive.

Because both products are positive, their average must also be positive. That average is the covariance, and, as you'll see, the sign of the covariance must also be the sign of the correlation coefficient. As noted earlier in this chapter, when lower values on one variable are associated with lower values on the other variable (and, therefore, higher values with higher values), the result is a positive correlation coefficient. The slope of the regression line is from the lower left (low values on both the horizontal and the vertical axis) to upper right (high values on both axes).

That type of pairing is going on in Figure 2.5. Figure 2.6 shows the complementary effect.

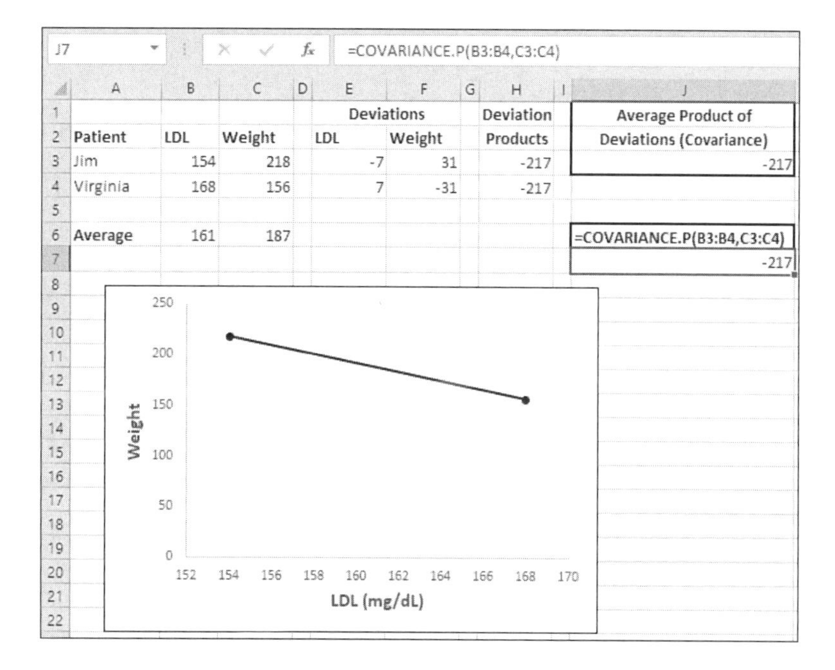

J7 =COVARIANCE.P(B3:B4,C3:C4)

	A	B	C	E	F	H	J
1				Deviations		Deviation	Average Product of
2	Patient	LDL	Weight	LDL	Weight	Products	Deviations (Covariance)
3	Jim	154	218	-7	31	-217	-217
4	Virginia	168	156	7	-31	-217	
5							
6	Average	161	187				=COVARIANCE.P(B3:B4,C3:C4)
7							-217

Figure 2.6
Changing just one value turns the covariance negative.

In Figure 2.6, Jim's LDL has been reduced from 174 in Figure 2.5 to 154. The result is to turn his LDL deviation negative. And because that change also reduces the mean LDL in cell B6, Virginia's LDL deviation is now positive because her observed value is now larger than the revised mean value.

The net of that one change in Jim's LDL is to give each patient one negative and one positive deviation score. In that case, both patients must have negative deviation products, as shown in cells H3 and H4 of Figure 2.6. When all the deviation products have negative values, their average—the covariance, and therefore the correlation coefficient—must also be negative.

When high values on one variable go with low values on the other variable, you get a negative correlation. The slope of the regression line is from the upper left (low values on the horizontal axis, high values on the vertical) to the lower right (high values on the horizontal, low values on the vertical).

From the Covariance to the Correlation Coefficient

The covariance always has the same sign as the correlation, so all we know so far is that with the two records as of Figure 2.5, the correlation between LDL level and body weight will be positive (but negative as of Figure 2.6). The meaning of the *value* of the covariance is still obscure. We don't learn much—in particular, whether the correlation is weak or strong—from the fact that the covariance is 93.

Here's how to get the correlation coefficient from the covariance:

$$r = S_{xy}/(S_x S_y)$$

where:

- r is the correlation coefficient.
- s_{xy} is the covariance.
- s_x and s_y are the standard deviations of variables X and Y, respectively.

The formula divides the covariance by the standard deviation of both the X and the Y variables, thus removing the effects of their scales from the covariance.

The standard deviation is itself built on the sum of *squared* deviations, and therefore cannot be a negative number. So whatever sign the covariance has carries forward to the correlation coefficient. The possible range of r includes both positive and negative values, but is restricted to the range –1.0 to +1.0 by the removal of the scales of the X and Y variables.

Figure 2.7 shows the calculation.

Cell J10 divides the covariance by the product of the standard deviations. The formula is given as text in cell J9. The CORREL() function gets you to the same place with a lot less work, but taking it step by step is the only way to see what's going on inside the black box.

So far we've looked only at situations in which the correlation is either –1.0 or +1.0. With just two records—here, Jim and Virginia—the correlation must be perfect. One way to think about that is to keep in mind that there's only one straight regression line that can be drawn through the charted points. Therefore, both points must lie directly on the regression line. In that case the relationship is perfect. Only when at least one charted point is off the regression line is the correlation greater than –1.0 and less than +1.0.

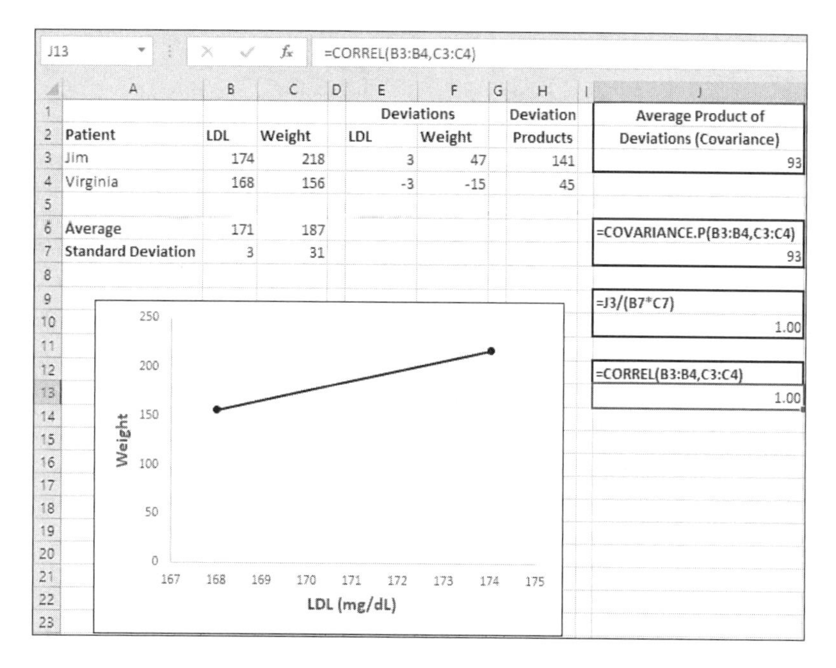

Figure 2.7
You can skip all the intermediate calculations by simply using the CORREL( ) function.

Let's have a look at the much more common situation, a less-than-perfect relationship. Although data sets that result in correlation coefficients of +1.0 or –1.0 aren't hard to assemble, they don't occur in the real world of medical, agricultural, economic, or any other meaningful field of research. There, correlation coefficients such as 0.36, 0.53, or 0.71 are the rule. So, because two records result in a perfect correlation, let's add a third patient to the two we've been looking at. See Figure 2.8.

In Figure 2.8, a third patient, Pat, shows up with an LDL value (172) that's just a smidgen above the new mean of 171.33, so Pat's LDL deviation score is positive. At 145, Pat's weight is below the mean of 173, so the Weight deviation score is negative. Multiplying a positive deviation by a negative deviation results in a negative product of –17.56. That negative product pulls Jim's and Virginia's positive products down, from the total that resulted in a correlation coefficient of 1.0 in Figure 2.7 to the 0.66 value shown in Figure 2.8.

What if Pat's LDL were a little lower, below the mean LDL value? Then Pat would have a negative deviation on LDL to go with a negative deviation on Weight, resulting in a positive product. In that case, all three deviation products would be positive, as shown in Figure 2.9.

A set of deviation products that are all positive (or all negative) does not necessarily mean that you have a perfect +1.0 or –1.0 correlation. Only when all data points lie directly on the regression line can the correlation be perfect. This issue will arise repeatedly throughout the remainder of this book because it bears directly on the accuracy of the predictions you can make using regression.

Figure 2.8
With three or more pairs of observations the correlation is no longer necessarily perfect.

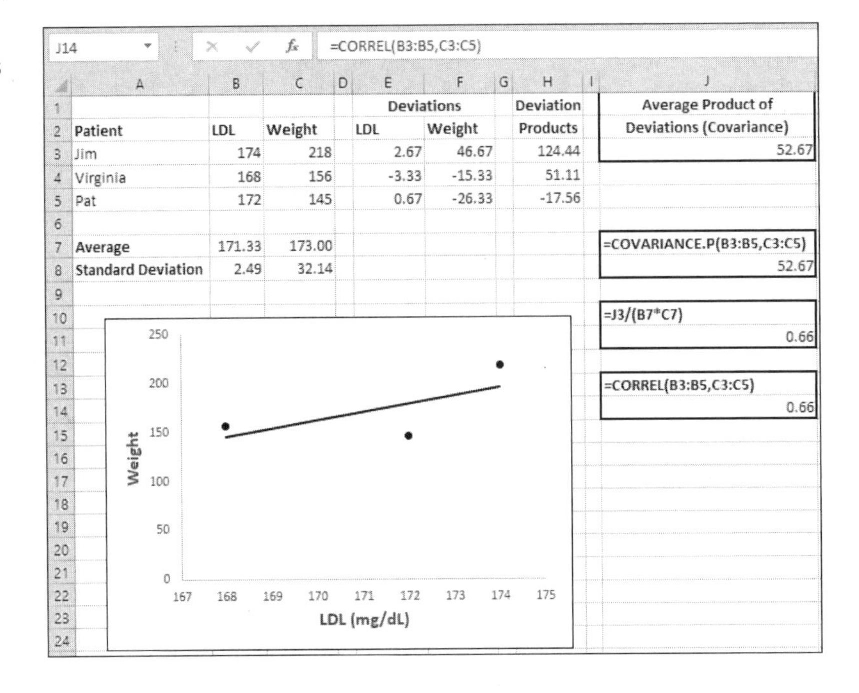

J14 =CORREL(B3:B5,C3:C5)

	A	B	C	D	E	F	G	H	I	J
1					Deviations			Deviation		Average Product of
2	Patient	LDL	Weight		LDL	Weight		Products		Deviations (Covariance)
3	Jim	174	218		2.67	46.67		124.44		52.67
4	Virginia	168	156		-3.33	-15.33		51.11		
5	Pat	172	145		0.67	-26.33		-17.56		
6										
7	Average	171.33	173.00							=COVARIANCE.P(B3:B5,C3:C5)
8	Standard Deviation	2.49	32.14							52.67
10										=J3/(B7*C7)
11										0.66
13										=CORREL(B3:B5,C3:C5)
14										0.66

Figure 2.9
The three charted data points are all off the regression line.

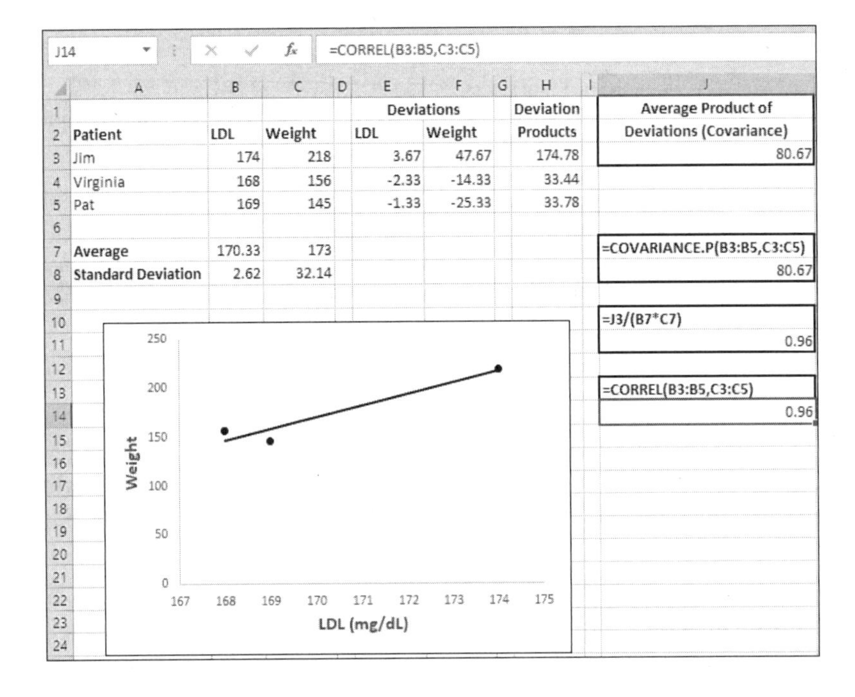

J14 =CORREL(B3:B5,C3:C5)

	A	B	C	D	E	F	G	H	I	J
1					Deviations			Deviation		Average Product of
2	Patient	LDL	Weight		LDL	Weight		Products		Deviations (Covariance)
3	Jim	174	218		3.67	47.67		174.78		80.67
4	Virginia	168	156		-2.33	-14.33		33.44		
5	Pat	169	145		-1.33	-25.33		33.78		
6										
7	Average	170.33	173							=COVARIANCE.P(B3:B5,C3:C5)
8	Standard Deviation	2.62	32.14							80.67
10										=J3/(B7*C7)
11										0.96
13										=CORREL(B3:B5,C3:C5)
14										0.96

Nevertheless, the correlation is now very strong at 0.96. If all your observations have positive deviation products (or if they all have negative products), you're going to wind up with a strong correlation coefficient.

Using the CORREL( ) Function

In the previous section I mentioned that the CORREL() function does all the work for you—it takes care of calculating the averages and standard deviations and covariances and it does all the multiplying, adding, and dividing. In Figure 2.9, for example, notice that the result shown in cell J14 returns the same value for the correlation coefficient as you see in cell J11, which is the end result of all those arithmetic calculations.

> **NOTE** Excel has another worksheet function, PEARSON( ), which takes the same arguments and returns the same results as CORREL( ). That's an unnecessary complication in Excel, of course, but it dates back at least to the mid-1990s and it's too late to discard it now. The PEARSON( ) function is named for Karl Pearson, the 19th and 20th century statistician who was most responsible for the development of the concepts and mathematics of the correlation coefficient. Use whichever function you prefer, but this book uses CORREL( ) because I find it quicker to type.

Understanding Bias in the Correlation

Notice that the CORREL() function has no P or S tag, as do the STDEV(), VAR(), and COVARIANCE() functions. The reason for the S tag in, for example, the VAR.S() function is to tell Excel to divide the sum of the squared deviations by the degrees of freedom, N – 1, rather than by the number of observations, N. So doing removes the negative bias in the sample variance and makes it an unbiased estimator of the variance in the population from which the sample was taken. In formulas rather than words, this calculation of the variance using a sample of values is an underestimate of the population variance and therefore biased:

$$S^2 = \Sigma_{i=1}^{N}(X_i - \overline{X})^2/N$$

But this calculation of the variance does not result in a biased estimator of the population variance:

$$S^2 = \Sigma_{i=1}^{N}(X_i - \overline{X})^2/(N - 1)$$

Is the sample correlation coefficient a biased estimator of the population correlation? Yes, it's biased, but the reason for its bias is different than for the bias in the variance or standard deviation. Those statistics are biased estimators when N is in the denominator because the sum of the squared deviations of individual values is smaller when you deviate them from the sample mean than when you deviate them from any other number, including the population mean. When the sum of the squared deviations is smaller than it should be, the sample variance underestimates the population variance. It turns out that dividing by the degrees of freedom rather than by the count of actual observations removes that bias from the sample variance.

In contrast, the bias in the correlation coefficient when it's based on a sample is due to the fact that its sampling distribution is skewed. When the population correlation coefficient (often symbolized as ρ or "rho," the Greek letter corresponding to the Roman "r") is positive, the sampling distribution of r is skewed negatively. When ρ is negative, r's sampling distribution is skewed positively. The amount of skew is in large part a function of the size of the sample. Small samples (say, 10 observations each) can skew substantially, but large samples with thousands of cases usually appear to be symmetrical.

For example, Figures 2.10 through 2.12 show the sampling distributions of r, based on samples of 10 observations each from a population in which the correlation between X and Y is 0.68, 0.05, and –0.62.

In Figure 2.10, the value of r in the population is 0.68. Most samples from that population will have sample values of r in the 0.6 to 0.8 range—as is evident from the chart. And there's plenty of room below 0.6 for sample correlations that go all the way down to –0.45, or 1.13 units below the population parameter. But because of the way that the correlation coefficient is calculated, its maximum value is 1.0 and samples with correlations higher than the 0.68 tend to bunch up in the right tail of the distribution.

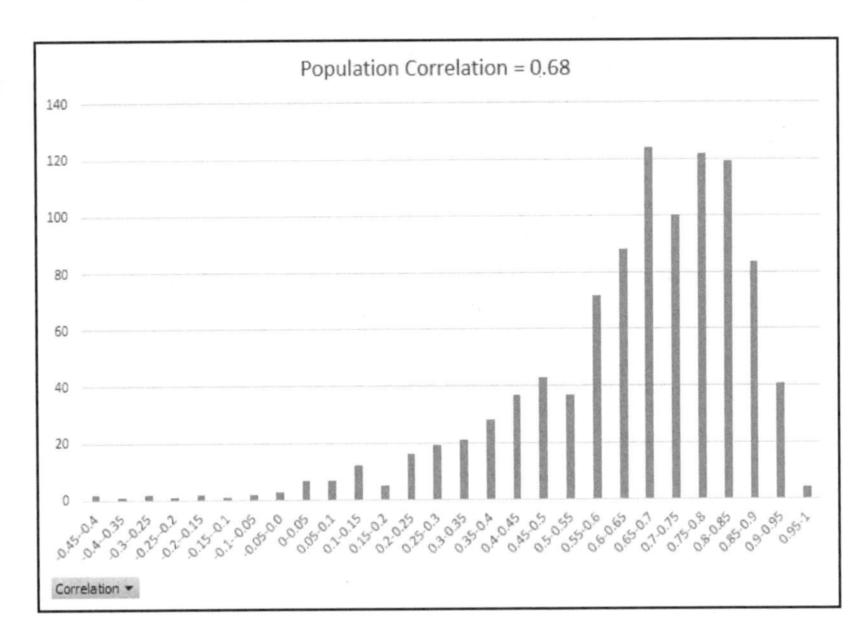

Figure 2.10
The sampling distribution of r when ρ is 0.68.

When the population parameter for the correlation is close to 0.0, as it is in Figure 2.11, there's plenty of room in each tail for the more extreme sample values (extreme given that the parameter is 0.0). The sampling distribution approaches a symmetric shape.

Figure 2.12 shows that when the population correlation is negative, you get a skewed sampling distribution just as in Figure 2.10, but it's skewed positive instead of negative.

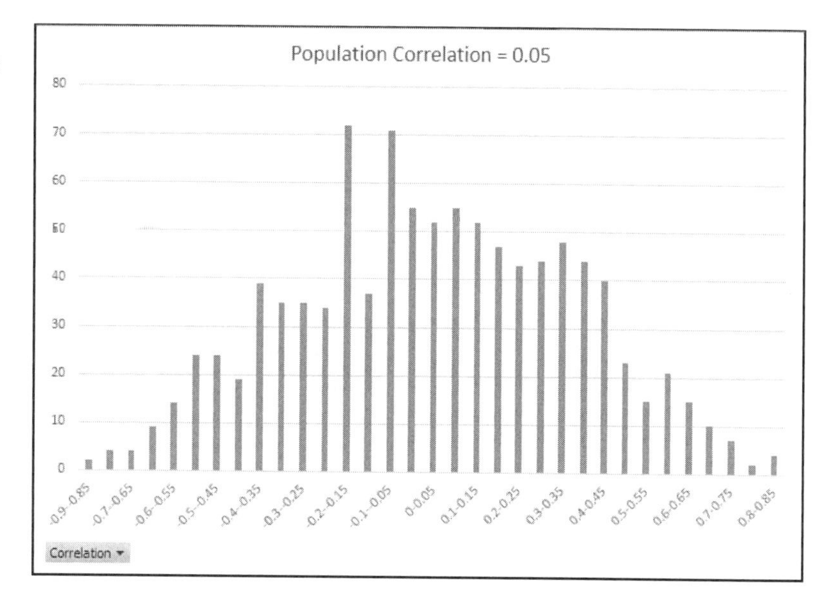

Figure 2.11 The sampling distribution of r when ρ is 0.05.

The effects shown in Figures 2.10 and 2.12 would be much milder with larger sample sizes. The skewness would be much less pronounced, and the range of the sample correlations would be much narrower. That's the basis for the adjustment for shrinkage in the *multiple* correlation coefficient, which I'll discuss in Chapter 5, "Multiple Regression."

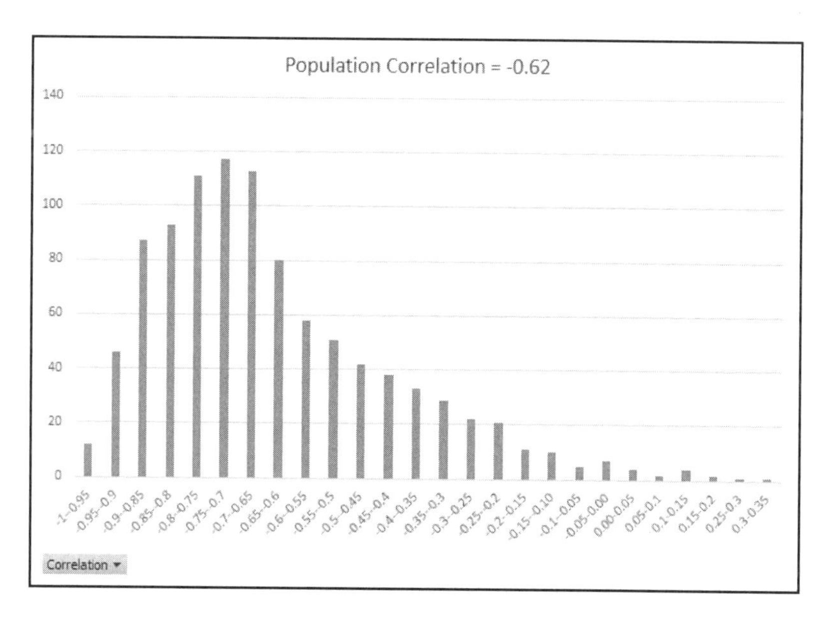

Figure 2.12 The sampling distribution of r when ρ is –0.62.

So, depending on the value of ρ, r is biased as an estimator, and the effect of the bias is greater with smaller sample sizes. Still, we don't try to correct for that bias using a method such as substituting the degrees of freedom for the number of observations in the formula for the standard deviation. In fact, the number of observations used to calculate the covariance has nothing to do with the calculated value of the correlation coefficient. This formula:

$$r = S_{xy} / (S_x S_y)$$

(which is given earlier in this chapter as a method of calculating r) is useful as a method of conceptualizing the relationships among the different ways of describing a data set (and, I suppose, a convenient way to calculate the correlation when you happen already to have the covariance and the standard deviations). But to see why the number of observations does not affect the outcome of the calculation, let's take it apart, as follows:

$$r = (\Sigma x_i y_i / N) / \sqrt{\Sigma x_i^2 / N} \sqrt{\Sigma y_i^2 / N}$$

The lowercase x's and y's represent the deviations of the observed X and Y values from their respective means. Move N from the covariance calculation to the denominator of the correlation calculation:

$$r = (\Sigma x_i y_i) / (N \sqrt{\Sigma x_i^2 / N} \sqrt{\Sigma y_i^2 / N})$$

Now, the product of the square roots of N in the denominators of the standard deviation calculations cancels out the N that was moved from the covariance:

$$r = \Sigma x_i y_i / \sqrt{\Sigma x_i^2} \sqrt{\Sigma y_i^2}$$

In words, the correlation can be calculated as the ratio of the sum of the products of the X and Y deviations to the square root of the sum of the squared X deviations, times the square root of the sum of the squared Y deviations. (You can see why it often pays to look at the formulas rather than the words.) Notice particularly that the number of observations is not used in the final formula for r. You could substitute (N – 1) for N and it would make no difference to the calculated value of r.

Checking for Linearity and Outliers in the Correlation

The standard correlation coefficient—the version developed by Karl Pearson around 1900 and the one that we typically mean when we refer to a correlation in a statistical context—is meant for use with variables that bear a linear relationship to one another. Figure 2.1 shows a good example of a linear relationship, with home prices rising according to a moderately strong relationship to family income.

Figure 2.13 has an example of the sort of relationship between variables that Pearson's r was not intended to handle.

Figure 2.13 shows a chart of the relationship between the length of time allotted to practice a manual task (on the horizontal axis) and the number of times the task was performed accurately in one minute (on the vertical axis). You can see that there's a diminishing returns aspect to the data: Beyond a certain amount of practice time, additional practice does not improve the score. The regression line in Figure 2.13 is therefore curved, not linear.

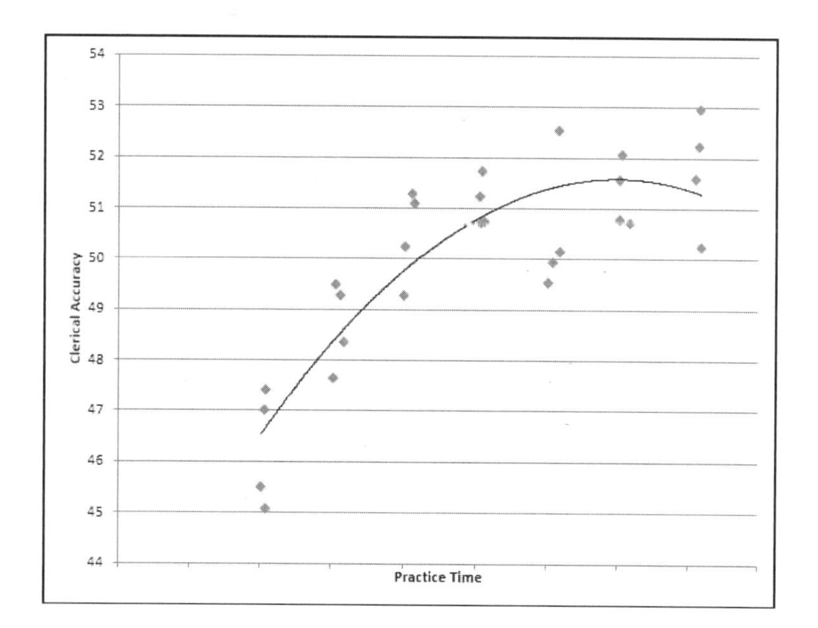

Figure 2.13
Pearson's *r* does not accurately quantify the strength of this relationship.

2

Using techniques that you'll read about in later chapters, you can calculate a measure of a curvilinear or nonlinear, rather than straight-line, relationship between these two variables. In this case the nonlinear correlation would be 0.87. The Pearson correlation would underestimate the strength of the relationship, returning 0.79. A difference of 0.08 in the strength of the correlation, especially in the upper range of possible correlation coefficients, is substantial.

You could easily have missed that nonlinear relationship if you had not charted the variables as shown in Figure 2.13. Figure 2.14 shows a related problem that jumps out at you if you chart the data.

Figure 2.14 is a slightly modified version of Figure 2.2. I added around $50,000 to the median household income for Alaska—a pretty good chunk of change, but exactly the sort of thing that can happen when there's an opportunity for transcription errors or a mistake in designing a database query.

A Pearson correlation based on the other 49 states returns –0.01, indicating no relationship between family income and the number of square miles within the state's borders. But if you include the one outlier state in the correlation, as you well might if it were buried halfway through a list returned by a database query, you would get a correlation of 0.58, a respectably strong relationship.

The reason for that startling increase in the value of the correlation coefficient is an increase in the value of the covariance, from a bit over 3 million to nearly 600 million. In turn, that increase is due to the distance of Alaska's median family income and its area in square miles (both fictional values in Figure 2.14, chosen to illustrate this point regarding outliers) from the means of the remaining states. When those deviations are squared, as they are when they enter the calculation, the effect on the correlation is pronounced.

Figure 2.14
It's easy to miss an outlier if you don't chart it.

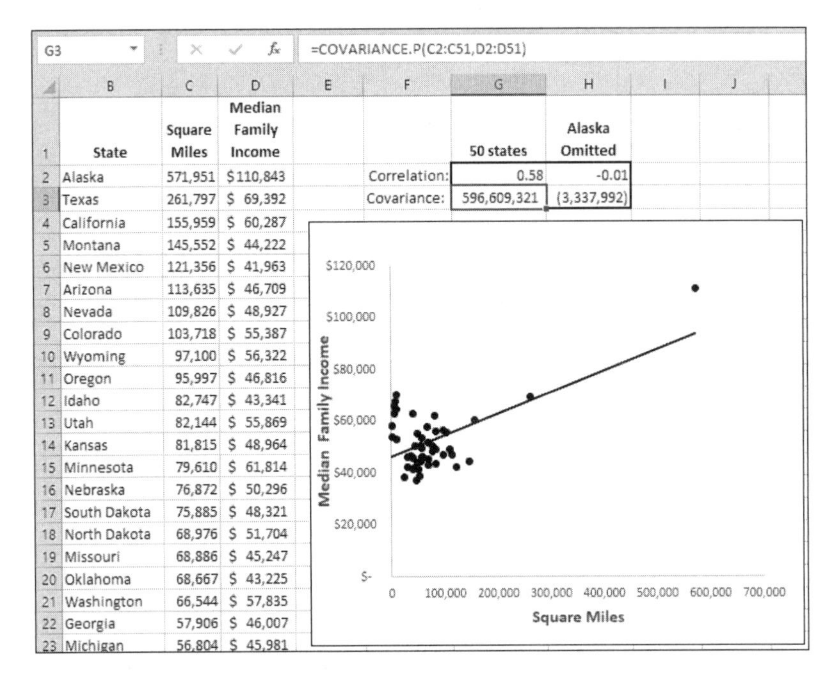

G3 =COVARIANCE.P(C2:C51,D2:D51)

	B	C	D	E	F	G	H
1	State	Square Miles	Median Family Income			50 states	Alaska Omitted
2	Alaska	571,951	$110,843		Correlation:	0.58	-0.01
3	Texas	261,797	$ 69,392		Covariance:	596,609,321	(3,337,992)
4	California	155,959	$ 60,287				
5	Montana	145,552	$ 44,222				
6	New Mexico	121,356	$ 41,963				
7	Arizona	113,635	$ 46,709				
8	Nevada	109,826	$ 48,927				
9	Colorado	103,718	$ 55,387				
10	Wyoming	97,100	$ 56,322				
11	Oregon	95,997	$ 46,816				
12	Idaho	82,747	$ 43,341				
13	Utah	82,144	$ 55,869				
14	Kansas	81,815	$ 48,964				
15	Minnesota	79,610	$ 61,814				
16	Nebraska	76,872	$ 50,296				
17	South Dakota	75,885	$ 48,321				
18	North Dakota	68,976	$ 51,704				
19	Missouri	68,886	$ 45,247				
20	Oklahoma	68,667	$ 43,225				
21	Washington	66,544	$ 57,835				
22	Georgia	57,906	$ 46,007				
23	Michigan	56,804	$ 45,981				

It's a good idea to look at your data in a chart, not only for a nonlinear relationship but for extreme outliers, to make sure you know what's going on with your data. Charting the variables, as in Figure 2.14, isn't the only way to diagnose a problem with the data that you want to analyze by means of a correlation. But it puts you in a position to evaluate for a nonlinear relationship, and to check for really egregious outliers, as well as other anomalies. Excel makes it so easy to create the charts that there's no good reason not to.

If you're new to Excel charts, here's a quick primer on how to get the chart in Figure 2.14:

1. Select the range C2:C51, which contains each state's area in square miles.
2. Hold down the Ctrl key and, while you're holding it down, select the range D2:D51, which contains each state's median family income (or, in the case of Alaska, a transcription error).

NOTE

If your variables occupy contiguous columns, as columns C and D in Figure 2.14, you can make one selection rather than two separate selections. In that figure, just click in cell C2 and drag through D51.

3. Click the Ribbon's Insert tab and locate the Charts group. Hover your mouse pointer over the different chart icons in that group until you locate the one with a popup that says "Insert Scatter (X,Y) or Bubble Chart." Click it.

You'll get an XY chart that looks substantially like the one in Figure 2.14. Right-click any data marker in the chart area and choose Add Trendline from the shortcut menu. Accept the default option of Linear Trendline and dismiss the Format Trendline pane.

A little tweaking, to do things like eliminating horizontal and vertical gridlines, will get you an exact replica of the chart.

Now, the reason I advise you to select a Scatter (X,Y) chart is that it's the only chart type, along with the Bubble chart, that treats both the horizontal and vertical axes as value axes. Excel distinguishes value axes from category axes. Value axes represent numeric quantities and preserve their quantitative differences. An 8 on a value axis is twice as far from 0 as a 4. On a category axis, though, consecutive values of 0, 8, and 9 would be equally far apart: Excel treats them as categories and there's no intrinsic reason not to make Pepsi, Coke, and Sprite equidistant from one another.

Figure 2.15 shows an example of what can happen if you choose, say, a Line chart instead of a Scatter chart.

The two charts in Figure 2.15 show the effect of using a Line chart instead of a Scatter chart to display the relationship between two variables—in this case, Height and Weight. The Scatter chart treats both the variable on the vertical axis and the variable on the horizontal axis as numeric variables. The observation on row 2 has the value 60 for its Height and 129 for its Weight.

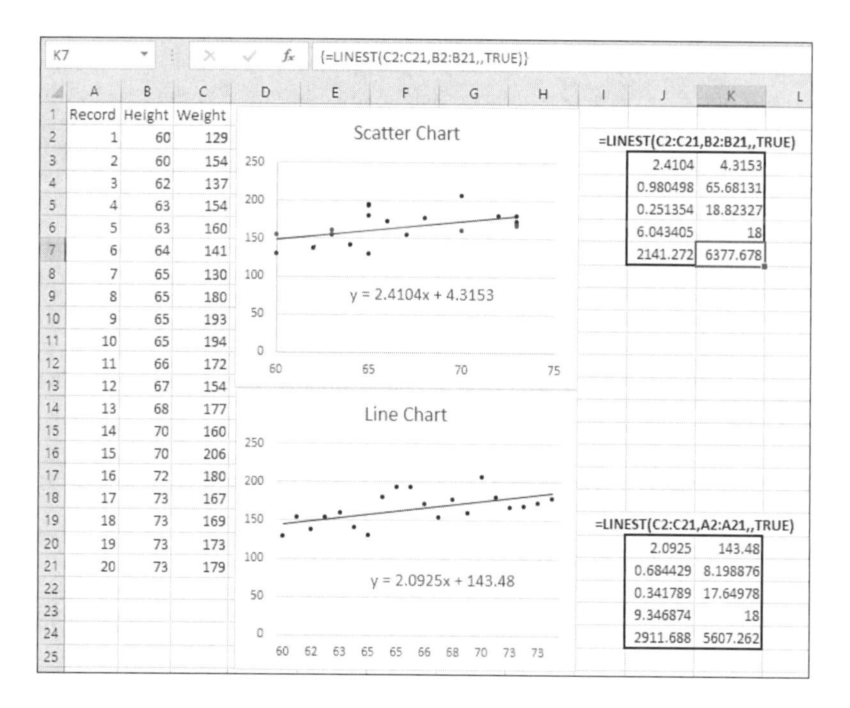

Record	Height	Weight
1	60	129
2	60	154
3	62	137
4	63	154
5	63	160
6	64	141
7	65	130
8	65	180
9	65	193
10	65	194
11	66	172
12	67	154
13	68	177
14	70	160
15	70	206
16	72	180
17	73	167
18	73	169
19	73	173
20	73	179

=LINEST(C2:C21,B2:B21,,TRUE)

J	K
2.4104	4.3153
0.980498	65.68131
0.251354	18.82327
6.043405	18
2141.272	6377.678

=LINEST(C2:C21,A2:A21,,TRUE)

J	K
2.0925	143.48
0.684429	8.198876
0.341789	17.64978
9.346874	18
2911.688	5607.262

Figure 2.15
The regression lines appear similar, but compare the regression equations.

In contrast, the Line chart treats the variable on its vertical axis as a numeric variable, but the variable on its horizontal axis as a *category* variable. As far as the Line chart is concerned, the first three Height values on its horizontal axis might as well be Alabama, Alaska, and Arizona instead of 60, 60, and 62. (I have formatted the data series in the Line chart to suppress the line that by default connects its markers. I did so to make it easier to compare the positions of the data markers on each chart.)

One result is that Excel has no quantitative basis to distinguish between observations on the horizontal axis. Notice on Figure 2.15's Line chart that the observations are equidistant on the horizontal axis. The values shown as labels on the horizontal axis are just that—labels, not numeric values. But in its Scatter chart, the distance between the observations depends on their relative Height values.

When you request a trendline with a Line chart active, Excel requires numeric values for the Height variable in order to calculate the regression equation. By choosing a Line chart and placing Height on its horizontal axis, you have told Excel that the Height values on the worksheet are really just category names. So Excel resorts to the number that identifies the record: It treats the Height value for the first record as 1, for the second record as 2, and so on.

To demonstrate this effect, have a look at the LINEST() results in the range J20:K24 of Figure 2.15. It assesses the relationship between the Weight values in column C with the Record Number values in column A. Notice that the regression coefficient and constant in cells J20 and K20 match those returned by the regression equation in the Line chart.

If you do the same with the Scatter chart and the LINEST() results in the range J3:K7, you find that the regression coefficient and constant returned by LINEST() are identical to those in the chart's regression equation. The relationship that LINEST() assesses in this case is between Weight and Height, not (as in the Line chart) between Weight and Record Number.

Clearly, then, the way that Excel treats numeric values and nominal categories on its chart axes has implications for how in-chart analyses work. Because a correlation coefficient assumes that both variables are measured on interval or ratio scales, be sure to chart them accordingly—on a Scatter (X,Y) chart, not on a chart type that might just look like a Scatter chart. (Bubble charts have their uses, but they only confuse the issue when it comes to basic correlation analysis.)

Avoiding a Trap in Charting

While I'm on the topic of charting data with Scatter charts, I want to alert you to a trap that cost me a couple of hours of head scratching before I got it figured out. To do so, it's necessary to anticipate some material on regression analysis that I won't start covering in detail until Chapter 5. But it's useful to discuss it here, in the context of charting correlated variables.

One important reason to analyze correlation is to prepare a regression analysis, which you can then use to predict the values of one variable given values from another. For example, you might use the correlation between height and weight to predict a person's weight from your knowledge of his height.

I don't recommend it as your sole means of predicting one variable's value from another variable, but one way to get the prediction equation is by way of a Scatter chart. You'll find one in Figure 2.16.

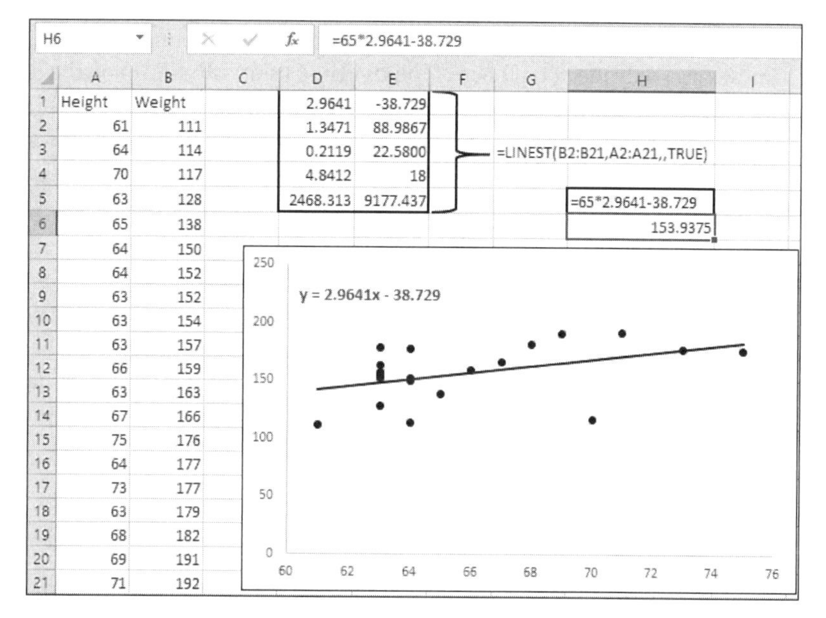

H6 =65*2.9641-38.729

	A	B	C	D	E	F	G	H
1	Height	Weight		2.9641	-38.729			
2	61	111		1.3471	88.9867			
3	64	114		0.2119	22.5800		=LINEST(B2:B21,A2:A21,,TRUE)	
4	70	117		4.8412	18			
5	63	128		2468.313	9177.437			=65*2.9641-38.729
6	65	138						153.9375
7	64	150						
8	64	152						
9	63	152						
10	63	154						
11	63	157						
12	66	159						
13	63	163						
14	67	166						
15	75	176						
16	64	177						
17	73	177						
18	63	179						
19	68	182						
20	69	191						
21	71	192						

Figure 2.16
If you request it, Excel calculates the equation and places it on the chart.

The chart in Figure 2.16 includes a *regression equation* that tells you how to predict variable Y from variable X. Excel derives the equation from the relationship—specifically, the correlation coefficient—between the Height and Weight variables provided in the ranges A2:A21 and B2:B21. I'll get much deeper into those weeds in Chapters 3 and 4, but for now it's enough to know that the equation tells you to do the following:

- Take some value such as 65 for height, which in this example is Height, the predictor variable.
- Multiply that value by 2.9641.
- Subtract 38.729.
- The result of 153.9 is the predicted weight for a height of 65.

Here's how to get that equation in the chart:

1. Create the Scatter chart.
2. Right-click any of the charted data markers.

3. Choose Add Trendline from the shortcut menu.
4. The Format Trendline pane appears. Accept the default Linear trendline. Scroll down if necessary to the checkbox labeled Show Equation on Chart, and fill the checkbox.
5. Dismiss the Format Trendline pane.

The regression equation is now on the chart. You might have to click and drag the equation to another place in the chart if a data marker or some other chart element obscures it.

There's another way, almost always a more useful one, to get that equation: the LINEST() function. I spend all of Chapter 4, "Using the LINEST() Function" on that function and its results and only touch on it here. The results of using LINEST() on the data in columns A and B appear in the range D1:E5 of Figure 2.16. Note the values in cells D1 and E1. They are identical to the values in the chart's regression equation. Like the charted equation, Excel's LINEST() function tells you to multiply a Height value by 2.9641 and subtract 38.729 (actually, it tells you to add a value of –38.729, an equivalent operation). So, if you want to predict the weight of a person whose height is 65 inches, using this data you could use this equation:

$$y = 2.9641x - 38.729$$

Now, how does Excel know which variable is the *x* or predictor variable in the Scatter chart, and which is the *y* or predicted variable? Excel's developers, doubtless for the best of reasons, decided long ago that the variable that appears to the *left* of the other variable on the worksheet is to be treated as the variable on the horizontal axis. The variable that appears to the *right* of the other variable on the worksheet is to be treated as the variable on the vertical axis.

You can enter the two variables in two noncontiguous columns and make a multiple selection of them (select the first variable's values, hold down Ctrl, and select the second variable's values). The result will be the same—the leftmost variable will be on the horizontal axis.

Furthermore, the Excel convention is that the variable on the chart's vertical axis is the one that is to be predicted in the regression equation that's displayed on the chart. The variable on the chart's horizontal axis is the predictor variable in the chart's regression equation. Therefore, given that Height is represented on the horizontal axis, the regression equation:

$$Y = 2.9641(X) - 38.729$$

means to predict a Y (weight) value by multiplying an X (height) value by 2.9641 and subtracting 38.729.

In sum, the predicted variable in a chart's regression equation is the variable found to the right of the other variable on the worksheet.

Still in Figure 2.16, have a look at the way that the arguments to the LINEST() function are entered in D1:E5:

```
=LINEST(B2:B21,A2:A21,,TRUE)
```

The LINEST() syntax makes the assignment of the variables' roles explicit in the arguments, rather than implicit in the variables' worksheet locations. In the LINEST() formula just given, the predicted variable is found in B2:B21. The predicted variable is always LINEST()'s first argument. The predictor variable is found in A2:A21. The predictor variable is always LINEST()'s second argument.

This arrangement—locating the predictor variable positionally for the purpose of the regression equation in a chart, but locating it as a specific argument in the worksheet function—is a trifle unfortunate because if it is consistent, it's consistent only by accident. But the decisions that led to the arrangement were probably well considered and we can live with them.

However, the Excel developers failed to reckon with human fallibility. Specifically, mine. See Figure 2.17.

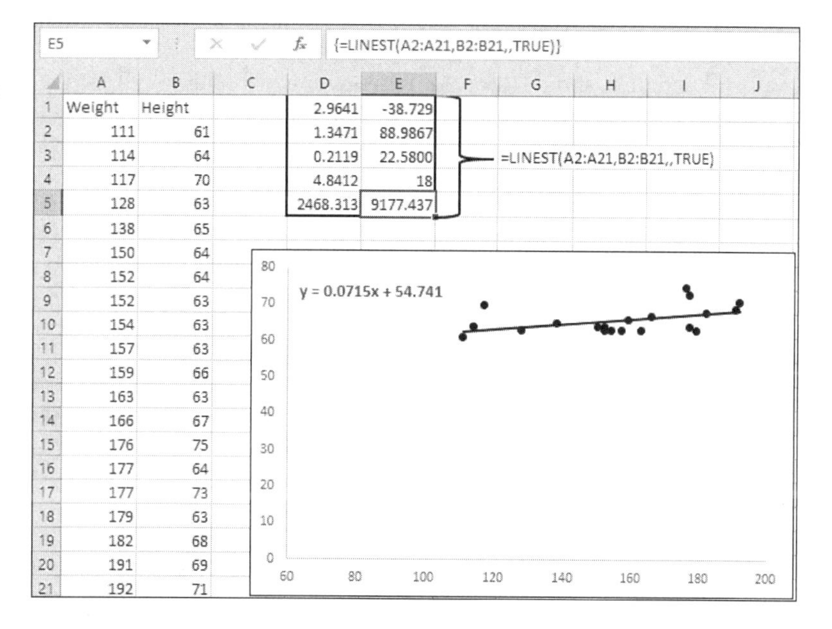

Figure 2.17
The LINEST() figures are no longer consistent with the charted regression equation.

The most obvious difference between Figures 2.16 and 2.17 is that in Figure 2.16 the predicted variable, Weight, is in column B. In Figure 2.17, it's in column A.

Now look at the first row of the LINEST() results in the range D1:E1. If you compare those values with the formula that shows up in the chart, you'll see that the coefficient (cell D1) and what's termed the *constant* (cell E1) no longer match with their values in the charted equation.

The most immediate reason is that the LINEST() equations in Figure 2.16 and Figure 2.17 both treat Weight as the predicted variable. The predicted variable is always the first argument to the LINEST() function. That argument is B2:B21 in Figure 2.16 and A2:A21 in Figure 2.17.

But because Figure 2.17 moves Weight to the right of Height, the chart puts Weight on the horizontal axis and Height on the vertical axis. Then, when it calculates the regression line and the regression formula, it treats Height, not Weight, as the predicted variable. The results are as follows:

- The data markers on the chart in Figure 2.17 are turned 90 degrees from their positions in Figure 2.16.
- The location of the regression line is altered.
- The regression equation on the chart is altered.

That's what threw me. To save myself I couldn't figure out why the coefficient and the constant in the LINEST() equation didn't match the coefficient and constant in the charted equation. I do a lot of work that depends in one way or another on regression analysis, so you would think that I would immediately understand the source of the problem, and therefore its fix. Probably I should have.

Okay, I definitely should have. All I can say is that I seldom call for the equation on the chart, so I seldom see it there. And because I seldom see it, I didn't have a chance to see the inconsistency years ago, figure it out, and put it in my hip pocket.

But why did it happen, as shown in Figure 2.17, that I put the predicted variable in column A and the predictor in column B? That has to do with how LINEST() works. And this is why I decided to come clean here and explain how easy it is to go wrong if you don't know where the traps are.

LINEST() is capable of predicting a single variable such as Weight from not just one but *several* predictor variables. (This book discusses that usage in Chapter 5.) If you want to analyze multiple predictors simultaneously, LINEST() requires that they occupy adjacent columns. So, if you want to allow room for additional predictors, the sensible layout is something like the one shown in the worksheet on the right in Figure 2.18.

Suppose your data is arranged as in the worksheet on the left in Figure 2.18. After looking at an analysis of the relationship between Height and Weight, you decide to shift to a multiple regression framework and add Age to the mix. To include Age, you would have to do something such as insert a new column between Height and Weight, so that the predictor variables, Height and Age, occupy adjacent columns.

Putting the predicted variable in the leftmost column is much more sensible, as is done in the worksheet on the right. That way, you can easily build out the set of contiguous columns for additional predictors as needed. But if you do that, the predicted variable winds up on the left, and an Excel chart turns it into a predictor variable.

I'm uncomfortably aware that I have spent several pages on this issue, which represents a trap that you might never fall into. Bear in mind, though, that you don't have to ask for the regression equation on the chart to be misled by which variable is treated as the predictor and which the predicted. The location and the slope of the regression line itself are determined by the regression equation. So be sure that you know exactly what's going on in your Scatter chart before you request a trendline.

Figure 2.18
Adding a predictor to the worksheet on the left requires a rearrangement of the data.

	A	B	C	D	E
1	Height	Weight			
2	61	111			
3	64	114			
4	70	117			
5	63	128			
6	65	138			
7	64	150			
8	64	152			
9	63	152			
10	63	154			
11	63	157			
12	66	159			
13	63	163			
14	67	166			
15	75	176			
16	64	177			
17	73	177			
18	63	179			
19	68	182			
20	69	191			
21	71	192			

	A	B	C	D
1	Weight	Height	Age	Minutes of Exercise Weekly
2	111	61	13	55
3	114	64	20	86
4	117	70	36	101
5	128	63	47	120
6	138	65	18	59
7	150	64	67	66
8	152	64	54	38
9	152	63	15	66
10	154	63	76	70
11	157	63	20	33
12	159	66	16	59
13	163	63	96	38
14	166	67	57	77
15	176	75	32	73
16	177	64	69	51
17	177	73	74	48
18	179	63	16	118
19	182	68	89	109
20	191	69	19	32
21	192	71	81	24

And if the ten minutes you've spent here reading about how I stubbed my toe on an Excel chart saves you two hours of head scratching, then that's a favorable rate of exchange.

Correlation and Causation

You've likely read or heard about the distinction between correlation and causation: Although correlation often implies the presence of causation, it does not necessarily ensure it.

Suppose your research finds that in a large number of locations along both the East and West coasts of the United States, a positive correlation exists between the mean ocean temperature up to one mile offshore and the number of whales sighted during migration season.

Some would say the organisms that whales feed on are attracted to the warmer waters, and that the whales swim to where the food is. And if the correlation is strong enough—0.80, perhaps—for your own criteria to rule out chance as an explanation for the strength of the relationship, you might decide that the explanation is correct.

Do not let yourself be misled by the empirical evidence and apparent logic of the argument. The correlation proves nothing. You don't use statistics to demonstrate whether a theory is true or false. You use *experimental design* to rule out competing explanations for an observation. You use statistics to summarize the information that's gathered by that experimental design, and to quantify the likelihood that you are making the wrong decision given the evidence you have.

Two kinds of competing explanations are commonly put forward when a causality is suggested as a reason for an observation. One is the *directionality* of the presumed causation: That is, the cause might be mistaken for the effect. The other is the presence of one or more other variables that exert a cause on both variables of interest.

Direction of Cause

It would be a stretch to claim that whales' body heat is causing the increase in ocean temperatures. However, an example of the directionality issue that's not too off-the-wall is the correlation between ownership of firearms and number of homicides by guns. Methodologies vary, but there appears to be a consistent correlation of roughly 0.30 between number of guns owned in a governmental unit (a municipality, a state, a country) and the number of homicides by gun.

If causality is involved in that relationship, which is the cause and which is the effect? Does the mere presence of more guns cause more homicides? Or do people who live in higher crime areas respond by buying more guns for protection?

Those questions aren't going to be answered here. In fact, they're not going to be answered anywhere by means of a correlational study. A correlation by itself cannot demonstrate either causation or, if present, the direction of the causation. The experimental approach would involve randomly selecting a large number of localities and then randomly assigning them to one of two groups. In one group, residents would be required to purchase more guns. In the other group, residents would be prohibited from purchasing more guns.

After some period of time, the gun-related homicides in each location would be counted and adjusted by covarying on number of guns owned prior to the experiment. If no meaningful difference in number of homicides existed between locales forced to acquire more guns and those prohibited from doing so, then you would have to reject the hypothesis that more guns cause more homicides.

Clearly, such an experiment could never take place, for reasons ranging from ethics to statutes to feasibility. Therefore, many researchers fall back on correlational studies as substitutes for true experimental designs. I frequently see reports of such research that state suggestively that links have been found between cell phone usage and cancer, vaccines and autism, distance from a residence to a high-tension power line and cancer (again), and so on. One characteristic of such studies tends to be the self reporting of the behaviors being studied (for example, frequency of cell phone usage). Another is the reliance on correlation coefficients (and their apparent statistical significance) to impute causality.

We can and should use correlation analysis to help us find promising questions to investigate via stronger experimental designs. If a correlation exists, there's reason to suspect that causation might be involved, perhaps even as a cause of the observed correlation. But the mere presence of a correlation is not in and of itself evidence of causality.

If reasons of cost, law, or research ethics prevent us from carrying out a true experiment, about all we can do is continue to observe the events in question and assume that the issue will clear up after sufficient time has passed. In the 1950s and 1960s, smoking was considered a possible cause of lung and oral cancers. The tobacco industry pointed out that no true experiment had ever conclusively demonstrated a causal relationship between smoking and cancer. It also pointed out that the various correlational studies couldn't fix the blame on tobacco: Correlation is not causation.

However, enough years eventually passed, with an accrual of more correlational studies, to make it impossible for anyone to seriously deny that the use of tobacco causes cancer. Evidently, although correlation does not necessarily mean causation, enough correlation can be pretty convincing.

Between 1930 and 1936 a spurt in population growth occurred in Oldenburg, Germany. There also occurred a concomitant jump in the number of reported stork sightings. The explanation, probably facetious, given at the time was that storks had come to Oldenburg and brought more babies with them. Of course, if the reported numbers are to be believed, the direction of cause ran the other way. A larger population has the opportunity to report more stork sightings.

A Third Variable

Sometimes a third variable, one other than the two involved with the correlation in question, is causally related to both variables in the correlation. For example, up until the mid-1990s, it was thought that no reliable relationship existed between the size of a police force relative to the population it served, and the crime rate in that population. Since that time, more sophisticated research designs and more robust statistical analysis has indicated that a relationship does in fact exist.

However, the correlations reported in the more recent studies do not necessarily mean that the relationship is in fact *causal*. A third variable, the socioeconomic level of the communities in question, tends to exert an impact on both the size of the police force and the crime rate. Wealthier communities tend to enjoy greater tax revenues, and have more dollars to spend on total salaries for their police forces. Those communities also tend to experience less crime, particularly crimes of violence, against persons and against property.

These studies suggest that to simply increase the size of a police force does not result in lower crime rates. The relationships among the variety of variables are complex and difficult to parse, but certainly it's not implausible that the wealth of a community impacts both the size of its police force and its crime rate.

Restriction of Range

An additional point to watch for is the loss of the full range of values in one of the variables.

Since the 1990s at least, standardized tests such as the SAT have been criticized for a variety of reasons, some pretty good, some less so. One of the critiques of college entrance exams has been that they do a poor job of predicting actual post-secondary scholastic performance.

If you were to take a sample of 50 college students and analyze their college grade point averages along with their SAT scores, you might get a result such as the one shown in Figure 2.19. Although lower SAT scores are certainly possible, the fact that these 50 students are in college means that in this fictitious sample their SAT scores have a floor of about 1100.

Figure 2.19
A correlation of 0.40 does not indicate a strong relationship.

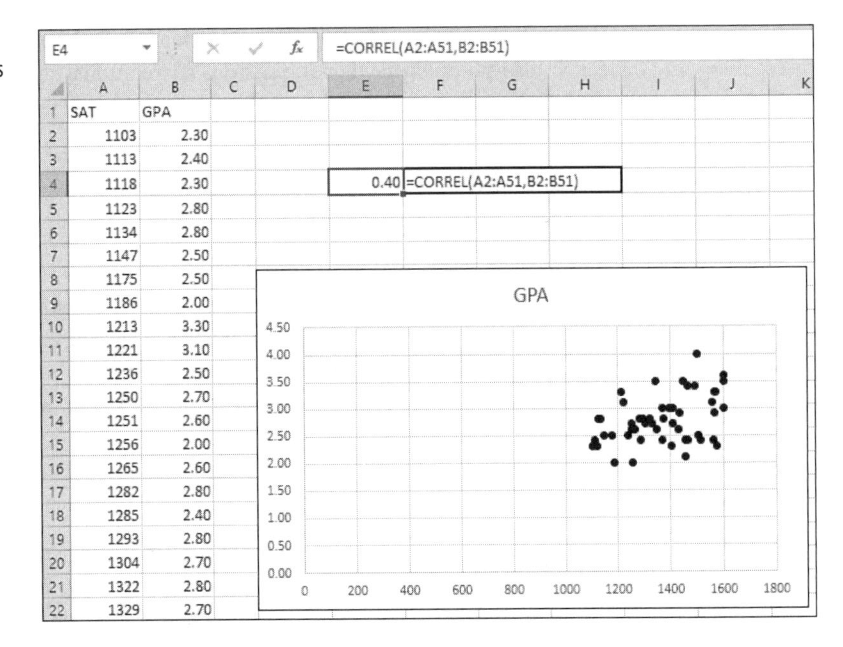

SAT	GPA
1103	2.30
1113	2.40
1118	2.30
1123	2.80
1134	2.80
1147	2.50
1175	2.50
1186	2.00
1213	3.30
1221	3.10
1236	2.50
1250	2.70
1251	2.60
1256	2.00
1265	2.60
1282	2.80
1285	2.40
1293	2.80
1304	2.70
1322	2.80
1329	2.70

The charted data points create what's basically a random scatter, but they do suggest a mild correlation between GPA and SAT score, echoed by the correlation coefficient of 0.40. If this were all you had to work with, you might well conclude that the SAT fails to do a credible job of predicting success in college.

Now suppose that you also had access to data on another 50 students who have dropped out of college before graduation because of low GPAs. Because they did not graduate, their data did not make it into the analysis in Figure 2.19. See Figure 2.20.

The pattern of the data markers on the chart in Figure 2.20 is similar to that in Figure 2.19. The range of both the SAT and GPA scores is lower in Figure 2.20. Again, though, there is little evidence that the SAT predicts performance in college.

But what if you put all 100 pairs of scores onto the same chart? That's been done in Figure 2.21.

In Figures 2.19 and 2.20, the restriction placed on the range of both variables, GPA and SAT score, artificially reduces the correlation, the apparent strength of relationship, between the variables. Although it might be correct to say that the correlation between SAT score and college GPA is rather weak *among those students who remain in college*, that's very different from saying that the correlation is rather weak in general—with the implication that the test is a poor tool for selection purposes. For good or ill, the SAT has been used to allocate college enrollment, a scarce resource. It's disingenuous to imply that the correlation is weak when the values that form the basis of the correlation have been, in effect, cherry-picked.

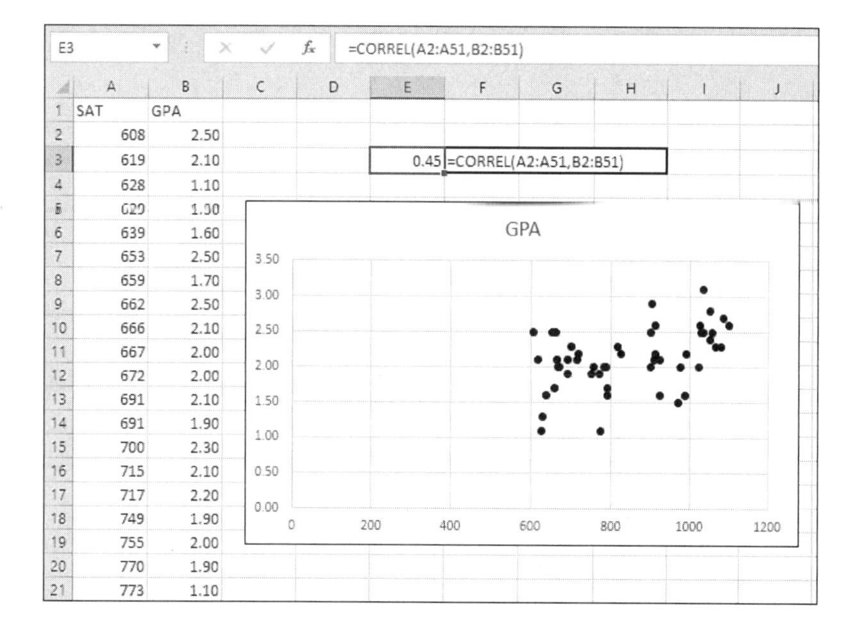

Figure 2.20
A correlation of 0.45 is just marginally stronger than the 0.40 shown in Figure 2.19.

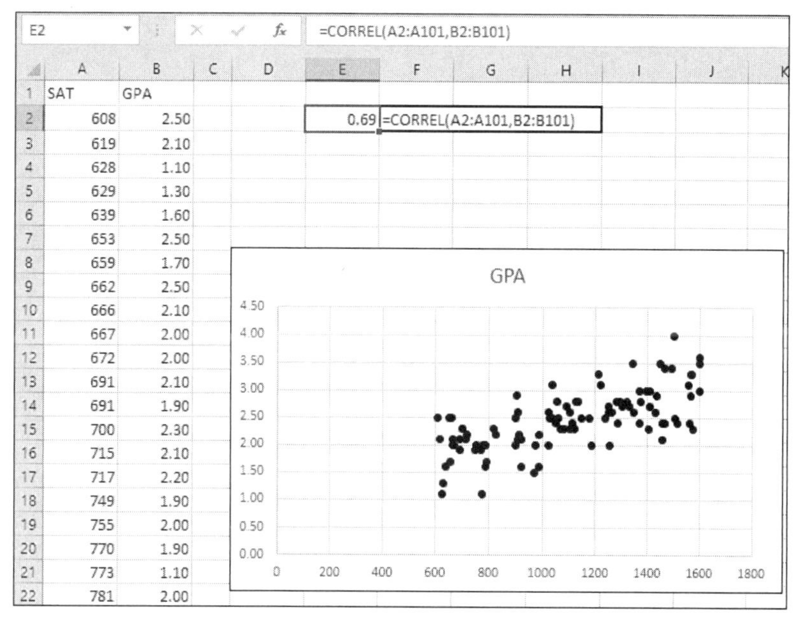

Figure 2.21
With the restrictions removed from the ranges, the correlation rises to nearly 0.70.

Chapter 1 discusses concepts and techniques that pertain to variance. This chapter has covered issues surrounding *co*variance—how two variables move in concert, or fail to do so. Chapter 3, "Simple Regression" on the subject of simple two-variable regression, combines the topics from the first two chapters to show how you can use them to understand more about the data you work with regularly. You'll also see how variability and correlation merge to form the foundations of more advanced techniques such as multiple regression, factorial analysis of variance, and the analysis of covariance.

Simple Regression

3

IN THIS CHAPTER

Predicting with Correlation and Standard Scores 60

Predicting with Regression Coefficient and Intercept 65

Shared Variance 71

The TREND( ) Function 82

Partial and Semipartial Correlations 90

As I write this chapter, a program at the University of Virginia, called the Reproducibility Project, is getting considerable publicity for its work. That work consists of replicating published experiments to determine whether the findings stand up when the studies are repeated. The project expends considerable effort to ensure that the conditions that were in place when the experiment was originally conducted are maintained when it's replicated. Even so, the findings in around 60% of the experiments are substantially weaker when repeated than they were as originally reported in refereed publications.

The popular media, such as *The New York Times*, report that the differences between the initial and the subsequent experimental findings appear to be largely due to issues of control: The experimenter's control of the subjects' behavior, the integrity of the differences between the experimental and the comparison treatments, and control of the underlying experimental design.

We'll have to wait and see the project's final report, if one is planned, to get information that's better articulated. Still, it's curious that one method of control, statistical control, wasn't mentioned as a source of the discrepant findings. Statistical control can be a powerful tool. It can never fully make up for shortcomings in a weak experimental design. A strong design with random selection and assignment, possibly including double-blind treatments, intelligently prepared and carefully executed, almost always results in a more satisfactory experiment and more credible findings than any statistical wizardry applied to the results of a grab-sample.

Nevertheless, the statistical control tools are available and useful if you use them well on data collected with due care. I'll start to explore them toward the end of this chapter. A fair amount of groundwork needs to be laid, so it's a slow build. But much of that groundwork has broad applicability, so I hope you'll stay with me. The next section starts the build, using the concepts developed in this book's first two chapters.

Predicting with Correlation and Standard Scores

When two variables have a moderate to strong quantitative relationship to one another—when their correlation is greater than .5, say—then it's possible to predict, with some accuracy, an unknown value of one variable from a known value of the other variable. This simple concept is the foundation of most of the interesting and useful applications of regression analysis.

You can study the process of prediction using concepts discussed in Chapter 1, "Measuring Variation: How Values Differ" and Chapter 2, "Correlation," correlation and variability. (If you're uncomfortable with the term *prediction*, you might think of it in terms of *estimation* instead.) At prediction's most basic level, you can forecast one variable from another by working with their correlation and the values as z-scores. Figure 3.1 starts a simple example.

In Figure 3.1 you see a random sample of ten houses, represented by their square footage and recent sales prices. Notice that the calculated correlation is .67. Because the correlation coefficient is positive, you can tell that the greater the measure of square feet, the costlier the house. It's a subjective assessment, but I would term it a moderately strong correlation. The accompanying chart shows no anomalies such as extreme outliers or a departure from a straight-line relationship, which might cause you to suspect that the Pearson correlation is the wrong measure.

Figure 3.2 extends the analysis to the prediction of home prices using the calculated correlation and the sales prices, converted to standard scores.

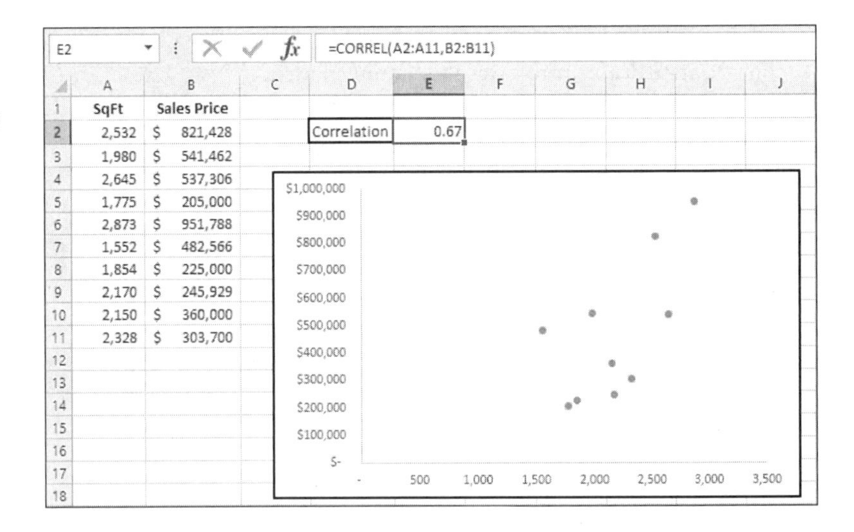

SqFt	Sales Price
2,532	$ 821,428
1,980	$ 541,462
2,645	$ 537,306
1,775	$ 205,000
2,873	$ 951,788
1,552	$ 482,566
1,854	$ 225,000
2,170	$ 245,929
2,150	$ 360,000
2,328	$ 303,700

Figure 3.1
Home prices are usually associated with the building's area in square feet.

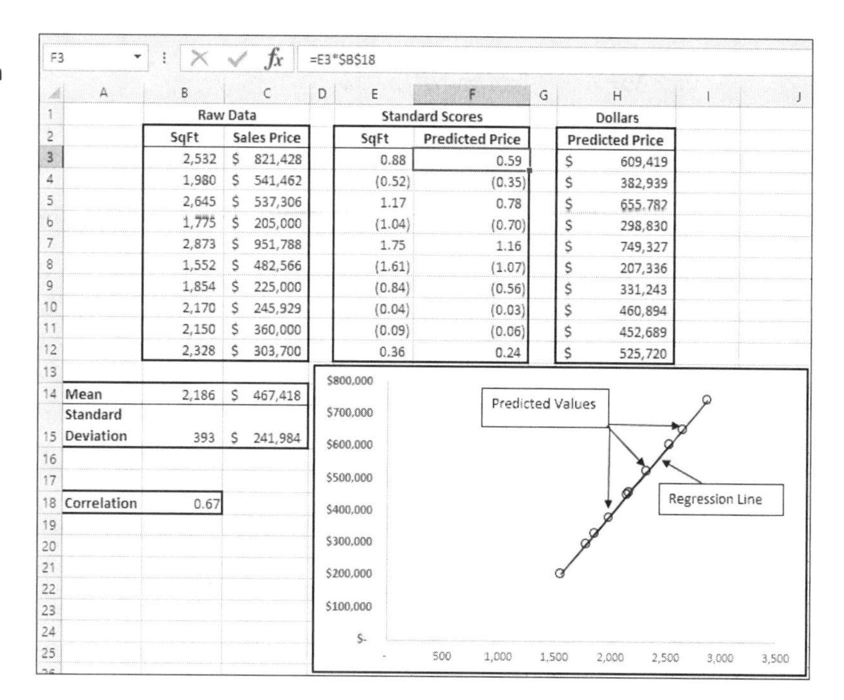

	A	B	C	D	E	F	G	H
1		Raw Data			Standard Scores			Dollars
2		SqFt	Sales Price		SqFt	Predicted Price		Predicted Price
3		2,532	$ 821,428		0.88	0.59		$ 609,419
4		1,980	$ 541,462		(0.52)	(0.35)		$ 382,939
5		2,645	$ 537,306		1.17	0.78		$ 655,782
6		1,775	$ 205,000		(1.04)	(0.70)		$ 298,830
7		2,873	$ 951,788		1.75	1.16		$ 749,327
8		1,552	$ 482,566		(1.61)	(1.07)		$ 207,336
9		1,854	$ 225,000		(0.84)	(0.56)		$ 331,243
10		2,170	$ 245,929		(0.04)	(0.03)		$ 460,894
11		2,150	$ 360,000		(0.09)	(0.06)		$ 452,689
12		2,328	$ 303,700		0.36	0.24		$ 525,720
14	Mean	2,186	$ 467,418					
15	Standard Deviation	393	$ 241,984					
18	Correlation	0.67						

Figure 3.2
The predicted prices form a straight line.

You could use any one of various ways to arrive at predicted sales prices given the data in Figures 3.1 and 3.2, and you would almost certainly *not* use the one shown in Figure 3.2 and described in the next section. However, it's worth looking at, once anyway, because no other method shows so clearly the relationship between correlation and prediction.

NOTE: The charted line representing the predicted sales prices in Figure 3.2 is termed a *regression line*. Excel uses the term *trendline* instead, but the two terms are synonymous.

Calculating the Predictions

In Figure 3.2, notice that the mean and the standard deviation of both square feet and sales price are calculated in the range B14:C15. For example, cell B14 contains this formula and uses the AVERAGE() function:

=AVERAGE(B3:B12)

This formula is in cell C15:

=STDEV.P(C3:C12)

Note that we are treating the observations in rows 3 through 12 as the population of interest, so we use the STDEV.P() function. If instead we used the STDEV.S() function, we would be treating the observations as a sample from which to estimate the standard deviation in a larger population. For more on this matter, see "Population and Sample Forms of the Variability Functions" in Chapter 1.

With the mean and the standard deviation of both square footage and sales price in hand, you can convert to and from z-scores. The first step takes place in the range E3:E12. The formula in cell E3 is:

=(B3–$B$14)/$B$15

In words, subtract the mean in cell B14 from the observation in cell B3. Then divide the difference by the standard deviation in cell B15. Because the formula places a dollar sign before both the row and the column of cells B14 and B15, you can copy and paste the formula from E3 into E4:E12. The reference to cell B3 adjusts to B4, B5, and so on, but the references to B14 and B15 remain constant as you copy and paste the formula down through cell E12.

Now you're ready to predict the sales price on the basis of square footage. Simply multiply the calculated correlation (in cell C18) by the z-score for square footage. In cell F3, the formula is:

=E3*$B$18

Copy and paste that formula into the range F4:F12 to get the remaining predictions. The formula results in another z-score: Each formula in F3:F14 is the z-score for sales price as estimated by the z-score for square footage, multiplied by the correlation between square footage and sales price. The next step will be to convert the predicted z-scores for sales price to dollars, but first note the following aspects of the predictions in F3:F14.

Correlation and Z-score Determine the Predicted Value

You multiply the z-score for square footage by the correlation to predict the z-score for sales price. The z-score for square footage expresses the number of standard deviations a given measure (such as 2,532 square feet in cell B3) falls from the mean: 2,532 square feet is .88 standard deviations above the mean square footage (2,186 in cell B14).

The correlation cannot be less than –1.0, nor greater than 1.0. Furthermore, the mean of a set of z-scores is always 0.0. Therefore, correlation limits how far the predicted z-score is from 0.0. If the z-score for square footage is 1.0 and the correlation is .67, then the predicted z-score is also .67. If the z-score for square footage is 2.0, a correlation of .67 results in a sales price z-score of 1.34.

Regression Toward the Mean

A correlation that's greater than –1.0 and less than 1.0 always pulls the predicted value back toward the mean of 0.0. The closer the correlation is to 0.0, the closer the predicted z-score is to 0.0.

Suppose a property's z-score for square footage is 1.0. If the correlation between square footage and sales price is .67, then the predicted sales price is 1.0 * .67, or .67. That's two-thirds of a standard deviation above the mean. But if the correlation is only .50, then the predicted sales price is 1.0 * .50, or .50. That's just half a standard deviation above the mean. Whenever the correlation is less than perfect, when it's greater than –1.0 and less than 1.0, the predicted score *must* be closer to its mean than the predictor score is to its mean.

This phenomenon is termed *regression toward the mean*, or simply *regression*. The term *regression* has come to encompass a variety of statistical techniques, as well as observed phenomena, that go well beyond simply predicting one variable from another by means of a correlation coefficient.

Returning to the Original Metric

At this point you have predictions of sales prices, but they are measured as z-scores rather than as dollars. Recall that to go from a measure such as dollars to z-scores, you subtract the mean from an observation and divide by the standard deviation. To go from z-scores to dollars, you reverse the process: Multiply by the standard deviation and add the mean.

Calculating the Predicted Values

In Figure 3.2 that process takes place in the range H3:H12. The formula in cell H3 is:

```
=F3*$C$15+$C$14
```

That is, multiply the predicted score for sales price in cell F3 by the standard deviation of sales prices in cell C15. Then add the mean sales price from cell C14. As usual, the dollar signs in $C$14 and $C$15 fix the references to those cells, so you can copy and paste the formula down through H12 without losing track of the mean and standard deviation.

Charting the Regression

It's easy, and often helpful, to chart the predicted values such as those in Figure 3.2, cells H3:H12. Here's one way:

1. Make a multiple selection of the predictor values in B3:B12 and the predicted values in H3:H12. That is, select B3:B12, hold down the Ctrl key, and select H3:H12.
2. Click the Ribbon's Insert tab.
3. In the Charts group, click the Scatter (XY) control.
4. Choose the Scatter with Straight Lines subtype.

This sequence of steps will get you the regression line itself and will confirm the location of the predictions with respect to the regression line. In regression problems, the predicted values always lie directly on the regression line—indeed, they define it.

> **TIP**
> It's customary in these charts to show the predicted variable on the chart's vertical (Y axis) and the predictor variable on the horizontal (X axis). To arrange for that, before you call for the chart, structure your worksheet so that the X values are in a column to the *left* of the Y values. It makes no difference whether you start by selecting the predictor or the predicted variable.

Generalizing the Predictions

Now, what's the point of all this? We have the actual square footage and the sales price of ten houses. Why predict their sales prices when we already know them?

There are quite a few reasons, although it's true that predicting the specific sales prices of these specific houses is not the principal reason. (However, the *differences* between the actual and the predicted sales prices is of major importance to your understanding of the relationship between the two variables.) The clearest reason to run the analyses that Figure 3.2 summarizes is to predict the sales price of *another* house, one that's not involved in establishing the regression equation. See Figure 3.3.

The data and calculations in columns A through C in Figure 3.3 are the same as in Figure 3.2. But suppose that there's another house whose square footage you know, but whose sales price has yet to be established. It's fairly straightforward to apply the knowledge you derive from the original ten houses to the new one. That's done in the range F3:F6 of Figure 3.3.

Figure 3.3
Here you apply the existing information to a new case.

C18 =CORREL(B3:B12,C3:C12)

	A	B	C	D	E	F	G
1		Raw Data					
2		SqFt	Sales Price		New House		
3		2,532	$ 821,428		SqFt	1,400	
4		1,980	$ 541,462		z-score (SqFt)	(2.00)	=(F3-$B$14)/$B$15
5		2,645	$ 537,306		Predicted z-score	(1.33)	=F4*$C$18
6		1,775	$ 205,000		Predicted sales price	$144,972	=F5*C15+C14
7		2,873	$ 951,788				
8		1,552	$ 482,566				
9		1,854	$ 225,000				
10		2,170	$ 245,929				
11		2,150	$ 360,000				
12		2,328	$ 303,700				
13							
14	Mean	2,186	$ 467,418				
15	Standard Deviation	393	$ 241,984				
16							
17							
18		Correlation	0.67				

Cell F3 contains the area, 1,400 square feet, of the new house. The remaining calculations in F4:F6 predict the house's sales price (the formulas in F4:F6 appear in G4:G6). The process is as follows:

1. Cell F4: Convert the 1,400 square feet in cell F3 to a z-score. Subtract the mean square footage in cell B14 from the value in cell F3, and divide the difference by the standard deviation in cell B15.
2. Cell F5: Get the predicted z-score for sales price by multiplying the area z-score in cell F4 by the correlation in cell C18.
3. Cell F6: Convert the predicted z-score for sales price to dollars. Multiply the z-score in cell F5 by the standard deviation of sales dollars in cell C15, and add the mean sales dollars in cell C14.

Excel offers several simpler ways to get to the same result. I'll start discussing those next, on the assumption that you would rather calculate predictions such as those made in this section with less effort. However, understanding how the simpler methods work depends on understanding how the interplay of correlations, standard deviations, and mean values results in a quantitative prediction. You'll probably never again derive a prediction in the way just described, but keep those relationships in mind.

Predicting with Regression Coefficient and Intercept

Much of this book is concerned with a form of regression analysis that's termed *multiple regression*. Using multiple regression, you have more than just one predictor variable, and exactly one predicted variable. (Yet another form of correlation analysis, *canonical correlation*, allows for both multiple predictor as well as multiple predicted variables.)

Excel offers a variety of worksheet functions that are designed for use in multiple regression. I'll discuss the use of those functions repeatedly throughout the remaining chapters. Although most of these functions are able to handle multiple predictor variables, they can also be used to deal with situations that have one predictor only—such as the prior section's example regarding the square footage and the sale prices of housing.

However, Excel also has some worksheet functions designed specifically for the single-predictor situation. Because you deal with multiple predictors much more frequently than with single predictors, you are likely to have much less use for the single-predictor functions than for their multiple-predictor counterparts.

Still, the single-predictor functions provide a useful introduction to the multiple-predictor functions. Counterintuitively, the single-predictors can be valuable in sussing out some multiple regression models. So let's take a closer look at them in this section.

The SLOPE( ) Function

Think back to your middle school algebra or geometry class. While you were learning about Cartesian coordinates, your teacher probably discussed the mathematics behind straight lines on charts. In particular, you might have heard the line's slope defined as "the rise over the run." Your teacher continued:

"Pick any two points on a straight line. Count how many units up (or down) you have to go to get from one point to the other. That's the rise. Count how many units left (or right) to get from one to the other. That's the run. The slope is the rise over the run. Divide the rise by the run."

Excel's SLOPE() function does all the counting and dividing for you. See Figure 3.4.

Figure 3.4
The SLOPE( ) function tells you how steep a straight line is.

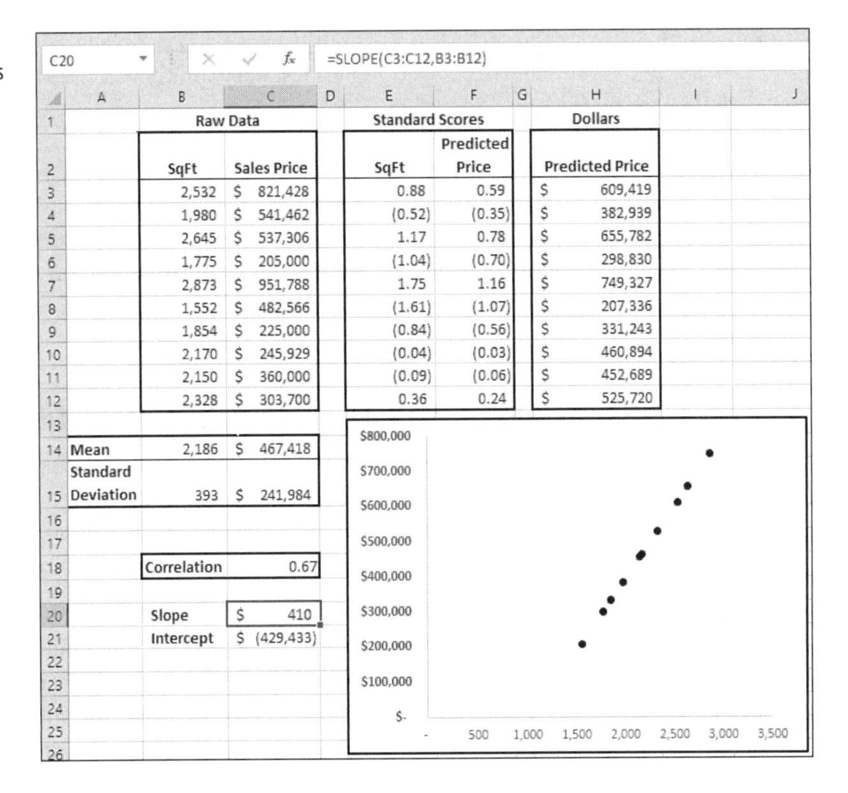

C20 =SLOPE(C3:C12,B3:B12)

	A	B	C	E	F	H
1		Raw Data		Standard Scores		Dollars
2		SqFt	Sales Price	SqFt	Predicted Price	Predicted Price
3		2,532	$ 821,428	0.88	0.59	$ 609,419
4		1,980	$ 541,462	(0.52)	(0.35)	$ 382,939
5		2,645	$ 537,306	1.17	0.78	$ 655,782
6		1,775	$ 205,000	(1.04)	(0.70)	$ 298,830
7		2,873	$ 951,788	1.75	1.16	$ 749,327
8		1,552	$ 482,566	(1.61)	(1.07)	$ 207,336
9		1,854	$ 225,000	(0.84)	(0.56)	$ 331,243
10		2,170	$ 245,929	(0.04)	(0.03)	$ 460,894
11		2,150	$ 360,000	(0.09)	(0.06)	$ 452,689
12		2,328	$ 303,700	0.36	0.24	$ 525,720
14	Mean	2,186	$ 467,418			
15	Standard Deviation	393	$ 241,984			
18		Correlation	0.67			
20		Slope	$ 410			
21		Intercept	$ (429,433)			

Figure 3.4 has the same data as you find in Figures 3.1 through 3.3. The SLOPE() function appears in cell C20. Here's how it's used there:

=SLOPE(C3:C12,B3:B12)

Excel terms the first argument in SLOPE() the *known y's*, so the y-values in this case are the sales prices found in C3:C12. They represent the variable that you want to predict. The second argument to SLOPE() is the *known x's*—the variable you want to predict from—and here it's represented by the values in the range B3:B12. Generally, then, the syntax for the SLOPE() function is as follows:

=SLOPE(*known y's*, *known x's*)

Be sure to keep that syntax in mind as you use SLOPE() or any of the other worksheet functions that are used to predict results. They are conceptually and computationally different from CORREL(). For example, the correlation between square footage and sales price is the same as the correlation between sales price and square footage. The order in which you supply the arguments to CORREL() doesn't matter. However, the order matters in SLOPE() and other prediction functions. I'll mention this issue, when it matters, as we cover other functions in this family.

The slope, the correlation, the standard deviations of the variables, and the covariance are closely related to one another. Figure 3.5 shows how they work together.

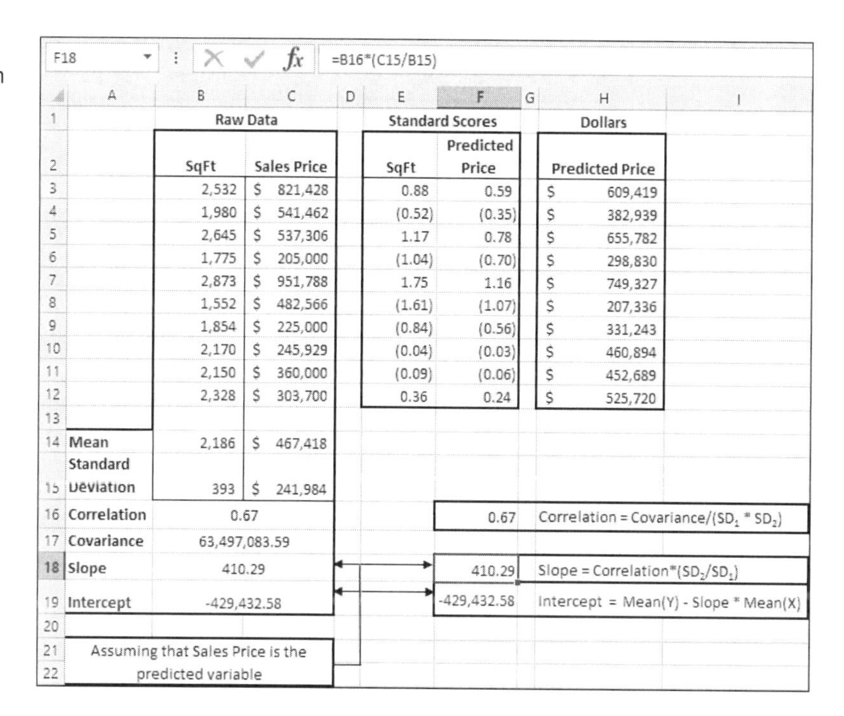

F18 =B16*(C15/B15)

	A	B	C	D	E	F	G	H
1		Raw Data			Standard Scores			Dollars
2		SqFt	Sales Price		SqFt	Predicted Price		Predicted Price
3		2,532	$ 821,428		0.88	0.59		$ 609,419
4		1,980	$ 541,462		(0.52)	(0.35)		$ 382,939
5		2,645	$ 537,306		1.17	0.78		$ 655,782
6		1,775	$ 205,000		(1.04)	(0.70)		$ 298,830
7		2,873	$ 951,788		1.75	1.16		$ 749,327
8		1,552	$ 482,566		(1.61)	(1.07)		$ 207,336
9		1,854	$ 225,000		(0.84)	(0.56)		$ 331,243
10		2,170	$ 245,929		(0.04)	(0.03)		$ 460,894
11		2,150	$ 360,000		(0.09)	(0.06)		$ 452,689
12		2,328	$ 303,700		0.36	0.24		$ 525,720
13								
14	Mean	2,186	$ 467,418					
15	Standard Deviation	393	$ 241,984					
16	Correlation	0.67				0.67	Correlation = Covariance/(SD_1 * SD_2)	
17	Covariance	63,497,083.59						
18	Slope	410.29				410.29	Slope = Correlation*(SD_2/SD_1)	
19	Intercept	-429,432.58				-429,432.58	Intercept = Mean(Y) - Slope * Mean(X)	
20								
21	Assuming that Sales Price is the							
22	predicted variable							

Figure 3.5 The relationship between the standard deviations converts the covariance to the correlation and to the slope.

The covariance is one measure of the degree to which two variables vary jointly, or *covary*. This book looks into the covariance more deeply in Chapter 2, where I note that the size of the covariance is largely dependent on the scale of measurement used by each of the two variables. Suppose that you were looking into the relationship between the annual salaries of two groups of people. The covariance could easily turn out in the hundreds of millions of dollars, as shown in Figure 3.6.

In Figure 3.6, the range B3:C12 contains the annual salaries of two groups of paired employees. The covariance for the two groups, in cell C14, is more than three quarters of a billion dollars. The range E3:F12 contains the number of inches that ten pairs of siblings, each consisting of a brother and a sister, grew between their twelfth and thirteenth birthdays. The covariance is less than a quarter of an inch. Despite the huge discrepancy in the covariances, the correlations in C16 and F16 are precisely equal.

Figure 3.6
The covariances are far apart but the correlations are identical.

C14 =COVARIANCE.P(B3:B12,C3:C12)

	A	B	C	D	E	F
1		Salary			Annual Growth (inches)	
2		Group 1	Group 2		Brother	Sister
3	Pair 1	$ 185,378	$ 162,528		2.78	2.45
4	Pair 2	$ 133,811	$ 164,473		2.01	2.47
5	Pair 3	$ 93,476	$ 144,774		1.40	2.17
6	Pair 4	$ 75,376	$ 138,832		1.13	2.08
7	Pair 5	$ 54,241	$ 119,607		0.81	1.79
8	Pair 6	$ 85,177	$ 106,747		1.28	1.60
9	Pair 7	$ 177,032	$ 95,360		2.66	1.43
10	Pair 8	$ 153,229	$ 118,930		2.30	1.78
11	Pair 9	$ 185,595	$ 194,457		2.78	2.92
12	Pair 10	$ 47,932	$ 96,379		0.72	1.45
13						
14	Covariance		765,437,033.31			0.173
15						
16	Correlation		0.479			0.479
17						
18	Standard deviation	51,519	31,019		0.8	0.47

The fact that the salary covariance is so different from the height covariance is due to the different measurement scales. The standard deviation of the salaries is in the tens of thousands, but for the heights the standard deviation is less than 0.5. If the scales are familiar and meaningful, as they often are in engineering, physics, and biology, the covariance itself can be meaningful. However, if the scales aren't constant—and the value of any unit of currency changes constantly—then the covariance is probably not as informative as you might want.

In a case of that sort, the correlation is a more useful measure of the strength of the relationship between two variables than the covariance. If you divide the covariance by the product of the standard deviations of the two variables, you get the correlation. The act of dividing by the standard deviations removes the influence of the scales of measurement and you're left with an estimate of the strength of the relationship, unencumbered by the effect of the original measures. Refer back to Figure 3.5, cell F16, to see how the covariance is converted to the correlation in the context of the example for square footage and sales prices. Note that Excel's CORREL() function returns precisely the same value in cell B16.

But when it's time to estimate the slope of the line that shows the relationship of two variables, the scale of measurement has to be re-introduced. In Figure 3.5, you'll find the slope calculated in cell F18, where the correlation is multiplied by the ratio of one variable's standard deviation to the other's. Cell F18 multiplies the correlation by the ratio of SD_2 to SD_1: the standard deviation of the second, predicted variable divided by the standard deviation of the first, predictor variable.

Back to the rise over the run: The ratio tells you how many units of rise are associated with one unit of run. Put differently, it tells you how many standard deviations of the predicted variable are associated with one standard deviation of the predictor variable.

Multiply that ratio by the correlation—which carries the information about the strength of the relationship and about its direction—and you have the slope. The SLOPE() function does not by itself illuminate the relationship between the slope and the correlation, but it's handy.

Note, by the way, that the covariance divided by the variance of one of the variables results in the slope used to predict the other variable. It's a useful exercise to derive that result from the equation that returns the slope from the correlation and the standard deviations.

The INTERCEPT( ) Function

The *intercept* is the point where the regression line intercepts the chart's vertical axis. See Figure 3.7.

The regression line's slope tells you how far the line moves on the vertical axis, given its movement on the horizontal axis. The regression line's intercept tells you the elevation above (or below) the zero-point on the vertical axis.

So, in Figure 3.7, the chart plots the brothers' actual values against the sisters' actual values. Each plotted point represents the intersection of a brother's value with his sister's value. The plotted points are accompanied by the regression line (or, as Excel terms it, the trendline). You can see that the charted regression line crosses the vertical axis at 1.5. That's the intercept, often called the *constant*. In a simple two-variable regression, the slope and the intercept together are sufficient to draw the regression line.

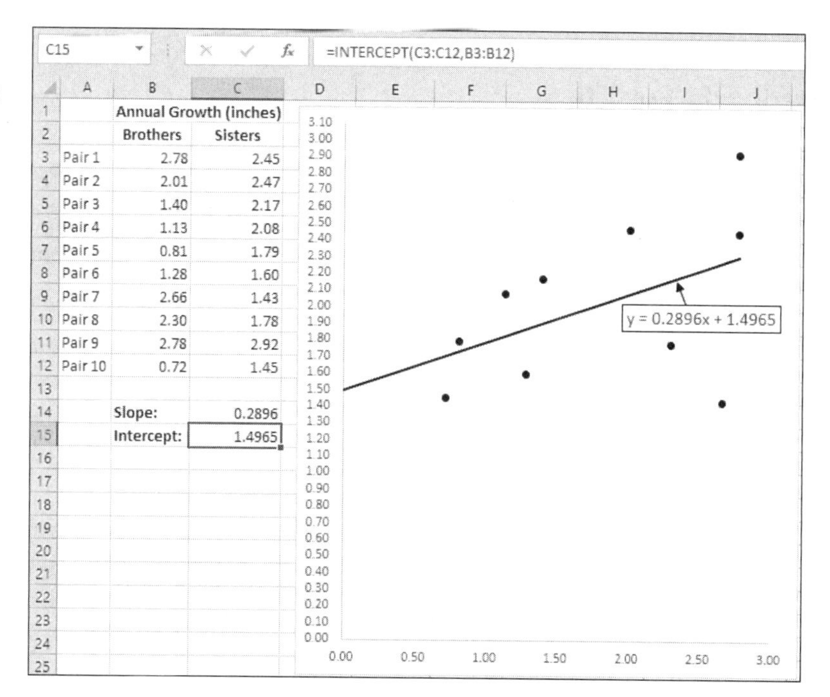

Figure 3.7
The regression line crosses the vertical axis at the *intercept*.

Generally, the equation for the intercept in the two-variable case is as follows:

$$Intercept = \overline{Y} - Slope * \overline{X}$$

where:

- $\overline{Y}$ is the mean of the predicted variable (in this example, that's the variable labeled *Sisters* in Figure 3.7).
- $\overline{X}$ is the mean of the predictor variable (in this example, that's the variable labeled *Brothers*).
- *Slope* is, of course, the slope that is calculated as described in the prior section.

Figure 3.7 also shows Excel's INTERCEPT() function in cell C15. Its arguments are identical to those for the SLOPE() function, given in the prior section:

=INTERCEPT(known y's, known x's)

or, as used in cell C15:

=INTERCEPT(C3:C12,B3:B12)

NOTE

It's probably clear to you, but just to be explicit I might as well point out that there must be as many known x values as there are known y values in the arguments to both SLOPE( ) and INTERCEPT( ). Also, as a practical matter you'll normally find that the records in the two ranges of values pair up in some way. In the present example, the first known x belongs to a brother, and the first known y belongs to his sister.

Charting the Predictions

In the case of simple regression—that is, one predictor and one predicted variable—Excel makes it very easy to chart the values that your regression equation predicts.

One method is simply to apply the result of the SLOPE() function and the INTERCEPT() function to the values of your predictor variable. In Figure 3.7, for example, you could enter this formula in cell D3:

=$C$14*B3+$C$15

Then copy and paste that formula down into D4:D12. Include the range D3:D12 along with the observed values in B3:B12 and C3:C12 when you create an XY chart. The plotted predictions will all line up directly where a trendline would be.

NOTE

I recommend an XY chart—also known as an XY (Scatter) chart—because other than bubble charts it's the only chart type in Excel that has value axes as both its horizontal and its vertical axis. All other two-axis chart types have a category axis. For bivariate analysis of the sort discussed in this book, you need two value axes.

If you don't care about getting the actual predictions plotted on your chart as distinct points, you can get a trendline as shown in Figure 3.7 even more easily. Just create an XY chart of the actual observations. Open the chart and right-click on one of the charted points. Choose Add Trendline from the shortcut menu and specify a Linear trendline. You'll get a trendline that shows where the predicted values would fall if you plotted them explicitly.

Shared Variance

Let's take a closer look at what's going on inside the equations that the previous section discusses. We're dealing with the quantitative relationship between two sets of values: a predictor variable and a predicted variable. They're called *variables*, of course, because their values vary. Two well-known ways to measure how much variability exists in a set of values are the standard deviation and the variance.

Both Chapter 1 and Chapter 2 discuss the standard deviation and the variance. But before introducing the notion of shared variance, it's useful to revisit these two measures of variability in the context of regression.

The Standard Deviation, Reviewed

You were probably familiar with the notion of a standard deviation before you opened this book. Its main function is to give you a sense of how far individual values stray from an average value. In a symmetric distribution such as the one shown in Figure 3.8, you can reasonably expect that 68% of the values will fall between one standard deviation below the mean and one standard deviation above the mean. Another 28% fall within two standard deviations from the mean, and another 4% are found more than two standard deviations from the mean.

In the late 1970s, according to tables prepared by the National Center for Health Statistics, the mean height of adult American males was 69.1 inches, with a standard deviation of 2.8 inches. From what we know about how standard deviations behave in a normal distribution, we can create the chart in Figure 3.8 using Excel.

The standard deviation provides a handy yardstick. Suppose that in the late 1970s your height was 71.9 inches. If someone had told you then that the mean height among males was 69.1 inches, and the standard deviation was 2.8 inches, you would know very quickly that you were taller than 84% of adult men.

In that example, the standard deviation expressed how men's heights vary. Using the standard deviation in that *sort* of way—to express the amount of variation in a set of raw numbers—is doubtless the most typical reason to calculate a standard deviation. But there are many other quantities that vary in interesting and useful ways. Those other quantities are usually data that has been processed in some way, and then the measure of variation is usually termed a *standard error* rather than a standard deviation.

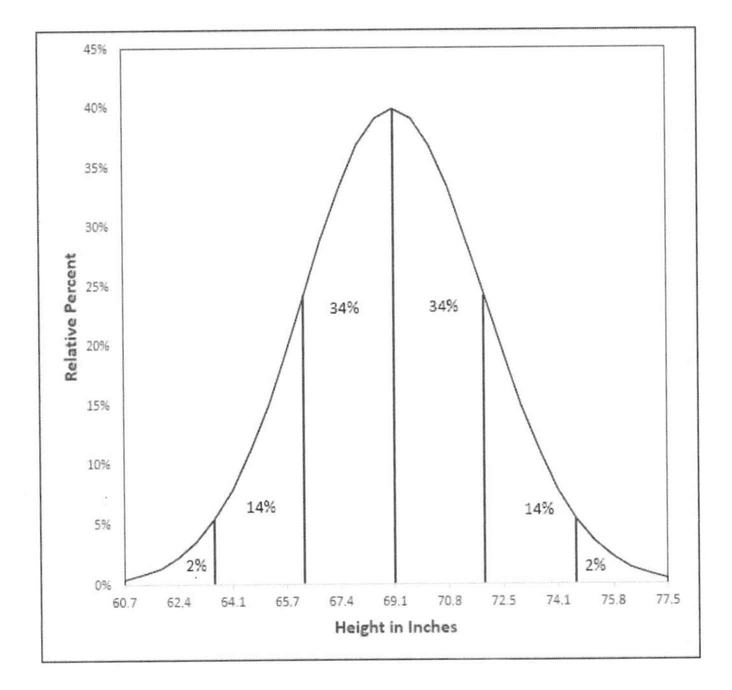

Figure 3.8
This is how the standard deviation divides up values that follow a normal distribution.

The term *error* in this context doesn't imply that anyone has done something wrong. It just refers to the fact that we frequently need to estimate a value that belongs to a population with a fallible sample estimate. For example:

- The *standard error of the mean* is the standard deviation of the means of two or more groups. You might have three groups of subjects who differ in some way—perhaps each group receives a different treatment. You calculate the mean of each group and find the standard deviation of those means. This procedure is often used to test whether any differences in the means of the groups are reliable and can be expected to recur if you repeat the experiment. If the treatments have no differential effects, then for your purposes the subjects all came from the same population with just one mean, and all three sample means are (probably erroneous) estimates of that population value.
- The *standard error of estimate* is the standard deviation of the errors made in estimating, via regression, values of a predicted variable from values in a predictor variable.
- A close conceptual relative of the standard error of estimate is the *standard error of measurement*, which is the standard deviation of the differences between an attribute as measured by some instrument, and the person or thing's "true score" on that attribute. The true score cannot be measured directly, and so the standard error of measurement must be inferred, usually from the instrument's reliability coefficient.

So the standard deviation has broad applicability across the fields of descriptive and inferential statistics. One of the reasons that it's such a widely used measure is that it's expressed in the original metric. In the prior example on the height of adult males, the research reports that the standard deviation is 2.8 inches. The original metric is also inches: When the researchers measured an individual's height, they reported it as some number of inches. That characteristic makes it easier to understand what a standard deviation is all about.

More About Sums of Squares

The term *sum of squares* is the source of some confusion. In statistics, it is used to mean the sum of the squared deviations of individual values from their mean. The confusion comes about because many readers understandably think that the phrase refers to the sum of the squares of the original values, not their deviations from the mean. The confusion gets reinforced by the *computational formula* for the variance. That formula isn't much used anymore because we routinely use computers instead of paper and pencil. Back in the day, though, students were often taught this formula for the variance:

$$(\Sigma_i X_i^2 - n\overline{X}^2)/n$$

where:

- X_i is the ith original observation.
- $\overline{X}$ is the mean of the original observations.
- n is the number of original observations.

If you're calculating the variance with paper and pencil, or even using a hand calculator, that computational formula is easier to use than its definitional counterpart:

$$\Sigma_i (X_i - \overline{X})^2/n$$

because the computational formula does not require that you subtract the mean from each observation and square the differences.

By now, though, personal computers have almost completely supplanted paper, pencil, and the calculator in statistical work, and applications such as Excel—not to mention such purely statistical applications as SAS, R, and Minitab—have made it unnecessary for texts to teach computational formulas for computational purposes.

Nevertheless, some confusion persists, and you'll do yourself a favor if you keep in mind that "sum of squares" in statistical writing almost always refers to the sum of the squared deviations from the mean.

Excel's choice of names for its associated functions is, fortunately, sometimes more precise than our standard English usage. The DEVSQ() function returns the sum of squares, and the "DEV" portion of the function name acts as a reminder that it's the squared *deviations* that are being summed. As an example, this formula:

```
=DEVSQ(A1:A20)
```

returns the sum of the squared deviations of the values in the range A1:A20 from their mean value. (The range occupied by the values in question is the function's only argument.) I'd be a little happier if the function's name were SUMDEVSQ() or even SUMSQDEV(), but Excel already has eight functions beginning with "SUM" and including yet another would start to make the dropdowns unwieldy.

Another function, SUMSQ(), does what its name implies: It sums the squares of its arguments. So this formula:

=SUMSQ(A1:A20)

squares each value in A1:A20 and returns the sum of those twenty squared values. The SUMSQ() function can be very useful, but not usually in the context of regression analysis, and this book has little more to say about it.

Sums of Squares Are Additive

To say that sums of squares are additive might sound mysterious, but—at least in the context of sums of squares—additivity is really a straightforward concept. All it means is that it's possible to add one sum of squares to another and get a meaningful result.

Whether or not you've ever studied the analysis of variance (*ANOVA*), it provides a useful example of the additivity of sums of squares. See Figure 3.9.

The basic idea behind this simple analysis of variance is to divide the total sum of squares into two components:

- Variability around the grand mean attributable to differences among the group means.
- Variability around the group means attributable to differences among the individuals.

Figure 3.9
ANOVA divides a total sum of squares into additive sets.

	A	B	C	D	E	F	G	H	I	J	K	L	M	N
1			Group 1	Group 2	Group 3									
2			15	4	17			Anova: Single Factor						
3			12	19	13									
4			19	13	12			SUMMARY						
5			15	12	8			*Groups*	*Count*	*Sum*	*Average*	*Variance*		
6			6	18	1			Group 1	10	116	11.6	30.71		
7			16	12	5			Group 2	10	114	11.4	21.16		
8			2	11	2			Group 3	10	86	8.6	26.27		
9			5	9	10									
10			11	7	6									
11			15	9	12			ANOVA						
12								*Source of Variation*	*SS*	*df*	*MS*	*F*	*P-value*	*F crit*
13								Between Groups	56.27	2	28.13	1.08	0.35	3.35
14			Group 1	Group 2	Group 3	Total		Within Groups	703.20	27	26.04			
15		Average	11.6	11.4	8.6	10.53								
16		Sum of Squares	276.4	190.4	236.4	759.47		Total	759.47	29				
17														
18								SS Between	56.27	=10*DEVSQ(C15:E15)				
19								SS Within	703.20	=SUM(C16:E16)				
20								SS Total	759.47	=DEVSQ(C2:E11)				

The Group 1 mean of 11.6 (see cell K6 or C15) is higher than the grand mean of 10.53 (see cell F15). Any members of Group 1 will tend to have a higher value than otherwise on the measure being used. Members of Group 3, which has a lower mean (8.6) than the grand mean (again, 10.53), will tend to have a lower value on the measure than otherwise. In this way, differences in group means exert an influence on how individual values vary. The sum of squares *between groups* quantifies the degree of that influence.

But apart from the variation in individual values that's attributable to, or associated with, differences in group means, there's the variation that's attributable to differences among individuals. We calculate that variation by measuring the deviations between each individual measurement and the mean of the group to which the individual belongs. The sum of squares *within groups* quantifies that variation.

Added together, the sum of squares between groups (often abbreviated as Sum of Squares $_B$) and the sum of squares within (Sum of Squares $_W$) account for the total sum of squares. Figure 3.9 demonstrates this additivity. Notice that the Sum of Squares $_B$ in cell I13 and the Sum of Squares $_W$ in cell I14 add up to the total Sum of Squares in cell I16.

The Data Analysis add-in has a tool called *ANOVA: Single Factor*. I used that tool to create the analysis shown in the range H2:N16 in Figure 3.9, based on the data in the range C2:E11. One of the drawbacks to the add-in's tools is that so many of them return not formulas that help tell you what's going on, but static values that tell you nothing more than the numeric results of the analysis. (At the risk of alienating any R fans who are still reading this, I should point out that the same drawback is typical of R results.)

We can substantiate the results of the add-in's tool by supplying the proper worksheet formulas. Still in Figure 3.9, I have supplied the following formulas to clarify what the ANOVA: Single Factor tool has done.

Sum of Squares Between

The sum of squares between is shown in cell I18. The formula in cell I18 is shown as text in cell J18, and is as follows:

```
=10*DEVSQ(C15:E15)
```

Notice that the result is identical to that returned by the ANOVA: Single Factor tool in cell I13. The formula uses the DEVSQ() function to calculate the sum of the squared deviations of the three group means from the grand mean. That value is then multiplied by 10, the number of observations in each group, because each of the squared deviations occurs ten times, once for each observation in a given group.

Sum of Squares Within

The sum of squares within is shown in cell I19, and the formula is shown as text in cell J19 as follows:

```
=SUM(C16:E16)
```

3

So, the formula merely totals the values in C16, D16, and E16. Those values are the sums of the squared deviations *within* each group: the squared difference between each individual value in the group and the group's mean. Those three sums of squares are obtained using these formulas:

Cell C16: =DEVSQ(C2:C11)

Cell D16: =DEVSQ(D2:D11)

Cell E16: =DEVSQ(E2:E11)

So these three sums of squares express the degree of variability within each group, irrespective of that group's distance from the grand mean. By totaling them, as is done in cell I19, you can calculate the total variability in the full set of scores as measured from each group's mean.

Sum of Squares Total

The total sum of squared deviations is easier to calculate, in Excel anyway, than either the sum of squares between or the sum of squares within. It's just the sum of the squared deviations of all the observations from their grand mean. The ANOVA tool supplies it as a static value in cell I16, and the formula in cell I20 returns the same value. The formula as text in cell J20 is as follows:

=DEVSQ(C2:E11)

As you'll see in subsequent chapters, you can convert the sums of squares to variances (termed *mean squares* in ANOVA parlance) and, by comparing the variances to one another, make inferences about the reliability of the differences in the group means. You make those conversions by dividing a sum of squares by its degrees of freedom, and the result is the mean square. For example, in Figure 3.9 the mean square between in cell K13 is the sum of squares between, 56.27, divided by the degrees of freedom between, 2. Notice that although the sum of squares between plus the sum of squares within equals the total sum of squares, the mean squares don't sum to anything meaningful. The sums of squares are additive and the mean squares aren't.

But that's not the main point of this discussion. The important takeaway here comes in three parts:

- A sum of squares (between groups) has been calculated using only the grand mean, the group means, and the number of observations in each group. This sum of squares ignores the variation between individual observations.
- A sum of squares (within groups) has been calculated using only the values in each group and their group means. This sum of squares ignores the variation of group means around the grand mean.
- A sum of squares (total) has been calculated using only the individual observations and their grand mean. This sum of squares ignores information about group membership and, therefore, about group means.

Although three different procedures are used to calculate the three sums of squared deviations, the sum of squares between plus the sum of squares within add up precisely to the sum of squares total. The procedure *partitions* the total sum of squares into a portion due to differences in the group means and another portion due to differences around the group means.

Also, because the sum of squares between plus the sum of squares within constitute the total sum of squares, you can divide the sum of squares between by the sum of squares total to calculate the percent of the sum of squares associated with the differences in the group means. In this case, that would be 56.27 divided by 759.47, or 7.41%.

NOTE

You don't see the ratio discussed in the immediately prior paragraph discussed very often in the context of a traditional analysis of variance such as that returned by the Data Analysis add-in's ANOVA tools. However, the ratio goes by several names including eta-squared and omega-squared, which depend on whether your interest centers on the sample or on the population from which you drew the sample.

In regression analysis, which is a different way of reaching the same conclusions available from the analysis of variance, the ratio is termed *R-squared* or R^2, and it's an indispensable part of regression analysis. The next section takes up the topic of R^2 in greater detail.

R^2 in Simple Linear Regression

A prior section of this chapter, "Predicting with Regression Coefficient and Intercept," discussed how to estimate or predict a value on one variable from a value on another variable. First, though, you need a set of values for both variables. Then you can use two Excel worksheet functions, SLOPE() and INTERCEPT(), to characterize the relationship between the two variables.

Figure 3.10 shows another example of the same sort of analysis, one that bears more directly on the concept of R^2 than do earlier examples of regression.

Figure 3.10 has fictional data, in the range B3:C13, on the height in inches and weight in pounds of eleven children. You would like to estimate their weights from their heights, to gauge whether any given child's weight is very high or very low given his or her height and the relationship between the two variables. (This analysis is, of course, close to inadequate as a diagnostic tool, but it serves the purpose of demonstrating more about R^2.)

Begin by calculating the slope and the intercept for the regression of weight on height. The formula for slope, used in cell E3, is:

```
=SLOPE(C3:C13,B3:B13)
```

The formula for the intercept, used in cell F3, is:

```
=INTERCEPT(C3:C13,B3:B13)
```

Figure 3.10
The sum of squares regression is analogous to the sum of squares between in ANOVA.

	A	B	C	D	E	F	G	H	I
1									
2		Height	Weight		Slope	Intercept		Predicted Weight	
3		27	76		2.35	6.22		69.79	
4		64	154					156.90	
5		37	91					93.33	
6		37	87					93.33	
7		49	115					121.58	
8		34	82					86.27	
9		60	175					147.48	
10		34	88					86.27	
11		61	125					149.83	
12		50	114					123.94	
13		51	148					126.29	
14									
15		SS Total:	11000.91	=DEVSQ(C3:C13)			SS Regression:	8899.76	=DEVSQ(H3:H13)
16									
17							SS Regression/SS Total:	0.81	=H15/C15
18									
19							Pearson's *r* between Height and Weight:	0.90	=CORREL(B3:B13,C3:C13)
20									
21							R^2 via worksheet function:	0.81	=RSQ(B3:B13,C3:C13)

As you saw earlier in this chapter, you can estimate the predicted variable Weight from the predictor variable Height by multiplying a Height value by the slope and adding the intercept. So the formula to estimate the weight of the first person, from his height in cell B3, is:

=($E$3*B3)+$F$3

That formula is used in cell H3 to return the value 69.79. There's an easier way, and I'll get to it, but for the time being let's keep the focus on R^2. We need the remaining ten Weight estimates. To get them, we want to multiply the slope in E3 times each value of Height, and add the result to the intercept in F3.

When the cell addresses $E$3 and $F$3 have been made absolute references by means of the dollar signs, you simply copy the formula in cell H3 and paste it down into H4:H13. If cell H3 is active, you can also click the selection handle (the square box in the lower-right corner of the active cell) and drag down through H13.

The raw materials are now in place, and calculating the sums of squares is easy. Cell C15 returns the total sum of squares for the original Weight variable, using the DEVSQ() function as described in the prior section. The formula is given as text in cell D15. Notice that this sum of squares is calculated in precisely the same way as the total sums of squares in cells C16, D16, and E16 in Figure 3.9.

Cell H15 returns what is termed the *Sum of Squares Regression*. The formula, also given as text in cell I15, is as follows:

=DEVSQ(H3:H13)

It is the sum of the squared deviations of the predicted values in H3:H13 from their own mean. The sum of squares regression is analogous to the Sum of Squares Between in a traditional analysis of variance.

Those two sums of squares, Sum of Squares Regression and Sum of Squares Between, appear to be calculated differently. In Figure 3.9, cell I18, the Sum of Squares Between is calculated by squaring the deviation of each group mean from the grand mean, and multiplying by the number of observations per group.

In Figure 3.10, the formula for Sum of Squares Regression simply totals the squared deviations of each predicted value from their mean. But conceptualize the data as eleven groups, with each value constituting its own group. Then we have a calculation that's identical to that used for Sum of Squares Between: the group means less the grand mean, the differences squared and the squares summed. With one observation per group, that observation *is* the group's mean. That also means that when we multiply the squared deviation by the number of observations in the group, we're multiplying by 1.

So, both conceptually and arithmetically, ANOVA's Sum of Squares Between is equivalent to regression's Sum of Squares Regression. Furthermore, the Sum of Squares Total in ANOVA has the same meaning and is calculated in the same way as the Sum of Squares Total in regression. In that case, you can determine the proportion of the total sum of squares that's associated with the regression of Weight on Height: simply the Sum of Squares Regression divided by the Sum of Squares Total. That ratio, 0.81 or 81%, shows up in cell H17 of Figure 3.10, and is termed R^2.

The reason for that name is that it's the square of *r*, the correlation coefficient. The correlation between these values for Height and Weight is in cell H19: 0.90. The square of that correlation, 0.81, is R^2, the proportion of the sum of squares in predicted Weight that is associated via regression with the sum of squares in Height.

NOTE

There's not a lot of logic to the symbols used in correlation and regression analysis. When you're dealing with the correlation between just two variables, it's referred to using a lowercase *r*. It's normal to use a capitalized *R* when the context is multiple correlation or regression, when you investigate the relationship between one variable (such as Weight) and the combination of two or more variables (such as Height and Age). Regardless of whether the context is simple or multiple regression, you usually see R^2 rather than r^2.

It happens that Excel has a worksheet function, RSQ(), which returns the R^2 value directly. Its usefulness is limited, though, because it works with two variables only—such as Height and Weight in this section's examples—and situations involving more than two variables are much more common. Nevertheless, you should be aware that the function exists. Figure 3.10 uses RSQ() in cell H21, where it returns the same value for R^2 as do the ratio of the sum of squares in cell H17 and the square of the Pearson correlation in cell H19.

Notice that the RSQ() function takes exactly the same arguments as does the CORREL() function: for example, RSQ(B3:B13,C3:C13) is simply the square of CORREL(B3:B13,C3:C13). Just as the correlation of Variable A with Variable B is identical to the correlation of Variable B with Variable A, RSQ(B3:B13,C3:C13) is precisely equivalent to RSQ(C3:C13,B3:B13).

Let's turn things around. See Figure 3.11.

Figure 3.11
The R^2 is the same regardless of which variable is treated as the predictor.

	A	B	C	D	E	F	G	H	I
1									
2		Weight	Height		Slope	Intercept		Predicted Height	
3		76	27		0.34	6.61		32.73	
4		154	64					59.53	
5		91	37					37.88	
6		87	37					36.51	
7		115	49					46.13	
8		82	34					34.79	
9		175	60					66.75	
10		88	34					36.85	
11		125	61					49.57	
12		114	50					45.79	
13		148	51					57.47	
14									
15		SS Total:	1605.64	=DEVSQ(C3:C13)			SS Regression:	1298.96	=DEVSQ(H3:H13)
16									
17							SS Regression/SS Total:	0.81	=H15/C15
18									
19							Pearson's *r* between Height and Weight:	0.90	=CORREL(B3:B13,C3:C13)
20									
21							R^2 via worksheet function:	0.81	=RSQ(B3:B13,C3:C13)

With simple correlation or regression, no order is implied by measures such as Pearson's *r*, which express the strength of the relationship between the two variables. The correlation between Height and Weight is the same as the correlation between Weight and Height. So it's not surprising that the order of the variables is likewise irrelevant to the *square* of the correlation.

Compare cell H17 in Figure 3.10 with cell H17 in Figure 3.11. The two cells contain identical values for R^2, even though the Sum of Squares Regression and the Sum of Squares Total are different in the two figures (because each figure estimates a different variable: Weight in Figure 3.10 and Height in Figure 3.11).

We say that two variables that have a non-zero correlation *share variance*, or *have variance in common*. It's usual to express that shared variance as a percent. So, in the current example, one could say that Height and Weight have 81% of their variance in common. This concept extends to multiple regression, where it's of critical importance to interpreting the analyses.

Sum of Squares Residual versus Sum of Squares Within

In this chapter's section titled "R^2 in Simple Linear Regression," I showed how the Sum of Squares Regression is equivalent to the Sum of Squares Between in a traditional analysis of variance. I also showed that the ANOVA's Sum of Squares Between and Sum of Squares Within together total to its Sum of Squares Total.

Where's the quantity in regression that's equivalent to the ANOVA's Sum of Squares Within? It's there, but it takes just a little more work to find it. The problem is that in the traditional ANOVA discussed earlier in "Sums of Squares Are Additive," you arrive at a Sum of Squares Within by calculating the sum of squares within each group, accumulating the squared deviations between the individual observations and their group's mean. Then you total the results.

In the discussion of the regression example, I noted that for the purpose of calculating the Sum of Squares Between, you can consider each observation as its own group. But that doesn't work well for calculating a quantity in regression analysis that's analogous to ANOVA's Sum of Squares Within: If each observation constitutes its own group, there's no within-group variability to add up. A group with one observation only has zero as its sum of squares (and, incidentally, zero as its variance).

A different approach is used in regression analysis to get the quantity that's analogous to ANOVA's Sum of Squares Within. It gets a different name, too: Sum of Squares Residual instead of Sum of Squares Within. See Figure 3.12.

Figure 3.12
Residuals are what's left over after removing predicted values from observed values.

B18 {=LINEST(C3:C13,B3:B13,,TRUE)}

	A	B	C	D	E	F	G	H	I	J	K
1											
2		Height	Weight		Slope	Intercept		Predicted Weight			Residuals
3		27	76		2.35	6.22		69.79			6.21
4		64	154					156.90			-2.90
5		37	91					93.33			-2.33
6		37	87					93.33			-6.33
7		49	115					121.58			-6.58
8		34	82					86.27			-4.27
9		60	175					147.48			27.52
10		34	88					86.27			1.73
11		61	125					149.83			-24.83
12		50	114					123.94			-9.94
13		51	148					126.29			21.71
14											
15		SS Total:			SS Regression:					SS Residual:	
16		=DEVSQ(C3:C13)	11000.91		=DEVSQ(H3:H13)			8899.76		=DEVSQ(K3:K13)	2101.15
17											
18		2.35	6.22								
19		0.38	18.07								
20		0.81	15.28		{=LINEST(C3:C13,B3:B13,,TRUE)}						
21		38.12	9								
22		8899.76	2101.15								

One way to get the Sum of Squares Residual value is to calculate the individual residuals and then get the sum of their squared deviations from their own mean. In Figure 3.12, the predicted values (calculated by the SLOPE() and the INTERCEPT() functions) appear in H3:H13. To get the residual values, just subtract the predicted results in H3:H13 from the observed values in C3:C13. The results—the residuals—appear in K3:K13. They are what's left over after subtracting the results of the regression from the original measures of the children's weights.

Now, by applying Excel's DEVSQ() function once again, you can calculate the Sum of Squares Residual. That's done with this formula in cell K16 of Figure 3.12:

=DEVSQ(K3:K13)

which returns the value 2101.15.

I just mentioned that one way to get the Sum of Squares Residual is by subtracting the predicted values from the observed values and applying DEVSQ() to the results. That implies there is another way, and there is. Notice that the Sum of Squares Regression, 8899.76 in cell H16, plus the Sum of Squares Residual, 2101.15 in K16, add up to the Sum of Squares Total of 11000.91 in cell C16. This is just as the Sum of Squares Between plus the Sum of Squares Within add up to the Sum of Squares Total in an ANOVA.

So, you could easily subtract the Sum of Squares Regression from the Sum of Squares Total in a regression analysis to get the Sum of Squares Residual. Or, for that matter, you could reverse the process and subtract the Sum of Squares Residual from the Sum of Squares Total to get the Sum of Squares Regression. That subtraction approach is precisely what Microsoft used prior to Excel 2003.

The problem was that Microsoft failed to appreciate the effect of forcing the intercept to equal zero, as the *const* argument to LINEST() can do. Combined with an intercept that's been forced to equal zero, determining either Sum of Squares Regression or Sum of Squares Residual by simple subtraction can result in impossible results such as negative values for R^2. If you're interested in the grisly history of this bug (which has been fixed since Excel 2003), I discuss it in Chapter 5, "Multiple Regression."

So, if you think you might like to deconstruct the LINEST() function as a way of understanding it better, or as a way of exploring how its various results work together, or just for fun, I urge you to avoid shortcuts. In the context of the present discussion, that means calculating the Sum of Squares Regression and the Sum of Squares Residual independently of each other. Don't get one of them via the shortcut of subtracting the other from the total sum of squares. That's a recipe not for diagnosing an error but for compounding one.

The TREND( ) Function

Excel has a worksheet function named TREND(), which relieves you of the hassle of calculating a regression's slope and intercept, then of applying the slope and intercept to the existing *known x* values so as to get the predicted *y values*. In effect, the TREND() function does the following:

- It takes the *known x's* and the *known y's* with which you supply it.
- It calculates, in the background, the slope and intercept.
- It returns on the worksheet the predicted values that result from applying the slope and intercept to the *known x's*.

Figure 3.13 shows how to use the TREND() function on the worksheet.

Figure 3.13
Using TREND( ) is quicker and easier than using SLOPE( ) and INTERCEPT( ).

G3 {=TREND(C3:C12,B3:B12)}

	A	B	C	D	E	F	G
1		Annual Growth (inches)			Annual Growth (inches)		Annual Growth (inches)
2		Brother	Sister		via SLOPE() and INTERCEPT()		via TREND()
3	Pair 1	2.78	2.45		2.30		2.30
4	Pair 2	2.01	2.47		2.08		2.08
5	Pair 3	1.40	2.17		1.90		1.90
6	Pair 4	1.13	2.08		1.82		1.82
7	Pair 5	0.81	1.79		1.73		1.73
8	Pair 6	1.28	1.60		1.87		1.87
9	Pair 7	2.66	1.43		2.27		2.27
10	Pair 8	2.30	1.78		2.16		2.16
11	Pair 9	2.78	2.92		2.30		2.30
12	Pair 10	0.72	1.45		1.71		1.71
13							
14		Slope:	0.29				
15		Intercept:	1.50				

In Figure 3.13, the range E3:E12 contains the values for the Sisters as predicted using SLOPE() and INTERCEPT() in combination with the values for their Brothers. For example, cell E3 contains this formula:

```
=$C$14*B3+$C$15
```

The slope in cell C14 times the value in cell B3, plus the intercept in cell C15, equals the predicted value in cell E3. Of course, before you can make the prediction, you have to calculate the slope and intercept using their respective functions. If you wanted to, you could combine the functions in this fashion:

```
=SLOPE(C3:C12,B3:B12)*B3+INTERCEPT(C3:C12,B3:B12)
```

It's often useful to combine or nest Excel functions, but this probably isn't one of those occasions. It's more trouble than it's worth. If you're looking for a better way to get the predictions, consider using the TREND() function, shown in the range G3 to G12. The TREND() function used there is as follows:

```
=TREND(C3:C12,B3:B12)
```

Notice that the *known y's* and the *known x's* are called out in TREND()'s arguments just as they are in SLOPE() and INTERCEPT(). This is as simple as TREND() gets. The TREND() function can accept two more arguments, optional ones, and I'll get to them shortly.

Array-entering TREND( )

One more point about TREND(): Unless you're using it to predict one value only, you have to *array enter* it. Put another way, to correctly use the TREND() function to return more than one predicted value, you must enter it in an *array formula*.

Array formulas have broad applicability in Excel, and you can use them to solve problems that are otherwise intractable. I'll discuss a subset of the reasons to use array formulas later in this book. But if you're new to array formulas, I urge you to take a close look at them in one of the excellent books that specifically concern Excel tools and techniques.

Here are two of the reasons, pertinent to the TREND() function, to use array formulas:

- A function that you want to use requires that you enter it as part of an array formula.
- You want the formula's results to occupy more than one cell.

Both of those reasons apply to the TREND() function. They also apply to the LINEST() function. These two, TREND() and LINEST(), are unquestionably the two most important regression functions in Excel's toolkit. So it's also important that you be familiar with at least the mechanics of array entering formulas.

NOTE

If you're already an array formula maven, you might as well skip the remainder of this section.

Here's how to get the TREND() function entered into the array formula in the range G3:G12 of Figure 3.13:

1. Select the range G3:G12. Press Delete to remove the existing array formula.
2. Type, but don't yet enter, this formula:

   =TREND(C3:C12,B3:B12)
3. Hold down the Ctrl and the Shift keys, and only then press Enter. Release the Ctrl and the Shift keys as you release the Enter key.

Done correctly, your array entry should return the same values as shown in Figure 3.13.

Notice the array formula as it appears in the Formula Bar shown in Figure 3.13. It is enclosed in curly brackets: { and }. That's Excel's way of telling you that it has accepted your formula as an array formula. Leave those brackets to Excel: If you try to enter them yourself as part of the formula, Excel treats what you enter as a simple text string.

A (very) few issues that you should be aware of if you're new to array formulas:

- There's no requirement that an array formula occupy a range of cells, as it does in Figure 3.13. Many useful array formulas are designed to occupy one cell only. In those cases, the array part of the formula is one (or more) of its arguments, rather than its results.
- After a multi-cell array formula has been successfully entered on the worksheet, you cannot edit (or delete) just one of the cells it occupies. Your edit must apply to all the cells that the array formula's results occupy, so it's often necessary to begin by selecting the formula's full range if you want to edit or delete an existing array formula.
- Knowing the dimensions of the range that you start by selecting is usually a combination of experience and familiarity with whatever function you might include in the formula. For example, the LINEST() function can never return more than five populated rows, but it can return as many as 64 columns.

TREND( )'s *new x's* Argument

As I mentioned earlier in this chapter, the TREND() function can take two arguments in addition to the *known y's* and the *known x's*. Those arguments are termed *new x's* and *const*.

The *new x's* argument must be TREND()'s third argument if you're going to use it. Its use arises most often in forecasting situations, when your predictor variable is some measure of a time period (such as the day of the month) and a variable that's to be forecast, such as month-to-date revenues. Then, you'll have a baseline of observations to use in calculating the slope and intercept. You'll also have the number of the day, still in the future, for which you want to estimate the month-to-date revenue.

Figure 3.14 provides an example.

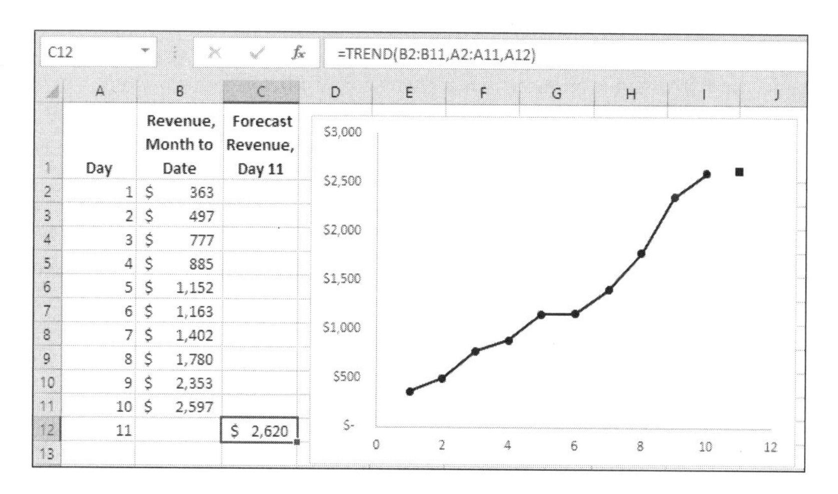

Figure 3.14
Regression is one popular way to forecast on the basis of time period.

In Figure 3.14, cell C12 contains the TREND() function. Because it occupies one cell only, the formula need not be array-entered. (It *could* be array-entered, though, and for consistency it probably makes sense to do so.) The formula in C12 is:

=TREND(B2:B11,A2:A11,A12)

The first two arguments, the *known y's* and the *known x's*, inform Excel that it is to calculate the regression equation, consisting of the slope and the intercept, based on the values in B2:B11 and A2:A11. The third argument informs Excel that it is to apply that regression equation to the value in cell A12. I have shown the known x's and known y's with circular markers in the chart. The predicted cumulative revenue as of Day 11 is shown with a square marker.

You can verify this, if you want, by using the SLOPE() and the INTERCEPT() functions on the ranges B2:B11 and A2:A11, and then using the results along with the value in cell A12. Done correctly, you will get the same result as appears in the figure.

(By the way, there are plenty of situations in which this sort of forecast, based on regression and using date or time as a predictor, is both sensible and reasonably accurate. But there are also many situations in which its use is unwise and potentially misleading. You can get what I think is pretty good guidance for telling the difference in *More Predictive Analytics*, published by Que in 2015.)

TREND( )'s *const* Argument

The fourth and final argument that TREND() takes is termed *const*, short for *constant* and referring to the regression equation's intercept. The LINEST() function also has a *const* argument and it has the same effect as it does in TREND(). Unhelpfully, however, *const* is the third of four arguments in LINEST() but is the fourth of four arguments in TREND().

The *const* argument can take on one of two values: TRUE or FALSE (equivalently, you can represent TRUE with 1 and FALSE with 0). If you specify TRUE, Excel uses the intercept in the regression equation that returns the predicted values. If you specify FALSE, Excel omits the intercept from the equation and recalculates the slope in order to minimize the errors using a regression line that intercepts the vertical axis at its zero point.

Many thoughtful statisticians believe that there are situations in which it is useful and informative to omit the intercept in the regression equation. Others, equally thoughtful, believe that it is *not* helpful, and potentially misleading, to do so. I discuss the matter in greater detail in Chapter 5. For the time being, I'll just say that I've long agreed with Leland Wilkinson, the principal developer of Systat, who wrote in the 1985 manual for that application: "If your constant really *is* near zero, then leaving CONSTANT in or out of the MODEL statement will make little difference in the output."

Here's an example to illustrate what I believe Wilkinson had in mind. Suppose that you're studying the relationship between acreage (the predictor x-variable), and crop yield (the predicted y-variable). The reality of the situation is that a plot of zero acres yields zero bushels of crop. The intercept—the constant—equals zero when the x-value (number of acres) is zero. You *expect* that to be the case. When the plot contains zero acres, you expect zero bushels of crop yield.

Therefore, given accurate measures of acreage and crop yield, you also expect the intercept to come very close to zero even when the constant is calculated normally. The intercept hews close to zero by the nature of the variables you're studying. In that case it cannot help matters to force the intercept to equal zero, and as you'll see in Chapter 5, it can hurt.

It may make it easier to understand what's going on with forcing the constant to equal zero (again, *intercept* is synonymous with *constant* in this context) if you have a look at an example. Figure 3.15 has one.

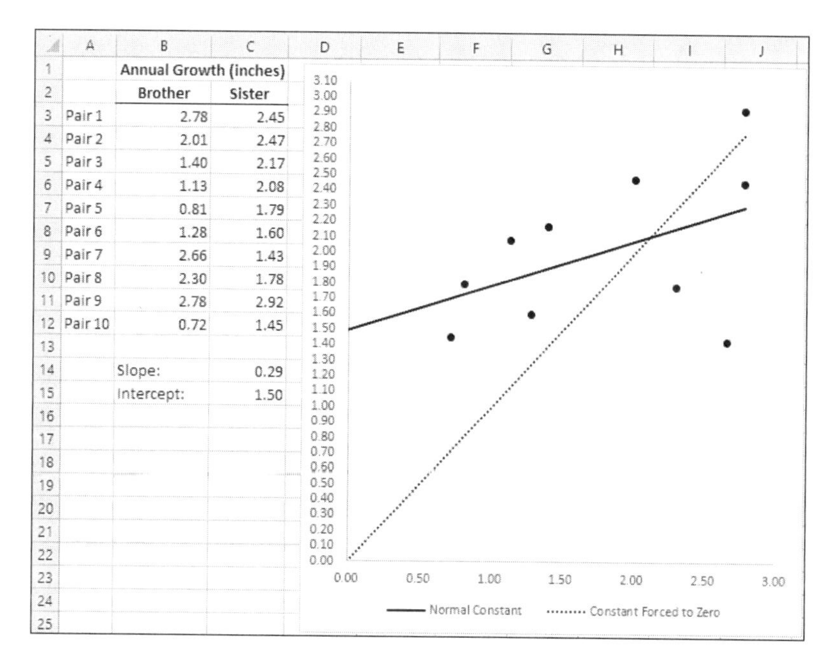

	Annual Growth (inches)	
	Brother	Sister
Pair 1	2.78	2.45
Pair 2	2.01	2.47
Pair 3	1.40	2.17
Pair 4	1.13	2.08
Pair 5	0.81	1.79
Pair 6	1.28	1.60
Pair 7	2.66	1.43
Pair 8	2.30	1.78
Pair 9	2.78	2.92
Pair 10	0.72	1.45
	Slope:	0.29
	Intercept:	1.50

Figure 3.15
Forcing a zero constant can mangle the meaning of the regression statistics.

Two regression lines appear in Figure 3.15, along with the ten individual plotted points. The solid line represents the situation in which the constant is calculated normally. It intercepts the vertical axis at 1.5, just as the INTERCEPT() function in cell C15 tells you it should.

The dotted line depicts the regression of the sisters' annual growth on the brothers' annual growth when the constant is forced to zero. Note that the dotted regression line intercepts the vertical axis at 0, at the point that the x values—the brothers' annual growth—also equals zero.

Suppose you use SLOPE() and INTERCEPT() to predict the sisters' values from the brothers' values, as shown in column E of Figure 3.13, and then plot those predicted values against the brothers' values in an XY (Scatter) chart. The results would show up exactly where the solid regression line appears in Figure 3.15. Therefore, you can estimate the accuracy of the prediction by measuring the distance between the plotted actuals (the circular markers in Figure 3.15) and the regression line. Figure 3.16 gives an example.

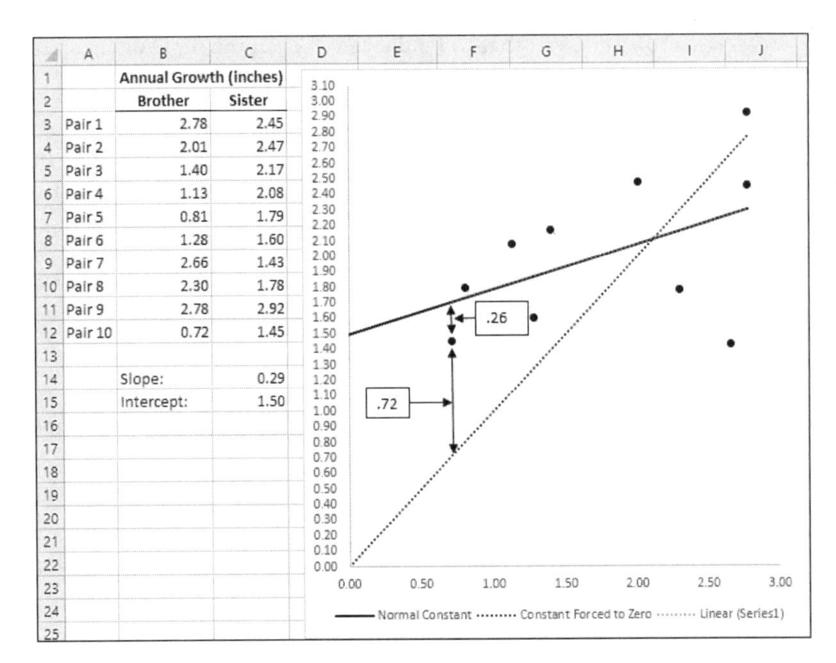

	A	B	C
1		Annual Growth (inches)	
2		Brother	Sister
3	Pair 1	2.78	2.45
4	Pair 2	2.01	2.47
5	Pair 3	1.40	2.17
6	Pair 4	1.13	2.08
7	Pair 5	0.81	1.79
8	Pair 6	1.28	1.60
9	Pair 7	2.66	1.43
10	Pair 8	2.30	1.78
11	Pair 9	2.78	2.92
12	Pair 10	0.72	1.45
13			
14		Slope:	0.29
15		Intercept:	1.50

Figure 3.16
The closer the observations to the regression line, the better the prediction.

Notice that Figure 3.16 shows the distance between the leftmost observation (at 0.72 on the horizontal axis and 1.45 on the vertical axis) and each of the two regression lines. That observation is .26 units from the solid regression line, for which the constant is calculated normally. But it's .72 units from the dotted regression line, for which the constant has been forced to equal zero. For that specific observation, the normal-constant regression is a more accurate predictor of the actual value than is the zero-constant regression.

If you do the same analysis for the remaining nine observations, you'll find that only one of the ten observations—the one in the upper-right corner of the chart—is closer to the zero-constant regression line than to the normal-constant regression line. Chapter 4, "Using the LINEST() Function," has much more to say about how to evaluate the accuracy of a regression, but for now it's pretty clear, even prima facie, that with this data set the regression line based on a normal constant is more accurate than the one based on a constant that has been forced to equal zero.

Calculating the Zero-constant Regression

Excel offers several ways to calculate a regression in which the constant is zero. One way is by means of chart options. Starting with an XY (Scatter) chart, right-click one of the markers in the charted data series. When you click Add Trendline in the shortcut menu, the Format Trendline pane appears. Choose Linear Trendline if necessary. Scroll down—again, if necessary—until you see the Set Intercept checkbox. Fill it, make sure that the associated edit box contains 0.0, and then dismiss the Format Trendline pane. That's how Figure 3.16 came by its dotted regression line.

Although this method provides one way to put a regression line (no matter how the constant is calculated) into a chart, it doesn't provide the predicted y-value for each x-value, and that's one principal purpose of a regression analysis. The method simply draws the trendline on the chart, and enables a couple of additional options such as showing the regression equation and the R^2 value as labels on the chart itself. To get the predicted y-values onto the worksheet, you need to use one of the worksheet functions, briefly discussed next.

The SLOPE( ) and INTERCEPT( ) Functions

Suppose you have a simple two-variable regression, where one variable represents the known x's and the other represents the known y's. In that case you can calculate the predicted y-values by applying the SLOPE() and INTERCEPT() functions to the known x's, as shown most recently in Figure 3.13. However, you can't use this method to force a zero intercept. The SLOPE() function assumes that the intercept is being calculated normally by the INTERCEPT() function, and SLOPE() therefore returns the slope of a regression line whose intercept is not forced to a particular value.

The TREND( ) Function

Using the TREND() function, it's possible to force the constant to zero *and* get the regression line slope that allows for a constant of zero. See Figure 3.17.

The range E3:E12 contains this array formula:

=TREND(C3:C12,B3:B12,,FALSE)

Notice that the third argument to the function is absent, as indicated by a pair of consecutive commas. You can supply a worksheet address for new x's to predict from, using the relationship between the known x's and the known y's. Here, though, no third argument is supplied, and therefore Excel resorts to the default, which is to predict using only the known x's.

Figure 3.17
Both TREND( ) and LINEST( ) support forcing the constant to zero.

I3 =B3*$G$3

	A	B	C	D	E	F	G	H	I
1		Annual Growth (inches)							
2		Brother	Sister		TREND() Function		LINEST() Function		Predicted Values from LINEST()
3	Pair 1	2.78	2.45		2.77		0.99	0	2.77
4	Pair 2	2.01	2.47		2.00		0.12	#N/A	2.00
5	Pair 3	1.40	2.17		1.39		0.88	0.76	1.39
6	Pair 4	1.13	2.08		1.12		64.79	9	1.12
7	Pair 5	0.81	1.79		0.81		37.53	5.21	0.81
8	Pair 6	1.28	1.60		1.27				1.27
9	Pair 7	2.66	1.43		2.65				2.65
10	Pair 8	2.30	1.78		2.29				2.29
11	Pair 9	2.78	2.92		2.77				2.77
12	Pair 10	0.72	1.45		0.72				0.72

The fourth argument is set to FALSE. That instructs the TREND() function to force the constant to zero, and to adjust the value of the slope accordingly. In Figure 3.17, then, the range E3:E12 returns the predicted values given a constant of zero. If you charted them against the known x's in an XY (Scatter) chart, you would get a series of points that fall directly on the dotted trendline shown in Figure 3.16.

The LINEST( ) Function

The TREND() function returns predicted values, based on known values and on the relationship between known x's and known y's. The LINEST() function does not return predicted values, but instead returns the equation used to obtain the predicted values, along with selected inferential statistics that bear on the reliability of the equation. (For example, given another sample of data that was collected in the same way, what's the probability of getting a similar set of results?)

The range G3:H7 in Figure 3.17 contains this array formula:

=LINEST(C3:C12,B3:B12,FALSE,TRUE)

The known y's and known x's are the same as in the TREND() example. The third argument, FALSE, tells LINEST() *not* to calculate the constant normally—that is, to force the constant to zero. The fourth argument tells LINEST() to return all the ancillary statistics, not just the regression coefficient (in a simple two-variable regression, the regression coefficient is the same as the slope) and the intercept. Note that the intercept, shown in cell H3, is zero. Because the intercept has been forced to zero, and is not an estimate based on the data, the value in cell H4 is #N/A, an error value. Normally it would show a measure of the amount of uncertainty in the intercept estimate, but there is no uncertainty when you call for a specific value, zero.

To get the predicted values from LINEST(), you multiply the known x's by the regression coefficient. There's no need to add the intercept, because it's been set to zero. So the formula to return the first predicted value in cell I3 is:

=B3*$G$3

It can be copied and pasted down through I12. The reference to the cell that contains the regression coefficient, $G$3, remains the same wherever it's pasted, because the dollar signs render it an absolute reference.

Notice that the predicted values according to LINEST() in column I are identical to those returned by TREND() in column E.

Partial and Semipartial Correlations

I wish there were some way I could hide the title of this section. I've been working with its concepts and methods for years and years, and I can promise you that they're straightforward, logical, and useful. So it makes me nuts that they have names like "partial correlation" and "semipartial correlation." They *sound* like they were coined for the purpose of intimidating people. But the publisher won't let me hide them. Those are the names we're stuck with and we might as well learn to live with them.

Partial Correlation

You use a partial correlation when you have three or more variables. Three correlation coefficients quantify the strength of the relationships between the three pairings of the variables. There are times when you would like to focus on the relationship between two of the variables as if they had *no* relationship with the third. It's entirely possible to estimate the strength of that relationship, which is termed a *partial* correlation.

Suppose that you're interested in the relationship between the percent of residents in a city who graduated from college and the number of books in the cities' libraries. You sample 50 cities and find that the correlation between the number of books and the percent of college graduates in those cities is a respectable 0.60. As you're compiling the data, it occurs to you that you should have accounted for at least one additional and important variable: the community's wealth.

The problem is that both the number of books in the library, and the percent of residents who graduate from college, are strongly related to the dollars that a community's residents have available to support institutions such as libraries, and to send young people to college. You consider the possibility that if you could hold community wealth constant—if you could somehow remove its effects on financial support for libraries and on the affordability of higher education—you might find a relationship between number of books and number of graduates quite different from the observed correlation of 0.60.

Of course, it would be very difficult to find another 50 cities, each of which has precisely the same wealth as the other 49, however that variable is measured. That's one way to hold wealth constant, but it just isn't feasible.

However, by *statistically* removing the effect of wealth from both the financial support of libraries and from the affordability of college, you might get a more accurate estimate of the quantitative relationship between the variables you're really interested in, number of books, and number of graduates. This statistical technique *partials out* wealth from both book acquisitions and college costs.

One way to calculate these partial correlations is by means of the following equation:

$$r_{CB.W} = \frac{r_{CB} - r_{CW}r_{BW}}{\sqrt{1 - r_{CW}^2}\sqrt{1 - r_{BW}^2}}$$

where:

- $r_{CB.W}$ is the correlation between College and Books with Wealth partialled out.
- r_{CB} is the correlation between College and Books.
- r_{CW} is the correlation between College and Wealth.
- r_{BW} is the correlation between Books and Wealth.

Although it's correct, there are two problems with that formula. One is that you can stare at it as long as you like and it still won't provide a flash of insight into what's going on inside it. (At any rate, it never did that for me.) The other problem is that it's fairly clumsy and prone to errors in placing actual correlations in the right places.

Another way of thinking about the problem of partial correlation is to cast it in the form of the analysis *residuals*, the differences between predicted values and the associated actual observations. Continuing with the previous example, if you can calculate the correlation between, say, College and Wealth, you can predict college from wealth using any of various techniques: standard scores and the correlation coefficient, functions such as SLOPE() and INTERCEPT() that return the regression coefficient and intercept, or a function such as TREND() that returns predicted values directly to the worksheet.

Further, if you can get the predicted values, you can subtract them from the actual observations to get the residual values. With the residual values for percent of college graduates in your sampled communities, you have information about college attendance rates with the effect of community wealth removed—or, in statistical jargon, partialled out.

Let's change the example a bit and see how it works out in the context of an Excel worksheet. See Figure 3.18.

Figure 3.18
Age is closely related to both Height and Weight.

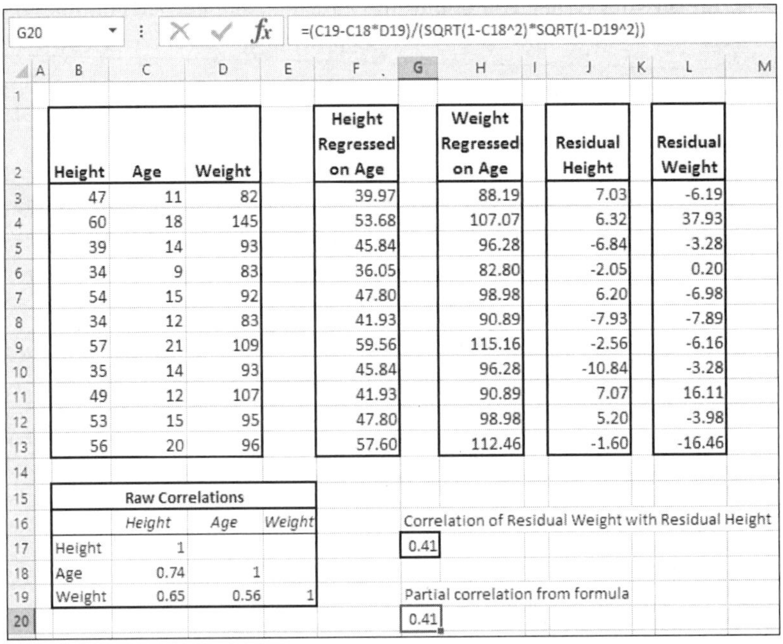

G20 =(C19-C18*D19)/(SQRT(1-C18^2)*SQRT(1-D19^2))

Height	Age	Weight	Height Regressed on Age	Weight Regressed on Age	Residual Height	Residual Weight
47	11	82	39.97	88.19	7.03	-6.19
60	18	145	53.68	107.07	6.32	37.93
39	14	93	45.84	96.28	-6.84	-3.28
34	9	83	36.05	82.80	-2.05	0.20
54	15	92	47.80	98.98	6.20	-6.98
34	12	83	41.93	90.89	-7.93	-7.89
57	21	109	59.56	115.16	-2.56	-6.16
35	14	93	45.84	96.28	-10.84	-3.28
49	12	107	41.93	90.89	7.07	16.11
53	15	95	47.80	98.98	5.20	-3.98
56	20	96	57.60	112.46	-1.60	-16.46

Raw Correlations

	Height	Age	Weight
Height	1		
Age	0.74	1	
Weight	0.65	0.56	1

Correlation of Residual Weight with Residual Height
0.41

Partial correlation from formula
0.41

Suppose that you're interested in investigating the relationship between height and weight in adolescents and young adults. Many variables such as nutrition and physical activity mediate that relationship, but surely one of strongest is age: The tendency is to grow taller and heavier as the years pass until a person reaches adulthood. You wonder what might happen to the relationship between height and weight if the effect of age on both variables were removed.

You collect information on height, weight, and age from a small sample of subjects and record it as shown in the range B3:D13 of Figure 3.18. A good early step with a problem such as this one is to get the correlations between the pairs of variables. I've done that in the range B16:E19. Notice that the correlations range from moderate (0.56) through fairly strong (0.74).

NOTE

I don't have much use for most of the tools in Excel's Data Analysis add-in. The analyses are underpowered and idiosyncratic. They are neither of production quality nor useful learning devices. But the Correlation tool can save you time and errors. (The same is true of the Covariance tool if you prefer to work with covariances instead of correlations.) The Correlation tool returns a lower-diagonal matrix of correlations (see B16:E19 in Figure 3.18), given a list or table of raw data such as that in B3:D13. Such a matrix is a useful way to describe how related variables behave—even more so when it's populated above the main diagonal as well as below it, so that it can be inverted or used in matrix multiplication. With five or more variables, populating it by hand with multiple instances of the CORREL( ) function is tedious and error prone. But the Correlation tool automates things, and it's one of the Data Analysis tools that's actually valuable.

The next step is to regress Height onto Age and then Weight onto Age. That's done in the ranges F3:F13 and H3:H13 of Figure 3.18. The values in those two ranges represent all the information about this sample's height and weight that can be estimated using knowledge of the subjects' ages. The array formula used in F3:F13 is:

```
=TREND(B3:B13,C3:C13)
```

And in H3:H13 it's this:

```
=TREND(D3:D13,C3:C13)
```

With the predicted values in F3:F13 and H3:H13, it's easy to get the residuals. Those are shown in the ranges J3:J13 and L3:L13. They're obtained simply by subtracting each predicted value from each actual observation. So, the residual in cell J3 is calculated with:

```
=B3–F3
```

to get the residual Height for the first subject, and this:

```
=D3–H3
```

to get the same subject's residual Weight.

We now have residual measures of Height and Weight, with all influence of Age removed from both sets of values, in J3:J13 and L3:L13. It only remains to get the correlation of those two sets of values, which is the partial correlation of Height and Weight with Age partialled out. That value is 0.41, as shown in cell G17, using this formula:

=CORREL(J3:J13,L3:L13)

Confirming that result, cell G20 uses this formula:

=(C19–C18*D19)/(SQRT(1–C18^2)*SQRT(1–D19^2))

to return the partial correlation by means of the formula given earlier in this section:

$$r_{HW.A} = \frac{r_{HW} - r_{HA}r_{WA}}{\sqrt{1 - r_{HA}^2}\sqrt{1 - r_{WA}^2}}$$

(replacing the subscripts with H for Height, W for Weight, and A for Age).

One more point about partial correlation before we move on to its close cousin, semipartial correlation. Figure 3.19 stresses the relationship between Age, the variable that's partialled out, and the predicted and residual values for both Height and Weight.

Figure 3.19 In simple regression, the predictor variable correlates perfectly with the predicted values.

L15 =CORREL(C3:C13,F3:F13)

	A	B	C	D	E	F	G	H	I	J	K	L
1												
2		Height	Age	Weight		Height Regressed on Age		Weight Regressed on Age		Residual Height		Residual Weight
3		47	11	82		39.97		88.19		7.03		-6.19
4		60	18	145		53.68		107.07		6.32		37.93
5		39	14	93		45.84		96.28		-6.84		-3.28
6		34	9	83		36.05		82.80		-2.05		0.20
7		54	15	92		47.80		98.98		6.20		-6.98
8		34	12	83		41.93		90.89		-7.93		-7.89
9		57	21	109		59.56		115.16		-2.56		-6.16
10		35	14	93		45.84		96.28		-10.84		-3.28
11		49	12	107		41.93		90.89		7.07		16.11
12		53	15	95		47.80		98.98		5.20		-3.98
13		56	20	96		57.60		112.46		-1.60		-16.46
14												
15			Raw Correlations				Age with *Height regressed on Age*:					1.00
16			*Height*	*Age*	*Weight*		Age with *Weight regressed on Age*:					1.00
17		Height	1									
18		Age	0.74	1			Age with Residual Height					0.00
19		Weight	0.65	0.56	1		Age with Residual Weight					0.00

The original observations, the predicted values for Height and Weight, and the residual values for Height and Weight are the same as shown in Figure 3.18. Notice the four correlations in cells L15, L16, L18, and L19. The variable that's treated as the predictor, Age, has a perfect correlation with its predicted values for both Height and Weight.

Also, because of the way that the residuals are derived from the predicted values and the original observations, the predictor variable has a 0.0 correlation with the residual values for Height and Weight. Another way of putting this is to note that Age shares *no* variance with either the Height residuals or the Weight residuals.

This effect is not due to anything peculiar to partial correlations. It's due to the nature of predicted values in simple, single-predictor regression. Recall that Excel's TREND() function simply calculates the regression equation's slope and intercept, just as do the SLOPE() and the INTERCEPT() functions, and displays the results of applying them to the predictor variable. That's what TREND() is doing in the ranges F3:F13 and H3:H13 of Figures 3.18 and 3.19.

So all we're doing with SLOPE() and INTERCEPT(), or equivalently with TREND(), is to multiply the original predictor values by a constant, SLOPE(), and adding another constant, INTERCEPT(). When you multiply a series of values by a constant, SLOPE(), you change the scale, but that has no effect on the *relative positions* of each value in the new scale with respect to the remaining values in that scale.

When you add a constant, INTERCEPT(), to a series of values, you don't even change the scale. All you do is change the mean of those values, moving each value up or down the original scale.

Because the relative positions of each value remain the same, their z-scores must also remain the same, and the inevitable result is that the correlation between the predictor variable and the predicted values must be 1.0. If the correlation with the predicted values is 1.0, there is no variance left to share with the residual values—and the correlation between the predictor variable and the residuals must be 0.0, just as shown in Figure 3.19.

I belabor this point because it reinforces the fact that when it's applied to partial correlation, it helps to make it clear that the calculation of the partial correlation, between two sets of residuals, leaves the variable that is partialled out well and truly partialled out. It has no variance to share with those residuals.

Things are a little bit different with *semipartial* correlations, which become central to certain applications of multiple regression. I take those applications up in Chapter 5, but first let's have a look at semipartial correlations in the context of simple regression.

Understanding Semipartial Correlations

I'd like to get a bit of terminology out of the way at the outset. The prior section discussed *partial* correlation. This section discusses *semipartial* correlation, which is sometimes called *part* correlation. It has always seemed to me that the terms *part correlation* and *partial correlation* are synonymous, and that the terms fail to suggest the very real and important distinction between the two types of correlation. But any difference between the terms *part* and *partial* is arbitrary and unhelpful.

The other term used, instead of *part* correlation, is *semipartial* correlation, and I hold my nose but use it where needed in the remainder of this book in preference to *part* correlation.

Recall from the prior section that partial correlation removes the effect of one variable (in its example, Age) from two other variables (Height and Weight), and the correlation between the residuals is termed a *partial* correlation.

It turns out that it's frequently useful to remove the effect of one variable from *only one* of two other variables, instead of from both, in multiple regression analyses. Because the effect is removed from only one variable, leaving the effect in place in the other variable, the approach is often termed *semipartial correlation*. Figure 3.20 has an example.

Figure 3.20
Semipartial correlations remove the effect of a variable from only one of the remaining variables.

G18 =(C19-C18*D19)/SQRT(1-C18^2)

Row	Height	Age	Weight	Height Regressed on Age	Residual Height
3	47	11	82	39.97	7.03
4	60	18	145	53.68	6.32
5	39	14	93	45.84	-6.84
6	34	9	83	36.05	-2.05
7	54	15	92	47.80	6.20
8	34	12	83	41.93	-7.93
9	57	21	109	59.56	-2.56
10	35	14	93	45.84	-10.84
11	49	12	107	41.93	7.07
12	53	15	95	47.80	5.20
13	56	20	96	57.60	-1.60

Raw Correlations

	Height	Age	Weight
Height	1		
Age	0.74	1	
Weight	0.65	0.56	1

0.65 Correlation of Height with Weight

Semi-partial Age from Height
0.34 Semi-partial correlation of Height with Weight

Semi-partial Age from Height
0.34 Correlation of Residual Height with Weight

As is the case with partial correlations, there's a formula for the semipartial correlation:

$$r_{W(H.A)} = \frac{r_{WH} - r_{HA} r_{WA}}{\sqrt{1 - r_{HA}^2}}$$

This formula uses the same subscripts as were used in the prior section, H for Height, W for Weight, and A for Age. Notice these two changes from the partial correlation formula:

- The subscript for the semipartial correlation itself now uses parentheses to indicate which variable has its effect removed from which variable. Here, W(H.A) indicates that the effect of Age has been partialled out of Height but not out of Weight.
- The denominator of the ratio no longer includes the square root of the percent of variance in Weight that is not predicted by Age. In this case we are not partialling Age from Weight—only from Height—so the R^2 between Age and Weight does not appear in the denominator.

This formula is used in cell G18 of Figure 3.20 to show the correlation of Height with Weight, after Age has been partialled out of Height. The semipartial correlation is 0.34—a value that's considerably smaller than the simple correlation, shown in cell G15, of the original Height and Weight measures. Removing the effect of Age from Height dramatically reduces the amount of variance shared by Weight with the residual Height values.

The original Height and Weight values have a 0.65 correlation, or 42% shared variance. The semipartial correlation of the Height residuals with the original Weight values is 0.34, so the shared variance is 12%, or 0.34 squared. It's important to bear in mind that the 30% difference in shared variance hasn't been lost. It's just that the 30% is also shared with Age, where we'll leave it, and we've chosen to partial it out of Height. We could go the other direction (and we shortly will) and partial it out of Age instead of Height.

An alternate way to calculate the semipartial correlation is directly analogous to the alternate method for partial correlations, shown in Figure 3.19. To partial Age out of Height, we can use TREND() to predict Height from Age and then subtract the predicted values from the actual observed Heights to get the residual Heights. Then, we get the semipartial correlation by correlating the residual Height values with the original Weight values. This is exactly what was done for Age and Height in Figure 3.19. It also takes place in Figure 3.20, in cell G21. There, the standard correlation coefficient is calculated for Weight with the residual Height values. Notice that it's identical to the semipartial correlation calculated by formula in cell G18.

In Figure 3.19, the discussion of partial correlations required that the effect of Age be partialled out of both Height and Weight. In contrast to Figure 3.19, Figure 3.20 shows only one set of predicted values and their associated residuals. At this point we remove the effect of Age from Height only, and not from Weight. Why is that?

Suppose that we were interested in predicting the value of Weight on the basis of Age *and* Height. That's multiple regression, which I take up in detail in Chapter 5. Because Age and Height both share variance with Weight we want to use both Age and Height to help one another predict Weight, by using both predictors in the regression equation.

But there's a problem: Age and Height share variance with one another, and some of that shared variance might also be shared with Weight. If you begin by regressing Weight onto

Age, you'll account for 32% of the variance in Weight: the simple correlation between Age and Weight is 0.56, and the square of that correlation is 0.32 or 32%.

Having regressed Weight onto Age, you continue by regressing Weight onto Height. Their simple correlation is 0.65, so 0.42 (that's 0.65 squared) or 42% of the variance in Weight is associated with the variance in Height.

By now you have explained 32% of the variance in Weight as associated with Age, and 42% of the variance in Weight as associated with Height. Have you therefore explained 32% + 42% = 74% of the variance in Weight?

No, you probably haven't. The only situation in which that could be true is if Age and Height, your two predictor variables, are *themselves* unrelated and therefore uncorrelated. It's certainly true that two predictor variables can be uncorrelated, but it's almost always because you have designed and conducted an experiment in such a way that the predictor (or explanatory) variables share no variance. You'll see instances of that in later chapters when we take up special coding methods.

In a situation such as the one described in this example, though, it would be pure luck and wildly improbable to find that your predictors, Age and Height, are uncorrelated. You didn't assign your subjects to particular values of Age and Height: That's just how they showed up, with their own ages and heights. Because their correlation with one another is 0.74, Age and Height share 0.74^2, or 55% of their variance.

So when you regress Weight onto Age, you assign some of the variance in Weight to the predictor Age. When you continue the analysis by adding Height to the equation, the variance in Weight shared with Age ought not to be available to share with the predictor Height. To simply add the squared correlations of Age with Weight and Height with Weight would be to count some of the variance twice: the variance that is shared by all three variables.

The solution is to use a semipartial correlation, and thereby to adjust the values of one of the two predictor variables so that they're uncorrelated and share no variance. (Conceptually, anyway—a function such as LINEST() that performs multiple regression is not based on code that follows this sequence of events, but LINEST() emulates them.) If you use a semipartial correlation to remove the effect of one predictor from the other, but leave that effect in place in the predicted variable, you ensure that the variance shared by the predictor and the predicted variable is unique to those variables. You won't double-count any variance.

That last paragraph contains an extremely important point. It helps lay the foundation of much discussion, in later chapters, of how we assess the effects of adding variables to a multiple regression equation. So I repeat it, this time in the context of this section's Height-Age-Weight example.

The predictor variables are Height and Age, and the predicted variable is Weight. This issue would not arise if Height and Age were uncorrelated, but they *are* correlated. Therefore, the two predictors share variance with one another. Furthermore, they share variance with the predicted variable Weight—if they didn't, there would be no point to including them in the regression.

The correlation between Height and Weight is 0.65. So if we start out by putting Height into the equation, we account for 0.65^2, or 42% of the Weight variance. Because Height and Age also share variance, some of that 42% is likely shared with Age, along with Height and Weight. In that event, if we just added Age into the mix along with Height, some variance in Weight would be accounted for twice: once due to Height and once due to Age. That would incorrectly inflate the amount of variance in Weight that is explained by the combination of Height and Age.

It can get worse—what if the correlation between Height and Weight were 0.80, and the correlation between Age and Weight were also 0.80? Then the shared variance would be 64% for Height and Weight, and also 64% for Age and Weight. We would wind up explaining 128% of the variance of Weight, a ridiculous outcome.

However, if we apply the notion of semipartial correlations to the problem, we can wind up with *unique* variance, variance that's associated only with a given predictor. We can take the semipartial correlation of Age with Height, partialling Age out of Height (but not out of Weight). See Figure 3.21.

Figure 3.21
We use semipartial correlations to remove the effect of one predictor from the other predictor, but *not* from the predicted variable.

E16 =RSQ(D3:D13,C3:C13)

Height	Age	Weight	Height Regressed on Age	Residual Height	Age regressed on Height	Residual Age
47	11	82	39.97	7.03	14.61	-3.61
60	18	145	53.68	6.32	18.25	-0.25
39	14	93	45.84	-6.84	12.37	1.63
34	9	83	36.05	-2.05	10.98	-1.98
54	15	92	47.80	6.20	16.57	-1.57
34	12	83	41.93	-7.93	10.98	1.02
57	21	109	59.56	-2.56	17.41	3.59
35	14	93	45.84	-10.84	11.26	2.74
49	12	107	41.93	7.07	15.17	-3.17
53	15	95	47.80	5.20	16.29	-1.29
56	20	96	57.60	-1.60	17.13	2.87

	R^2
Weight & Age	0.318
Weight & Residual Height	0.116
Total	0.434
Weight & Original Height	0.418

Age	Height	Intercept
0.900	0.917	41.644
1.892	0.715	23.248
0.434	15.081	#N/A
3.070	8	#N/A
1396.578	1819.422	#N/A

=LINEST(D3:D13,B3:C13,,TRUE)

Figure 3.21 shows how to use the residuals of one predictor, having partialled out the effects of the *other* predictor. The residuals of Height, after removing from it the effects of Age, appear in the range G3:G13. These cells are of particular interest:

- Cell E16 shows the R^2, the shared variance, between Weight and Age. It's returned easily using Excel's RSQ() function:

  =RSQ(D3:D13,C3:C13)

  At this point we're interested in the R^2 between the actual observations and no variance has been partialled from the first predictor variable (or, for that matter, from the outcome variable Weight).

- Cell E17 shows the R^2 between Weight and the residual values of Height, having already partialled Age out of Height:

  =RSQ(D3:D13,G3:G13)

  We partial Age out of Height so that we can calculate the R^2 between the outcome variable Weight and the *residual* values of Height. We have already accounted for all the variance in Age that's associated with Weight. We don't want to double-count any of that shared variance, so we first partial variance shared by Age and Height out of Height, and then determine the percent of variance shared by Weight and the Height residuals.

- Cell E18 shows the total of the R^2 values for Weight with Age, and for Weight with the residual Height values. They total to 0.434.

- Cell E20 shows the R^2 between Weight and the original Height values. Notice that it is several times larger than the R^2 between Weight and the residual Height values in cell E17. The difference is due to partialling Age out of Height.

3

Also notice that the total of the R^2 values, in E18, is exactly equal to the value in cell G18. That value, in G18, is the R^2 for the full multiple regression equation, returned by LINEST() and making simultaneous use of Age and Height as predictors of Weight. Had we simply added the raw R^2 values for Age with Weight and Height with Weight, we would have come up with a total R^2 value of 0.318 + 0.418 or 0.736, a serious overestimate.

What if we had started with Height as the first predictor instead of Age? The results appear in Figure 3.22.

In contrast to Figure 3.21, Figure 3.22 begins by regressing Age onto Height instead of Height onto Age, in the range F3:F13. Then the residual Age values are calculated in G3:G13 by subtracting the predicted Age values from the actual observations in column B.

Then, in Figure 3.22, the unadjusted R^2 for Weight with Height appears in cell E16 (in Figure 3.21, E16 contains the unadjusted R^2 for Weight with Age). Figure 3.22 also supplies in cell E17 the R^2 for Weight with the residual values of Age in G3:G13.

Compare Figures 3.21 and 3.22, and note that the individual R^2 values in G16 and G17 differ. The difference is strictly due to which predictor variable we allowed to retain the variance shared with the other predictor variable: Age in Figure 3.21 and Height in

Figure 3.22. *The total variance explained by the two predictor variables together is the same in both cases.* But the amount of variance in Weight that's attributable to each predictor is a function of which predictor we allow to enter the equation first.

Figure 3.22
The order in which predictors enter the equation affects only the degree of their contribution to the total R^2.

E17 =RSQ(D3:D13,G3:G13)

	A	B	C	D	E	F	G	H	I	J
2		Age	Height	Weight		Age Regressed on Height	Residual Age		Height Regressed on Age	Residual Height
3		11	47	82		14.61	-3.61		39.97	7.03
4		18	60	145		18.25	-0.25		53.68	6.32
5		14	39	93		12.37	1.63		45.84	-6.84
6		9	34	83		10.98	-1.98		36.05	-2.05
7		15	54	92		16.57	-1.57		47.80	6.20
8		12	34	83		10.98	1.02		41.93	-7.93
9		21	57	109		17.41	3.59		59.56	-2.56
10		14	35	93		11.26	2.74		45.84	-10.84
11		12	49	107		15.17	-3.17		41.93	7.07
12		15	53	95		16.29	-1.29		47.80	5.20
13		20	56	96		17.13	2.87		57.60	-1.60
14										
15					R^2		Height	Age	Intercept	
16				Weight & Height	0.418		0.917	0.900	41.644	
17				Weight & Residual Age	0.016		0.715	1.892	23.248	
18				Total	0.434		0.434	15.081	#N/A	
19							3.070	8	#N/A	
20							1396.578	1819.422	#N/A	
21							=LINEST(D3:D13,B3:C13,,TRUE)			

At this point that might seem a trivial issue. What's important is how accurately the overall regression equation performs. The contribution of individual variables to the total explained variance is by comparison a relatively minor issue.

Except that it's not. When you begin to consider whether to even use a variable in a multiple regression equation, it's a relatively major issue. It can affect your assessment of whether you've chosen the right model for your analysis. I'll take those matters up in some detail in Chapter 5. First, though, it's necessary to add Excel's LINEST() function to this book's toolkit. The LINEST() function is critically important to regression analysis in Excel, and Chapter 4 discusses it in much greater detail than I have thus far.

3

Using the LINEST( ) Function

4

IN THIS CHAPTER

Array-Entering LINEST( ) 103
Comparing LINEST( ) to SLOPE( ) and INTERCEPT( ) 107
The Standard Error of a Regression Coefficient 109
The Squared Correlation, R^2 117
The Standard Error of Estimate 120
Understanding LINEST( )'s F-ratio 129
The General Linear Model, ANOVA, and Regression Analysis 146
Other Ancillary Statistics from LINEST( ) 149

The worksheet function LINEST() is the heart of regression analysis in Excel. You could cobble together a regression analysis using Excel's matrix analysis functions without resorting to LINEST(), but you would be working without your best tools. So here you are, three chapters into a book entirely about regression, and I haven't even said anything about how to enter Excel's most important regression function into a worksheet.

I'll get into that next, but first let me explain an apparent contradiction. Projected into its smallest compass, with just one predictor variable, LINEST() returns a maximum of 10 statistics. Yet you can use LINEST() to perform something as simple as a test of the difference between the means of two groups, or as complex as a factorial analysis of covariance, complete with factor-by-factor and factor-by-covariate interactions. You can use it to run what's termed *model comparison*, which enables you to assess the statistical effect of adding a new variable (or of removing an existing variable) from a regression equation. You can use LINEST() to perform curvilinear regression analysis and orthogonal contrasts.

It's all in the way that you arrange the raw data before you point LINEST() at it. I'll get into that issue in subsequent chapters. First, though, it's important to understand the mechanics of putting LINEST() on the worksheet, as well as what those ten statistics mean and how they interact with one another. That's the purpose of this chapter.

Let's start with the mechanics.

Array-Entering LINEST( )

Different functions in Excel react in different ways when you array-enter them, by way of Ctrl-Shift-Enter, compared to how they behave in a formula that's entered normally, using just the Enter key. For example, the TRANSPOSE() function, which swaps the rows and columns in a matrix, returns nothing more than #VALUE! if you enter it normally: You *must* use Ctrl+Shift+Enter to get TRANSPOSE() to return even the first value in the original matrix.

On the other hand, the TREND() function is perfectly accurate if you enter it normally, with Enter, if you begin by selecting one cell only. Admittedly, it would be rare to want the result of a regression equation for the first record only. My point is that, syntactically, it works.

The LINEST() function resembles TREND() in that respect. It will return an accurate result if you select one cell only and enter the function normally. But you won't get accurate results, or you won't get all the results you're entitled to, if you don't handle things as Excel expects.

So let's take a brief look at what problems to avoid and how to avoid them.

Understanding the Mechanics of Array Formulas

Chapter 3, "Simple Regression" in the section "Array-entering TREND()," discusses some of the purely mechanical aspects of entering array formulas in Excel. Because those aspects are pertinent to using LINEST() as well as TREND(), I repeat some of them here:

- It's the formula that gets array-entered, not the function. However, array formulas generally contain functions. You array-enter the formula in order to take full advantage of the function's capabilities.
- You array-enter a formula in Excel by typing it and then, instead of pressing Enter, you hold down the Ctrl and Shift keys and continue to hold them down as you press Enter.
- Excel won't complain if you use Enter by itself instead of Ctrl+Shift+Enter with LINEST(), but you won't get the full set of results that the function is capable of returning.
- You need a mix of experience with and knowledge of the function so that you'll know the dimensions of the range that the array formula will occupy. You need to select that range of cells before you begin to type the formula.
- Typically, TREND() requires that you begin by selecting a range that's one column wide and with as many rows as are in your range of observations. Also typically, LINEST() requires that you begin by selecting a range that has five rows and as many columns as are in your range of observations.

So if you're using LINEST() to return a simple regression analysis, with one predictor variable and one predicted variable, you would probably begin by selecting a range consisting of five rows and two columns. Five rows because LINEST() cannot return more than five rows of useful information—any sixth and subsequent row that you select will contain #N/A error values. Two columns because you need one for the predictor variable's regression coefficient, and one for the intercept (also termed the *constant*).

Inventorying the Mistakes

Figure 4.1 has a few examples of how you can go wrong mis-entering LINEST()—like I have, more times than I care to count.

Figure 4.1 contains predictor values (which Excel terms the *known x's*), the heights of 20 people, in column A. The predicted variable (in Excel terms, its values are the *known y's*), weight, is in column B. The idea is to determine the nature of the relationship between the two variables—possibly with the intention of estimating the weight of a 21st person from knowledge of his or her height.

Figure 4.1
Not all of these examples represent erroneous usage.

G15 fx {=LINEST(B2:B21,A2:A21)}

	A	B	C	D	E	F	G	H	I
1	Height (inches)	Weight (pounds)		Correct					
2	72	131		2.092	-3.591	=LINEST(B2:B21,A2:A21,TRUE,TRUE)			
3	58	97		0.818	54.216				
4	61	144		0.267	21.118				
5	60	120		6.546	18				
6	63	100		2919.471	8027.329		Correct, wrong range selected		
7	68	150					2.092	-3.591	#N/A
8	67	170		Correct, possibly insufficient			0.818	54.216	#N/A
9	58	130		2.092	-3.591		0.267	21.118	#N/A
10	68	146					6.546	18	#N/A
11	63	106		Correct, insufficient			2919.471	8027.329	#N/A
12	65	105		2.092			#N/A	#N/A	#N/A
13	58	150							
14	72	170		Incorrect			Incorrect		
15	72	156		2.092	#VALUE!		2.092	-3.591	
16	63	127		#VALUE!	#VALUE!		2.092	-3.591	
17	75	172		#VALUE!	#VALUE!		2.092	-3.591	
18	72	150		#VALUE!	#VALUE!		2.092	-3.591	
19	76	121		#VALUE!	#VALUE!		2.092	-3.591	

The Correct Version

The figure shows a correctly entered instance of LINEST() in the range D2:E6. The formula, array-entered in that range, is:

=LINEST(B2:B21,A2:A21,TRUE,TRUE)

The arguments are as follows:

- The address for the known y values (B2:B21), the variable to be predicted.
- The address for the known x values (A2:A21), the variable to predict from.
- TRUE, specifying that the constant is to be calculated normally. TRUE is the default here, so you could omit it: =LINEST(B2:B21,A2:A21,,TRUE)
- TRUE, requesting all the regression statistics in rows 2 through 5 of the results. FALSE is the default, so leaving its position blank returns only the first row of results.

If you were to select any cell in the array formula's range, D2:E6, and look in the formula box, you would see the array formula surrounded by curly braces that you did not type. Excel supplies those when it recognizes that you have array-entered the formula using the keyboard sequence Ctrl+Shift+Enter.

Not Enough Cells, Part I

The range D9:E9 shows what can happen if you don't select enough rows or columns prior to array-entering the formula. Probably, the user should have selected a range with five rows and two columns. The regression coefficient and the intercept are correct—the formula was typed correctly—but that's all you get. I can't remember the last time that all I wanted was the coefficient and intercept, other than as a means of showing how *not* to use LINEST().

Not Enough Cells, Part II

In cell D12, the user probably forgot to select a range of cells before starting to type the formula.

Wrong Keyboard Sequence

The range D15:E19 shows what can happen if you start by selecting the correct number of rows and columns, but forget to array-enter the formula. As shown here, the keyboard sequence Ctrl+Enter does fill the selected range, but it doesn't array-enter the formula. For that, you need Ctrl+*Shift*+Enter.

Too Many Rows and/or Columns

The range G7:I12 shows what happens if you start by selecting too many rows or columns before array-entering the formula. LINEST() will never return more than five rows of data, so row 6 will always have #N/A.

Omitting LINEST()'s Fourth Argument

The user began by selecting the range G15:H19, and so probably intended to get information about both the regression coefficient and the intercept, as well as the associated statistics in LINEST()'s third through fifth rows. But the array formula as entered is =LINEST(B2:B21,A2:A21), and so omits the fourth argument, which Excel terms *stats*. That argument defaults to FALSE, and its omission is equivalent to stating that you don't want the associated statistics. However, because the formula was array-entered in a five-row range, Excel responds by giving you the coefficient and the intercept five times.

So, a variety of ways exist for you to go wrong in the process of invoking Excel's LINEST() function. The sort of thing shown in Figure 4.1 happened to me often enough when I was first exploring the LINEST() function that I decided to *always* select all five rows and as many columns as called for by the data. I decided to always omit the third *const* argument and to always use TRUE for the fourth *stats* argument. Sticking with that sort of routine becomes automatic after just a little while and saves a lot of error correction over time.

Comparing LINEST( ) to SLOPE( ) and INTERCEPT( )

The SLOPE() and INTERCEPT() functions discussed in Chapter 3 are intended for much simpler, more restricted situations than what LINEST() is equipped for. You can get the same information from LINEST() as you do from SLOPE() and INTERCEPT() if you have that simple situation, and you get much more in the way of ancillary information from LINEST(). Figure 4.2 has a concrete example.

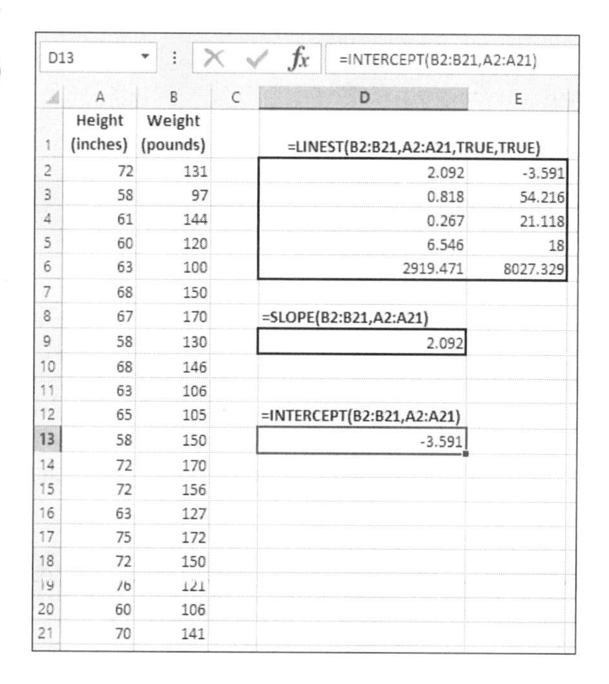

D13 =INTERCEPT(B2:B21,A2:A21)

	A	B	C	D	E
1	Height (inches)	Weight (pounds)		=LINEST(B2:B21,A2:A21,TRUE,TRUE)	
2	72	131		2.092	-3.591
3	58	97		0.818	54.216
4	61	144		0.267	21.118
5	60	120		6.546	18
6	63	100		2919.471	8027.329
7	68	150			
8	67	170		=SLOPE(B2:B21,A2:A21)	
9	58	130		2.092	
10	68	146			
11	63	106			
12	65	105		=INTERCEPT(B2:B21,A2:A21)	
13	58	150		-3.591	
14	72	170			
15	72	156			
16	63	127			
17	75	172			
18	72	150			
19	76	121			
20	60	106			
21	70	141			

Figure 4.2
SLOPE( ) and INTERCEPT( ) duplicate LINEST( )'s first row.

The range D2:E6 contains the result of pointing the LINEST() function at the data in the ranges A2:A21 and B2:B21. The formula as array-entered in D2:E6 is given in cell D1.

> **NOTE** In simple, single-predictor regression, the slope of the regression line *is* the regression coefficient: You use it to multiply each actual observed predictor value, and then add the intercept. In multiple regression, several predictors are combined and no single predictor by itself defines the slope of the regression line. Therefore, in the context of multiple regression, it's usual to refer to a predictor's regression coefficient rather than to its slope.

The SLOPE() function is entered (normally) in cell D9, and its formula in cell D8. Similarly, the INTERCEPT() function is used in cell D13, and its formula appears in cell D12. Both the SLOPE() and the INTERCEPT() functions point at the same ranges as LINEST() does.

Compare the result of the SLOPE() function in D9 with the LINEST() result in cell D2. They are identical. The INTERCEPT() function in D13 returns the same result as does LINEST() in cell E2.

In other words, in return for your selection of a range of cells instead of one cell and your willingness to hold down Ctrl and Shift as you press Enter, you get all this in addition to the regression coefficient and intercept:

- The standard errors of the regression coefficient and intercept (the range D3:E3)
- The R^2 for the regression (cell D4)
- The standard error of estimate (cell E4)
- The F-ratio for the full regression (cell D5)
- The degrees of freedom for the residual sum of squares (cell E5)
- The sum of squares for the regression (cell D6)
- The sum of squares for the residual (cell E6)

That's a favorable rate of exchange.

Those additional statistics are not just nice-to-haves. They represent critical information about the nature and the quality of the regression equation. There's absolutely no way that I would ever make an important decision or recommendation that makes use of a regression analysis without an opportunity to examine *at least* the standard errors, the R^2, the F-ratio, and the degrees of freedom for the residual.

Furthermore, as you'll see in Chapter 5, "Multiple Regression," using not just one but several predictor variables is possible and frequently desirable. In the context of the example shown in Figure 4.2, you might add Age to Height as a predictor of Weight. This approach is termed *multiple regression* because you're using multiple predictor variables. LINEST() is entirely capable of dealing with that situation, whereas SLOPE() and INTERCEPT() are not. The latter two functions are restricted to one predictor variable only.

Sometimes you must choose to show just a single figure in the LINEST() results, for some reason such as restrictions imposed by the layout of the rest of a worksheet. In that case you can resort to the INDEX() function. That function selects and displays values selected from a matrix such as the one returned by LINEST(). For example, the regression's R^2 is always in the third row, first column of the LINEST() results. To return only the R^2 for the regression shown in Figure 4.2, you could use this formula:

```
=INDEX(LINEST(B2:B21,A2:A21,TRUE,TRUE),3,1)
```

Here, you're presenting a matrix of values to the INDEX() function: That matrix is identical to the matrix that LINEST() returns to the worksheet when it's array-entered. The source of the matrix is provided by the first argument to INDEX(), which is the LINEST() function and *its* arguments.

The second argument to INDEX() is the row, and the third argument is the column, whose intersection is the cell of the matrix that you're interested in. The R^2 value is in the third row, first column, so the formula as shown returns that value.

By the way, if you use the INDEX() function to get at just one figure from the LINEST() results, you need not array-enter it. For example, you could select a single cell on the worksheet shown in Figure 4.2, type the formula just given, and press Enter to get the single value 0.267.

Now that I've called out the specific statistics that LINEST() offers you, in addition to the regression coefficient and the intercept, we might as well take a closer look at the meaning of each of those statistics, and at the ways that they interrelate.

The Standard Error of a Regression Coefficient

The range that's occupied by actual LINEST() results is generally asymmetric. When we get to multiple regression in Chapter 5, you'll see that the first two rows of the results extend to the right as the number of predictor variables increases. However, the portion of the range occupied by the ancillary statistics is fixed: It always occupies the first two columns of the final three rows. The number of predictors used has an effect on the *values* in that range, but not on the number of those ancillary statistics. You always get six of them, whether you're using 1 predictor or 64.

4

However, the first two rows of the LINEST() results have as many columns as you have predictor variables, plus 1. So if you're in a simple one-predictor situation, LINEST() returns two columns in its first two rows. The first row contains the regression coefficient in its first column and the intercept in the second column. This book has already examined the nature of the regression coefficient and the intercept in Chapter 3.

The second row's first and second columns contain the coefficient's standard error and the intercept's standard error, respectively. Let's take a closer look at the standard errors.

The Meaning of the Standard Error of a Regression Coefficient

Figure 4.3 shows the data and the LINEST() results that last appeared in Figure 4.2, but the focus here is on the standard errors in the second row of the LINEST() results.

A standard error is a standard deviation, and it takes errors as the underlying data whose variability is for some reason interesting. The *standard error of estimate* is one example. One way to calculate the standard error of estimate is to find the standard deviation of the errors made in predicting, say, weight from height. Other things being equal, a larger standard error of estimate results from a regression equation that's less accurate as a predictor. A more accurate regression equation results in a smaller standard error of estimate.

A regression coefficient is subject to a different sort of error. That error is due not to the accuracy of the regression equation—after all, the equation's accuracy depends heavily on the regression coefficient itself—but to sampling error.

Figure 4.3
You might regard the regression coefficient for Height as statistically significant.

D8 =D2/D3

	A	B	C	D	E
1	Height (inches)	Weight (pounds)		=LINEST(B2:B21,A2:A21,TRUE,TRUE)	
2	72	131		2.092	-3.591
3	58	97		0.818	54.216
4	61	144		0.267	21.118
5	60	120		6.546	18
6	63	100		2919.471	8027.329
7	68	150			
8	67	170		2.559	
9	58	130			
10	68	146			
11	63	106			
12	65	105			
13	58	150			
14	72	170			
15	72	156			
16	63	127			
17	75	172			
18	72	150			
19	76	121			
20	60	106			
21	70	141			

4

In Figure 4.3, we have a regression coefficient of almost 2.1 (cell D2) and an associated standard error of 0.818 (cell D3). If you divide the coefficient by the standard error, you get 2.1 / .818 = 2.559. In other words, the regression coefficient of 2.1 is more than two-and-a-half standard errors from zero. We'll look more carefully at how to evaluate that result later in this chapter, but for the moment: Why is it an important result?

A Regression Coefficient of Zero

Consider the possibility that the regression coefficient is really zero, not 2.1. In that case, the regression equation might be 0 * Height + 1325.16. That's the situation depicted in Figure 4.4.

Suppose that instead of height and weight you were investigating the relationship of height and street address. Now the regression equation, in effect, drops the person's height from consideration, by multiplying it by a zero regression coefficient. Then it adds 1325.16: the mean of the predicted variable.

In other words, when the regression coefficient is zero, you might as well drop it from consideration as a predictor and just use the mean of the predicted variable as its best estimate.

That's the reason you want to know whether a regression coefficient is a long way from zero, at least as measured by standard errors. (If the regression coefficient were calculated at 10,000 and the standard error of the coefficient were 1,000,000, then the coefficient would be greater than zero by only 1% of a standard error.)

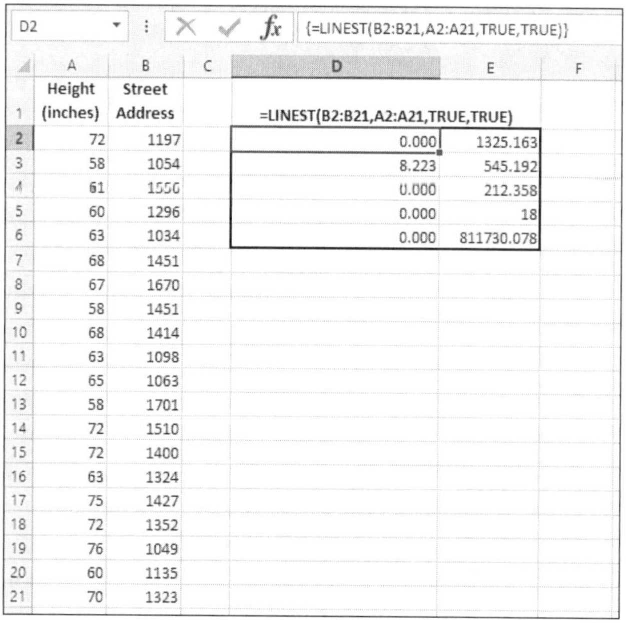

Figure 4.4
The regression reveals no relationship between the variables.

D2 fx {=LINEST(B2:B21,A2:A21,TRUE,TRUE)}

	A	B	C	D	E	F
1	Height (inches)	Street Address		=LINEST(B2:B21,A2:A21,TRUE,TRUE)		
2	72	1197		0.000	1325.163	
3	58	1054		8.223	545.192	
4	61	1556		0.000	212.358	
5	60	1296		0.000	18	
6	63	1034		0.000	811730.078	
7	68	1451				
8	67	1670				
9	58	1451				
10	68	1414				
11	63	1098				
12	65	1063				
13	58	1701				
14	72	1510				
15	72	1400				
16	63	1324				
17	75	1427				
18	72	1352				
19	76	1049				
20	60	1135				
21	70	1323				

In the case that's depicted in Figure 4.3, the regression coefficient for Height is 2.559 standard errors above zero (see cell D8). If you've been around statistical analysis for a while, your experience tells you that two and a half standard errors is quite a distance, and you would surely regard the population value of the coefficient as non-zero. A coefficient that's two-and-a-half standard errors from zero is very unlikely if the coefficient, calculated using the entire population, is in fact zero. Of course, that term "very unlikely" is a subjective judgment.

If we're after something more objective, perhaps we should resort to a test of statistical significance. A statistic, such as a regression coefficient in LINEST()'s first row, divided by the standard error of that statistic, found in LINEST()'s second row, results in a *t-ratio*. You can evaluate a t-ratio with a *t-test*—for example, with one of Excel's t-test functions. That will give you a probability and put you on firmer ground regarding the reliability of the regression coefficient.

Figure 4.5 shows how this process works. Figure 4.5 (and Figures 4.1 through 4.4, for that matter) displays data obtained from a sample in the range A2:B21. What if it's a bad sample? What if the coefficient for the regression of weight on height *in the full population* were 0.0? In other words, what if no relationship exists between the two variables in the population, and the luck of the sampling happened to result in a regression coefficient of 2.092? Is it even possible to determine the probability of getting a regression coefficient of 2.092 in a sample when, in the population, the regression coefficient is actually 0.0? Yes, it is. I wouldn't have asked the question otherwise.

Figure 4.5
The t-test quantifies the probability of getting so large a regression coefficient.

E14 =T.DIST.RT(E10,E11)

	A	B	C	D	E
1	Height (inches)	Weight (pounds)		=LINEST(B2:B21,A2:A21,TRUE,TRUE)	
2	72	131		2.092	-3.591
3	58	97		0.818	54.216
4	61	144		0.267	21.118
5	60	120		6.546	18
6	63	100		2919.471	8027.329
7	68	150			
8	67	170		Regression Coefficient	2.092
9	58	130		Standard Error of Coefficient	0.818
10	68	146		t-ratio	2.559
11	63	106		Degrees of Freedom	18
12	65	105		Probability of sample	
13	58	150		coefficient when population	
14	72	170		coefficient equals zero	1.0%
15	72	156			
16	63	127			
17	75	172			
18	72	150			
19	76	121			
20	60	106			
21	70	141			

4

Measuring the Probability That the Coefficient is Zero in the Population

You start by getting a t-ratio, which is the ratio of (in this case) the regression coefficient to its standard error. Figure 4.5 gets the coefficient from the LINEST() results in cell D2 and linked from there into cell E8. It also gets the standard error of the coefficient in cell D3 and from there to cell E9. Then the t-ratio is formed in cell E10 by dividing E9 into E8.

We also need the *degrees of freedom* for the t-ratio. The degrees of freedom for a t-ratio is closely related to the number of observations on which the ratio is based. In the case of the regression coefficient, it is $n - k - 1$, where n is the number of observations, and k is the number of predictor variables. The result in this example is 20 – 1 – 1, or 18. We can also pick up that figure from the LINEST() results, from cell E5.

Lastly, we point Excel's T.DIST.RT() function at the t-ratio and its degrees of freedom. Cell E14 contains this formula:

```
=T.DIST.RT(E10,E11)
```

The formula returns 0.00987 or, rounded, 1%. That tells us this: If the regression coefficient in the population is 0%, the probability of getting a sample of 20 people, for whom the regression coefficient is calculated at 2.092, is just shy of 1%. Is that a more objective way to evaluate the strength of the coefficient than just noticing that it's 2.5 standard errors greater than zero?

Well, yes, in a sense, but that really begs the question. When you convert a t-ratio to a measure of probability, you're arranging for a more sophisticated test, one which takes the number of sample observations into account, but you've accomplished little more than a change of the metric, from number of standard errors to a probability measured as a percentage. You're still left with the issue of the reliability of the finding. Is 2.5 standard errors a powerful enough finding that you can reasonably expect *another* sample to yield a non-zero regression coefficient? Is a probability of less than 1% sufficiently unusual that you can reject the notion that the population's regression coefficient is zero?

Statistical Inference as a Subjective Decision

Testing statistical significance increases the sophistication of your analysis and, as I noted earlier in this chapter, it puts you on somewhat firmer ground. But it doesn't change what is at root a subjective judgment to an objective one.

Figure 4.6 illustrates these concepts graphically.

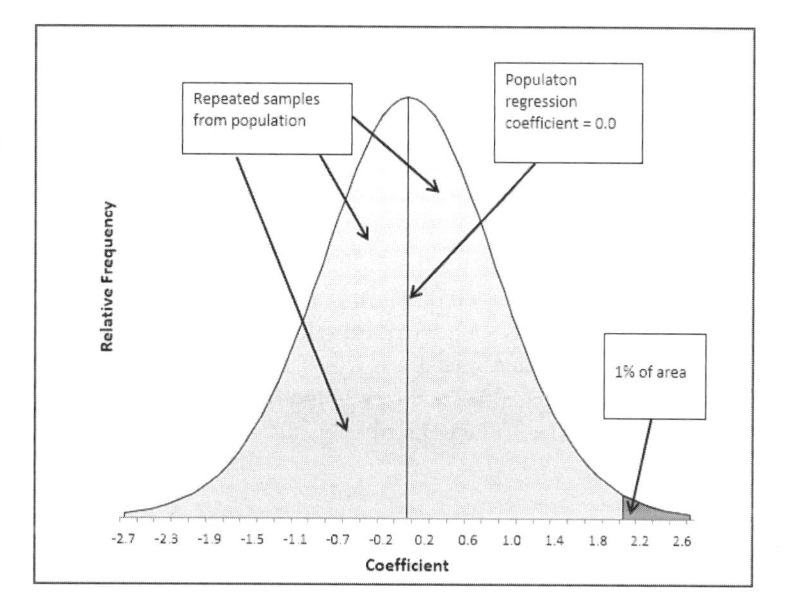

Figure 4.6
Locating the observed coefficient in a normal curve shows that it's rare if the coefficient is zero in the population.

Figure 4.6 shows, as a normal curve, the distribution of the regression coefficients calculated from thousands of possible samples of 20 people each. In the population, the regression coefficient is 0.0. Purely through the errors that are induced by imperfect sampling, some of the calculated regression coefficients miss the actual population value of 0.0. Many of them miss by just a little—say, 0.05 or –0.03—but some miss by a good bit more—as much as 2.0 or more. Of course, as the size of the error increases, the number of such errors drops. You have just a relative handful of samples that have a regression coefficient greater than 2.0.

> **NOTE** This sort of distribution, which shows the relative frequency of a statistic calculated on many samples, is called a *sampling distribution*.

Charted in this way, the regression coefficient follows a normal curve. We know enough about the normal curve to be able to state that only about 1% of these samples would return a regression coefficient as far from 0.0 as 2.092.

So we're confronted by a choice of one of these two possibilities:

- We continue to suspect that the regression coefficient is 0.0 in the population we sampled from. The luck of the draw just gave us an unrepresentative sample of observations.
- We decide that the obtained regression coefficient of 2.092 is so rare in the distribution around a population of 0.0 that it must come from a different population, one where the regression coefficient is not 0.0.

If you choose the first possibility, you shouldn't believe any of the regression statistics. A sample so unrepresentative of its population that it returns a regression coefficient occurring only 1% of the time can't be good enough to support the calculated R^2. Or the F-ratio. Or any other related statistic.

Or you might decide that the obtained regression coefficient is perfectly representative. It's just that you got it by sampling from a population where the regression coefficient *isn't* 0.0. It's larger than 0.0, and your best estimate of the population's actual regression coefficient is the one you obtained from your sample: 2.092.

If you decide on a criterion for the probability of the outcome before you collect and analyze the data, you're following the decision rules associated with traditional hypothesis testing, including concepts such as critical values and null hypotheses. That approach has much to recommend it, and it leads easily to the use of confidence intervals. This chapter looks into the topic of confidence intervals and regression, in a somewhat different context, in the later section titled "The t Distribution and Standard Errors."

WHEN ZERO DOESN'T REALLY MEAN ZERO

This book begins to examine *multiple* regression in the next chapter, but there is a situation that can arise (seldom, thank goodness) in multiple regression that bears on the present discussion and I would like to deal with it here.

When you have multiple predictors, as you do in multiple regression, it can happen that one predictor can be expressed as a combination of, or as a simple function of, other predictors. For example, it might turn out that Predictor 3 is the sum of Predictor 1 and Predictor 2. Or the values in Predictor 2 might be exactly 3.1416 times those in Predictor 1. This can happen by accident if, for example, one of the predictors is populated or coded incorrectly. It can also happen intentionally, if the user is not acquainted with what can result.

This condition is termed *collinearity* or *multicollinearity*. Traditional algorithms for solving what are called the *normal equations* to reach a least squares solution usually fail when collinearity is present in the data, because that makes it impossible to invert one of the important matrixes. That was the case with the LINEST() function from its introduction in Excel 3 in the mid-1990s through Excel 2002. The results provided by LINEST() with collinear data ranged from simply wrong to impossible.

Excel 2003 introduced a variety of modifications to the code that underlies the LINEST( ) function. Some of those modifications diagnose collinear data properly. I believe, though, that you should be warned about how LINEST( ) reports the condition to you.

LINEST( ) returns a regression coefficient of 0, *and* a standard error of the coefficient that also equals 0, when a predictor is collinear with another predictor. Figure 4.7 provides an example.

Figure 4.7
Notice that LINEST( ) returns 0 for both the coefficient and the standard error of Predictor 1.

F3 {=LINEST(D3:D22,A3:C22,,TRUE)}

	A	B	C	D	E	F	G	H	I
1						=LINEST(D2:D21,A2:C21,,TRUE)			
2	Predictor 1	Predictor 2	Predictor 3	Predicted Variable		Predictor 3	Predictor 2	Predictor 1	Intercept
3	2	4	16	48		-0.7953	-0.57824	0	38.60312
4	19	38	12	20		0.816408	0.32607	0	12.33422
5	5	10	11	4		0.162221	14.94431	#N/A	#N/A
6	3	6	10	15		1.645869	17	#N/A	#N/A
7	10	20	14	6		735.1515	3796.649	#N/A	#N/A
8	2	4	10	37					
9	11	22	9	23					
10	9	18	8	0					
11	8	16	18	32					
12	18	36	9	0					
13	13	26	5	39					
14	15	30	3	34					
15	17	34	7	8					
16	6	12	10	3					
17	5	10	15	25					
18	12	24	3	24					
19	15	30	4	19					
20	7	14	2	40					
21	12	24	12	2					
22	19	38	10	3					

Collinearity is a fairly rare event, and when the data submitted to LINEST( ) is assembled carefully it becomes extremely rare. However, it does occur. The combination of a coefficient of 0, as in cell H3 of Figure 4.7, with a standard error of 0, as in cell H4, is even less likely, but it can occur. If one of the predictors is a constant—if all of its values are the same number—you can get a coefficient and a standard error that both equal 0, and a constant predictor value is *not* an example of collinearity.

Now, this is a personal belief, and my own. I think that Microsoft should have arranged to supply an error value such as #NUM! instead of 0 when collinearity is present. You, as a user, should not be asked to shoulder the responsibility of diagnosing the reason for 0s such as those in cells H3 and H4 of Figure 4.7.

The t-ratio and the F-ratio

Before we leave the topic of regression coefficients and their standard errors, let me draw your attention once more to the t-ratio shown in Figure 4.5, as well as the F-ratio. The t-ratio appears in cell E10 in that figure, and the F-ratio appears in cell D5. The t-ratio, obtained by dividing the regression coefficient by its standard error, addresses the question of whether the regression coefficient is 0.0. If the t-ratio is large enough, given its degrees of freedom, to be sufficiently improbable as a sample value when the coefficient for the full populations is 0.0, then you might reject the hypothesis that the sample came from such a population. You're the judge of whether the associated probability statement makes the result "sufficiently improbable."

We'll get into the topic of the F-ratio later in this chapter, in the section titled "Interpreting the F-ratio." First, though, I should point out that the F-ratio serves a similar function for the full regression as does the t-ratio for a single regression coefficient. In the context of the full regression analysis, one way of expressing this concept is to say that the F-ratio is used to test whether the R^2 for the regression is large enough to reject the hypothesis that the R^2 for the regression in the full population is 0.0—that is, no variability shared by the predictor variable and the predicted variable.

4

In fact, when the regression analysis involves only one predictor variable, as is the case in the examples discussed in this chapter, the t-ratio and the F-ratio are equivalent. With just one predictor, the test that 0% of the variance is shared—in the population—is equivalent to the test that the slope of the regression line is 0.0—again, in the population.

So you might expect that in the single predictor situation, the F-ratio and the t-ratio return equivalent probability statements—and they do. (We have not yet arrived at the point of discussing the F-test as a stand-in for the t-test.) Nevertheless, you can test their equivalence. With just one predictor variable, the F-ratio is exactly the square of the t-ratio.

And so it is in Figure 4.5. The square of the t-ratio of 2.559 is exactly equal to the F-ratio of 6.546, out to the thirteenth decimal place. (Beyond that we get into rounding errors caused by differences in how the two ratios are calculated.)

Interval Scales and Nominal Scales

Bear that condition, "With just one predictor variable," in mind. In the present example, we have just one predictor: Height. It is what's termed an *interval variable*. Variables measured on an interval scale have numbers that indicate the size of differences: For example, the difference between 5'0" and 5'2" is the same as the difference between 6'0" and 6'2". We can represent the entire set of values assumed by the variable with just one column on an Excel worksheet.

But as you'll see in Chapter 7, "Using Regression to Test Differences Between Group Means," it's not only possible but desirable to represent a *nominal variable* numerically, using a coding approach. A nominal variable is one that assumes two or more values that are just names: For example, Male and Female, or Anglo, Asian, and Latin.

You could represent Male and Female by assigning them codes such as 0 and 1. Then a single variable—in worksheet terms, a column—would be sufficient to represent the difference between the two sexes, to distinguish a Male from a Female.

Using the same approach to three nominal values such as Anglo, Asian, and Latin would require not one but two variables or columns: one to distinguish Anglos from the other two groups, and one to distinguish Asians from the other two groups. (As you'll also see in Chapter 7, this approach automatically results in the distinction between Latins and the other two groups.)

The traditional way of analyzing this sort of data is the Analysis of Variance (*ANOVA*), and it's in that context that you might have encountered the numeric equivalence between a t-ratio and an F-ratio, where $t^2 = F$. It's often noted that, comparing a t-test to an ANOVA, t^2 equals F when there's only one variable that takes on only two values. (I discuss the F-ratio in much more detail later in this chapter, in "Understanding LINEST()'s F-ratio.")

When we code such a binomial variable, we need one worksheet column only, just as in this section's example regarding Height and Weight: Height, an interval variable, can be accounted for using just one column. So in those cases, t^2 equals F, but not when multiple columns are needed either for multiple interval variables, or for even just one nominal variable that has more than two values (and therefore requires at least two columns to distinguish among those values).

Finally, it's useful to bear in mind that the F-ratio for the full regression tests whether R^2 is different from 0.0 in the full population. That R^2 incorporates the effects of *all* the predictor variables on the regression. Thus far we have considered simple, single-predictor regression only. However, in multiple regression, where more than just one predictor is involved, the full regression reflects the relationship between each predictor and the predicted variable. It's that multiple predictor relationship to the predicted variable that the F-ratio assesses.

In contrast, the regression coefficient applies to one predictor only. Therefore, in the context of multiple regression, t^2 can equal F only when all the remaining regression coefficients equal 0.0—that is, when they have no observed relationship to the predicted variable. That happens only with made-up data.

In sum, t^2 and F express the same degree of relationship, but are equivalent only when there's one predictor variable in a regression, or, what's the same thing, just two categories in a nominal variable.

The Squared Correlation, R^2

Chapter 3 discusses the statistic R^2 in some detail, particularly the notion that R^2 expresses the proportion of the variability in the predicted variable that is associated with variability in the predictor variable. This chapter has much more to say about R^2—it's involved in one way or another with the standard error of estimate, the F-ratio, and both the regression sum of squares and the residual sum of squares. R^2, the squared correlation, shows up in just about everything you might do in the area of regression analysis.

Figure 4.8 summarizes how the squared correlation quantifies the degree of shared variance.

Figure 4.8
R^2 is easily calculated from the sums of squares.

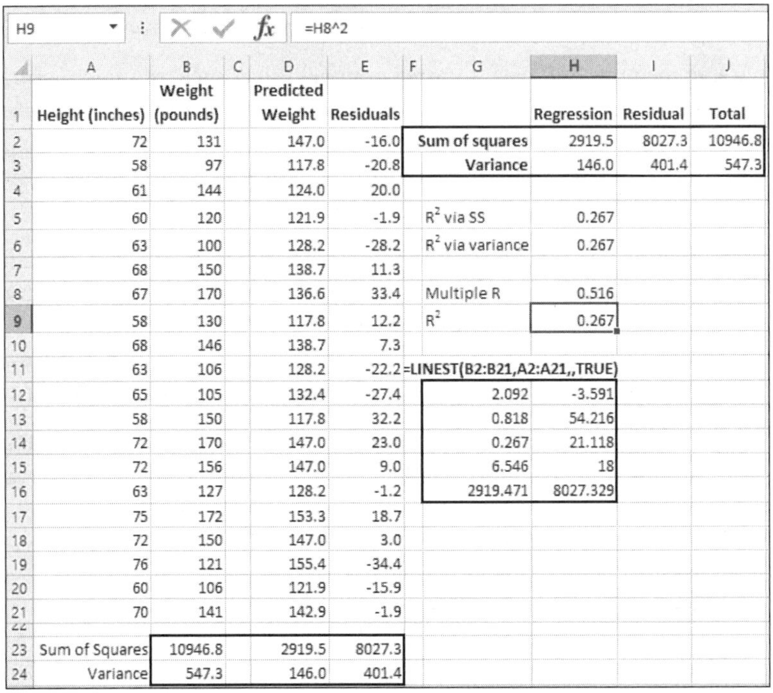

H9 =H8^2

	A	B	C	D	E	F	G	H	I	J
1	Height (inches)	Weight (pounds)		Predicted Weight	Residuals			Regression	Residual	Total
2	72	131		147.0	-16.0		Sum of squares	2919.5	8027.3	10946.8
3	58	97		117.8	-20.8		Variance	146.0	401.4	547.3
4	61	144		124.0	20.0					
5	60	120		121.9	-1.9		R² via SS	0.267		
6	63	100		128.2	-28.2		R² via variance	0.267		
7	68	150		138.7	11.3					
8	67	170		136.6	33.4		Multiple R	0.516		
9	58	130		117.8	12.2		R²	0.267		
10	68	146		138.7	7.3					
11	63	106		128.2	-22.2	=LINEST(B2:B21,A2:A21,,TRUE)				
12	65	105		132.4	-27.4		2.092	-3.591		
13	58	150		117.8	32.2		0.818	54.216		
14	72	170		147.0	23.0		0.267	21.118		
15	72	156		147.0	9.0		6.546	18		
16	63	127		128.2	-1.2		2919.471	8027.329		
17	75	172		153.3	18.7					
18	72	150		147.0	3.0					
19	76	121		155.4	-34.4					
20	60	106		121.9	-15.9					
21	70	141		142.9	-1.9					
22										
23	Sum of Squares	10946.8		2919.5	8027.3					
24	Variance	547.3		146.0	401.4					

In Figure 4.8, the raw data appears in A2:A21 (the Height variable) and B2:B21 (the Weight variable). Weight, predicted from Height, is in D2:D21, where the TREND() function is used to calculate the predicted values. The array formula in D2:D21 is:

=TREND(B2:B21,A2:A21)

We get the residual values in column E easily, by subtracting the predicted weights from the actual observations in column B. Now that we have three series of values—the actual observations, the predicted values, and the residuals—we're in a position to calculate and use the sums of squares.

The "sum of squares" is the sum of the squared *deviations* of the observations from their mean. So the sum of squares in cell B23 of Figure 4.8 is calculated by subtracting the mean of the weights from each weight, squaring the differences, and summing the squared differences (or *deviations*). The Excel worksheet function DEVSQ() does all that on your behalf. The formula in cell B23:

=DEVSQ(B2:B21)

returns the sum of the squared deviations of the observed weights from their mean value. The variance of the 20 weights can be easily calculated by this formula:

=VAR.P(B2:B21)

as it is in cell B24, or with a trifle more difficulty with this formula:

=B23/20

which stresses that the variance is the average of the squared deviations.

The same calculations appear in cells D23 (sum of squares) and D24 (variance) for the predicted values, and finally in E23 (sum of squares) and E24 (variance) for the residual values. For easier comparison I have repeated those sums of squares and variances in the range H2:J3 of Figure 4.8.

In that range, notice that the figure labeled "total" sum of squares in cell J2 is identical to the sum of squares for the original Weight observations in cell B23. Also note that the "regression" sum of squares in cell K2 is identical to the sum of squares for the predicted values in cell D23. Finally, notice that the "residual" sum of squares in cell I2 is equal to the sum of squares for the residuals in cell E23.

The sum of the regression sum of squares and the residual sum of squares is precisely equal to the total sum of squares as calculated directly on original observations. All this means that the regression has divided the total sum of squares into two completely separate portions: one sum of squares based on predicted values, and one based on unpredicted values. The fact that the two sums of squares add up to the total sum of squares means that we can divide the regression sum of squares by the total sum of squares, to obtain the percent of the total variability that can be predicted by the regression.

In this case that ratio is .267 (of course, expressed as a percentage, 26.7%). We call that value the R^2, the square of the correlation. That's apt enough, because you'll note that the correlation between the original Weight values and the predicted Weight values, obtained with this formula:

=CORREL(B2:B21,D2:D21)

in cell H8, is .516. That's the square root of R^2, as shown in cell H9.

(All these calculations relating the sums of squares to one another and to the value of R^2 can also be done with the variances.)

Thought of in this way, it's easier to feel comfortable with the notion of R^2 as a measure of shared variability. That sense of comfort helps later on, when we begin to examine multiple regression, the effect of adding new variables to a regression equation, and the effect of variance shared by multiple predictor variables.

Finally, notice in Figure 4.8 that the calculations of R^2 from the sums of squares (or from the variances) in cells H5, H6, and H9 all agree with the value in cell G14, the value for R^2 returned directly by the LINEST() worksheet function.

We next take a look at the standard error of estimate and its relationship to R^2.

4

The Standard Error of Estimate

Let's start an examination of the standard error of estimate by revisiting the Height and Weight data set in Figure 4.9.

As does Figure 4.8, Figure 4.9 shows the actual predictions of the Weight variable given its relationship to the Height variable. The residual values for Weight are in the range E2:E21. When you predict a variable such as Weight from a variable such as Height, one of the issues you're interested in is the accuracy of the predictions. One way of assessing their accuracy is to quantify the errors made by the predictions—that is, the residuals, the differences between the actually observed values (column B in Figure 4.9) and the predicted values (column D in Figure 4.9).

Figure 4.9
Both R^2 and the standard error of estimate express the accuracy of the predictions from the regression.

	A	B	C	D	E	F	G	H	I	J
1	Height (inches)	Weight (pounds)		Predicted Weight	Residuals			=LINEST(B2:B21,A2:A21,,TRUE)		
2	72	131		147.0	-16.0			2.092	-3.591	
3	58	97		117.8	-20.8			0.818	54.216	
4	61	144		124.0	20.0			0.267	21.118	
5	60	120		121.9	-1.9			6.546	18	
6	63	100		128.2	-28.2			2919.471	8027.329	
7	68	150		138.7	11.3					
8	67	170		136.6	33.4					
9	58	130		117.8	12.2					
10	68	146		138.7	7.3					
11	63	106		128.2	-22.2					
12	65	105		132.4	-27.4					
13	58	150		117.8	32.2					
14	72	170		147.0	23.0					
15	72	156		147.0	9.0					
16	63	127		128.2	-1.2					
17	75	172		153.3	18.7					
18	72	150		147.0	3.0					
19	76	121		155.4	-34.4					
20	60	106		121.9	-15.9					
21	70	141		142.9	-1.9					
22										
23	Sum of Squares	10946.800		2919.471	8027.329					
24	Standard deviation				20.555					

NOTE It's sometimes said, a bit carelessly, that the standard error of estimate is the standard deviation of the residuals. That's a useful way of thinking about the concept, but it's not quite right: For example, compare cells I4 (the standard error of estimate) and E24 (the standard deviation of the residuals) in Figure 4.9. I clarify this issue in this chapter's section titled "Standard Error as a Standard Deviation of Residuals."

Although we tend to use the relatively innocuous term *residuals* for those differences, it's perfectly accurate to term them *errors*—hence the term *standard error of estimate*. That statistic, which is returned in the third row, second column of the LINEST() results (cell I4 in Figure 4.9), helps you to gauge how far any given prediction might be from the associated actual.

Of course, given the data shown in Figure 4.9, you *know* what the actual observation is, and you know the value of the prediction. However, one important function of regression analysis is to estimate the value of a predicted variable when it has not yet been actually observed. In the context of the present example, you might wonder what weight it would predict for someone whose height is 78 inches. That's a value for Height that you have not yet observed, and therefore you have not observed a value for Weight that's associated with that Height.

So you might plug the Height value of 78 into the regression equation and get a predicted value of 159.6 for Weight. Now, if someone who stands 6'6" subsequently walks into your lab and steps on the scale, you would not expect that person to weigh exactly 159.6, but it is true that the more accurate your regression equation, the more closely you would expect to predict the actual weight.

The smaller the standard error of estimate, the more accurate the regression equation, and the more closely you expect any prediction from the equation to come to the actual observation. The standard error of estimate provides you a way to quantify that expectation. Because the standard error of estimate is a type of standard deviation, you can use it to establish brackets around the specific prediction—in the present example, around a prediction of 159.6 pounds for a person who stands 6'6".

Suppose that over time, a respectable number of people—say, 25— who stand 6'6" come to your lab. The *t* distribution tells us that 95% of them would weigh between 2.1 standard errors below, and 2.1 standard errors above, the predicted weight. So, if you want to predict the weight of a random person who is 78 inches tall, your best estimate would be 159.6 pounds, and you would expect that 95% of people that tall would weigh between 115.3 (159.6 – 2.1*21.1) and 203.9 (159.6 + 2.1*21.1). The boundaries around the predicted value of 159.6 are found as discussed in the next section.

The t Distribution and Standard Errors

The t distribution is similar to the standard normal curve, but its shape varies along with the number of observations that underlie it. An example appears in Figure 4.10, similar to the example shown in Chapter 1 in Figure 1.8.

Two conditions sometimes force you to resort to the t distribution rather than the normal curve as a benchmark:

- You have a relatively small sample size, and therefore a relatively small value for the degrees of freedom. In this example, we have 20 observations. We lose two degrees of freedom due to two constraints on the data: the mean of the 20 observations and the regression coefficient for Weight on Height. That's small enough, if just barely, to warrant using the t distribution instead of the normal curve: Note in Figure 4.10 that the t distribution for 18 degrees of freedom is just perceptibly different from the normal curve.
- You do not know the standard deviation of the variable in the population from which you're sampling. That's the case here: We use the standard deviation of the sample weights to estimate the standard deviation in the population.

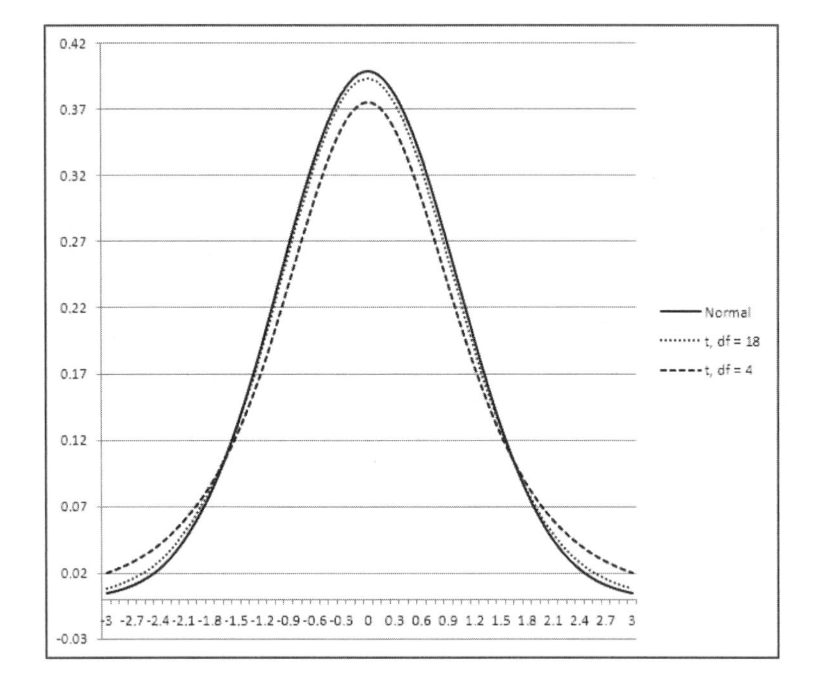

Figure 4.10
The smaller the degrees of freedom, the flatter the curve.

At this point we're back to the subjective part of this process. We're confronted by the definition of "improbable." Is an occurrence improbable if it takes place one time in 20 opportunities? One time in 100? One in 1,000? You'll have to decide that for yourself. For the purpose of the present example, I'll assume that you regard one time in 20 as improbable. Working your way through the example, it will be clear how to adjust the procedure so that you can be more (or less) picky than that.

HOW SIGNIFICANT IS IT?

The selection of an "improbability level" (in statistical jargon, that's often referred to as *alpha*) is one that has vexed quantitative research for decades. In some cases, it's possible to bring cost-benefit analysis to bear on the question. Then, you might choose a level (more jargon: *set alpha*) to a level that balances the costs of a false positive with those of a false negative. Of course, many decisions aren't susceptible to cost-benefit analysis: How do you measure the cost of administering an NSAID to reduce inflammation when doing so increases the probability of a stroke or heart attack?

Questions such as that are often difficult to answer and many researchers have just shrugged their shoulders and picked one of the conventional values for alpha. You've seen the footnotes in the reports: "p < .05" or "p < .01". Why those specific values?

Well, the principal reason is that for many years researchers had to rely on appendices in textbooks that called out, for a given t-statistic or F-ratio and the associated degrees of freedom, the value necessary so as to be regarded as "statistically significant" at the .05 or the .01 level. Those textbook appendices typically had room for only those two levels of significance of the F-ratios, when the tabled levels crossed the degrees of freedom for the numerator (usually in columns) and those for the denominator (usually in rows). The appendices supplied a few more levels of significance for t-ratios, which have only one value to tabulate for degrees of freedom. The researchers reported what they could, using the available criteria for the .05 and .01 levels of "statistical significance."

Today, it's easy to determine a critical t-statistic or F-ratio at a given level of improbability. Suppose the following:

- Your cost-benefit analysis tells you to limit the likelihood of a false positive to 3.5%.
- You have 18 degrees of freedom.
- You make a directional ("one-tailed") hypothesis and expect a negative t-value.

You're not out of luck—as you would have been if you had to rely on a textbook appendix. Just enter this in an Excel worksheet:

=T.INV(0.035,18)

The result will tell you that you need a t-value of −1.926, or even farther from 0.0, to regard the outcome as statistically significant with an alpha of 0.035.

To review: You would like to set the brackets around the predicted value of 159.6. That's the weight that your regression equation predicts for someone who is 78 inches tall, and you want to set brackets around that value. Those brackets should capture 95% of the weights you obtain when you measure more people whose height is 78 inches.

NOTE I'm using the predicted value of 159.6 pounds for someone who stands 78 inches tall solely as an example. In regression analysis, we assume that the variance of prediction errors is the same regardless of the value of the predictor. See the section titled "Homoscedasticity: Equal Spread," later in this chapter, for more information. For the time being, simply be aware that you can put the brackets around any value of weight predicted by the regression equation, whether that prediction is 100, 159.6, or 500 pounds.

Putting the brackets around a predicted weight of 159.6 pounds, the result is an interval in which 95% of actual weights would fall if more people who stand 6'6" are measured and weighed.

Now, how many standard errors from 159.6 pounds capture 95% of the possible observations of the weight of those who stand 6'6"? We resort to Excel's T.DIST() for the answer:

=T.INV.2T(.05,18)

and Excel returns 2.1. That means that 95% of the t distribution with 18 degrees of freedom lies between 2.1 standard errors above, and 2.1 standard errors below, the mean of the distribution. 2.5% lies more than 2.1 standard errors above, and another 2.5% lies more than 2.1 standard errors below. (That's the effect of the "2T" portion of the function's name: It tells Excel to divide the 5% equally between the two tails of the distribution.)

How big is one standard error? In this case, it's 21.1 pounds. So 2.1 standard errors is 2.1 * 21.1 or 44.3, and the interval is 88.6 pounds wide. Placed around a predicted weight of 159.6, we wind up with an interval from 115.2 to 203.9.

How do we interpret this information? Recall that we're looking for a context for the predicted weight of 159.6. Suppose that we replicated our 20-person sample hundreds of times, and put an interval of +/–44.3 pounds around the predicted weight associated with a height of 78 inches. Because the interval was constructed to capture 95% of the area under the t distribution, we can expect that the actual population weight, predicted from a height of 78", would fall within 95% of those hypothetically calculated intervals.

Which is more rational: to suppose that the interval we calculated is one of the 5% that will not contain the population value, or one of the 95% that will contain it? The latter, surely.

Now, that probably seems like a pretty wide interval, running as it does from 115.2 to 203.9. Bear in mind, though, that it's based on a regression equation whose R^2 is only .276. Just a little more than a quarter of the variability in height is shared with weight. A more accurate set of predicted weights would come about from a regression equation with a larger R^2, such as the one shown in Figure 4.11.

For example, if I change the Height measures so that the R^2 increases from .276 to .899, as I did in Figure 4.11, the standard error of estimate comes down from 21.1 to 7.8. Because that also changes the entire regression equation, the predicted weight for a height of 78 inches is 175.5. The bracket would still be based on 2.1 standard errors, but a standard error of 7.8 results in a bracket that's 32.9 pounds wide, running from 159.1 pounds to 192.0, instead of one that's 88.6 pounds wide. That context results in a much smaller scope for error, and you still have 95% confidence that you have captured the population weight for six-and-a-half footers.

4

Figure 4.11
The larger the R^2, the smaller the standard error of estimate.

	A	B	C	D	E	F	G	H	I	J
1	Height (inches)	Weight (pounds)		Predicted Weight	Residuals			=LINEST(B2:B21,A2:A21,,TRUE)		
2	67	131		138.5	-7.5			3.366	-87.071	
3	55	97		98.1	-1.1			0.266	17.589	
4	65	144		131.7	12.3			0.899	7.833	
5	60	120		114.9	5.1			160.414	18	
6	58	100		108.2	-8.2			9842.388	1104.412	
7	70	150		148.6	1.4					
8	74	170		162.0	8.0					
9	61	130		118.3	11.7					
10	69	146		145.2	0.8					
11	59	106		111.5	-5.5					
12	59	105		111.5	-6.5					
13	65	150		131.7	18.3					
14	77	170		172.1	-2.1					
15	73	156		158.7	-2.7					
16	63	127		125.0	2.0					
17	78	172		175.5	-3.5					
18	72	150		155.3	-5.3					
19	65	121		131.7	-10.7					
20	58	106		108.2	-2.2					
21	69	141		145.2	-4.2					
22										
23	Sum of Squares	10946.800		9842.388	1104.412					
24	Standard deviation				7.624					

Standard Error as a Standard Deviation of Residuals

If you read about the standard error of estimate in statistics textbooks, you find that it's sometimes referred to as the standard deviation of the residuals. That's a good, intuitively useful definition. It's a reminder that the less accurate the regression's predictions, the larger the residuals, and the larger the standard error of estimate.

The problem is that when we're confronted with the term *standard deviation*, we all think of dividing by n or by $n-1$. If we just want an expression of the degree of variability in a set of observations, we use a formula like this one:

$$s = \sqrt{\Sigma_{i=1}^{n}(X_i - \overline{X})^2/n}$$

In words, this standard deviation is the square root of the average squared deviation from the mean of the data set. In Excel, you use the STDEV.P() worksheet function. (The "P" at the end of the function name indicates that you are regarding the set of observations as a population rather than as a sample from a population.)

But if you want to use the sample standard deviation to estimate the standard deviation of the population from which the sample was taken, you use a formula like this one:

$$s = \sqrt{\Sigma_{i=1}^{n}(X_i - \overline{X})^2/\,(n-1)}$$

Notice that the divisor in the ratio inside the radical sign is now $n - 1$ instead of n. Instead of using the average squared deviation, you're using the sum of the squared deviations divided by the degrees of freedom, $n - 1$. (I discuss the rationale for this adjustment briefly in Chapter 1 of this book, and at some length in Chapter 3 of *Statistical Analysis: Microsoft Excel 2013*.) In Excel, you use the STDEV.S() worksheet function instead of the STDEV.P() worksheet function. The "S" at the end of the function name tells Excel that you are regarding the set of observations as a sample from a population.

Now, when you use the standard error of estimate to assess the amount of variability in the errors of your regression equation's predictions, you are generalizing from a sample to a population. In effect, you're asking how well you can expect your equation to perform when you hand it new observations from a population. So you need to employ the degrees of freedom correction in the formula.

One final twist and we're there. The term *degrees of freedom* implies the number of observations that are "free to vary." If all you've done is grab some observations *as a sample*, you lose one degree of freedom when you calculate their standard deviation.

That's because the measure of variability is based on deviations from the mean of the observations, which acts as a constraint. Suppose you have 10 values in your sample, and their mean is 7. With the mean fixed, you can freely change all but one of the original observations and keep the same mean. The tenth observation is constrained by the fixed mean to a particular value, some figure other than its original value. Figure 4.12 shows how this works in practice.

Figure 4.12
With one constraint, the degrees of freedom is $n-1$.

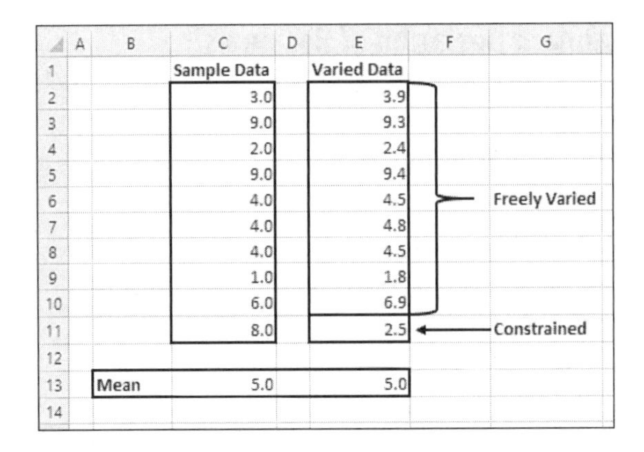

	A	B	C	D	E	F	G
1			Sample Data		Varied Data		
2			3.0		3.9		
3			9.0		9.3		
4			2.0		2.4		
5			9.0		9.4		
6			4.0		4.5		Freely Varied
7			4.0		4.8		
8			4.0		4.5		
9			1.0		1.8		
10			6.0		6.9		
11			8.0		2.5		Constrained
12							
13		Mean	5.0		5.0		
14							

In Figure 4.12, the range C2:C11 contains a random sample of ten data points from a population. Its mean is 5.0. The range E2:E10 shows the first nine values from column C, each varied slightly by adding a fractional value to the associated original observation.

The tenth observation in cell E11 is *constrained* by the need to keep the mean at 5.0. Given the changes to the values in E2:E10, the value in cell E11 must be 2.5. The other nine values are free to vary, and those nine values constitute the sample's degrees of freedom when you're making an estimate of the population's standard deviation.

When you use those ten observations to arrive at a regression coefficient, you are adding another constraint, besides the one imposed by the mean of the observations. So you lose another degree of freedom. Furthermore, as you add new predictor variables to the regression analysis, you lose an additional degree of freedom for each additional predictor.

That's the reason that the intuitively appealing definition of the standard error of estimate, the standard deviation of the residuals, is nevertheless somewhat misleading. A textbook invites you to calculate the residuals, point a function such as STDEV.S() at them, and expect that the result will equal the standard error of estimate.

That won't happen, because the standard deviation of the errors of estimate has more constraints, and fewer degrees of freedom, than the verbal definition leads you to expect. See Figure 4.13.

The proper formula for the standard error of estimate, as shown in Figure 4.13, is as follows:

$$\text{Standard Error of Estimate} = \sqrt{\Sigma_{i=1}^{n}(X_i - \bar{X})^2/(n-k-1)}$$

In that formula, k is the number of predictor variables in the regression equation. So in simple, single-predictor regression, the degrees of freedom equals the number of observations, n, less 1 for the regression coefficient, k, less 1 for the mean. With three predictor variables in a multiple regression equation, the standard error of estimate would be calculated using the number of observations minus 4.

Figure 4.13
Accounting properly for the constraints gets you the correct standard error of estimate.

D23 =DEVSQ(D2:D21)

	A	B	C	D	E	F	G	H	I
1	Height (inches)	Weight (pounds)		Predicted Weight	Residuals		=LINEST(B2:B21,A2:A21,,TRUE)		
2	72	131		147.0	-16.0		2.092	-3.591	
3	58	97		117.8	-20.8		0.818	54.216	
4	61	144		124.0	20.0		0.267	21.118	
5	60	120		121.9	-1.9		6.546	18	
6	63	100		128.2	-28.2		2919.471	8027.329	
7	68	150		138.7	11.3				
8	67	170		136.6	33.4				
9	58	130		117.8	12.2		SS Residual	8027.329	=E23
10	68	146		138.7	7.3		DF	18	=H5
11	63	106		128.2	-22.2		Variance	445.963	=H9/H10
12	65	105		132.4	-27.4		Std Error of Etimate	21.118	=SQRT(H11)
13	58	150		117.8	32.2				
14	72	170		147.0	23.0		$1 - R^2$	0.733	=1-G4
15	72	156		147.0	9.0		Total Variance	608.156	=B23/H5
16	63	127		128.2	-1.2		Variance of Estimate	445.963	=H14*H15
17	75	172		153.3	18.7		Std Error of Etimate	21.118	=SQRT(H16)
18	72	150		147.0	3.0				
19	76	121		155.4	-34.4		Square root of SS Resid/DF	21.118	=SQRT(H6/H5)
20	60	106		121.9	-15.9				
21	70	141		142.9	-1.9				
22									
23	Sum of Squares	10946.800		2919.471	8027.329				

All that is done in Figure 4.13:

- The sum of the squared deviations of the residuals, calculated in cell E23, is picked up in cell H9. It's also available in the LINEST() results.
- The degrees of freedom, $n - k - 1$, in cell H10, is picked up from the LINEST() results.
- The variance of the residuals, using the correct degrees of freedom in the denominator, is calculated in cell H11. You might see this statistic referred to as the *variance of estimate* in some sources.
- The standard error of estimate is calculated in cell H12 as the square root of the value in cell H11.

Notice that the value for the standard error of estimate in cell H12 is identical to that returned by LINEST() in cell H4.

Here's another way of expressing the standard error of estimate:

$$S_y\sqrt{1 - R^2}$$

In words, that's the standard deviation, using $n - k - 1$ as the degrees of freedom, of the *original* values for the predicted variable (in this example, weight in pounds) times the square root of 1 minus R^2.

I like that definition because it ties the measure of R^2 directly to the calculation of the standard error of estimate. Consider the expression $(1 - R^2)$. Because R^2 quantifies the percentage of variability in weight that's associated with variability in height, $(1 - R^2)$ must be the percentage of variability in weight that is *not* predictable by height.

So if you multiply the variance of the predicted variable, S_y^2, by the percentage of unpredicted variance:

$$S_y^2(1 - R^2)$$

you get the variance of estimate, and its square root is the standard error of estimate:

$$S_y\sqrt{1 - R^2}$$

Figure 4.13 also traces this calculation in the range H14:H17, as follows:

- Cell H14 contains the percentage of unpredictable variance.
- Cell H15 contains the variance of the Weight variable, using $n - k - 1$ as the denominator in the ratio.
- Cell H16 contains the product of the unpredictable percentage in H14 and the variance in H15. This gives a measure of the unpredictable variance, measured by weight in pounds: The variance of estimate.
- Cell H17 contains the standard error of estimate, the square root of the variance of estimate.

Finally, Figure 4.13 shows in cell H19 the fastest way to get the standard error of estimate, short of just looking at the LINEST() results. It is the square root of the ratio of the Sum of Squares Residual, returned by LINEST(), to $n - k - 1$, also returned by LINEST(). This formula is just a very slightly tweaked version of the calculations in the range H9:H12, but it helps to reinforce the notion that all the statistics in the final three rows of the LINEST() function are intimately interrelated.

Homoscedasticity: Equal Spread

One of the assumptions that's frequently touted as necessary to carry out a regression analysis is the assumption of equal spread. (It's also termed *homoscedasticity*, but I admit that I've never run into that word outside of a statistics text.)

Here's the idea: You assume (and sometimes test) that the variance of the variable to be predicted is the same for each value of the predictor variable. So, in the context of the Height-Weight example that this chapter has used, the assumption is that the variance of Weight for all the people who stand, say, 5'6" is the same as the variance of Weight for all the people who stand, say, 6'2". The variability of the Weight values is assumed to be equal for each value of Height.

This is one of those assumptions that's more of a nice-to-have than a sine qua non. If the assumption is violated, no violence is done to the predicted values, to the assessment of how replicable the findings are, to the values of the regression coefficients or their standard errors.

All you've lost if the assumption is violated is the ability to use the same error of estimate regardless of which value of the predictor variable you're interested in. The principal rationale for using the standard error of estimate in that way is to construct a confidence interval—that is, the brackets discussed earlier in this chapter, in the section, "The Standard Error of Estimate"—around the predicted value.

This issue can become more important when you get into the business of converting a nominal variable to numeric codes and using regression analysis to test the differences between group means. Then, the relationship between group variances and the number of observations per group can get to be a problem. Even then, that problem arises only when both the group variances *and* the group sizes are unequal.

By the way, don't be misled by the cigar-shaped appearance of a scatter chart of a predictor that correlates fairly well with a predicted variable. The ends of the cigar are typically narrower than the center of the distribution, suggesting that the variability is greater in the middle of the distribution than in its tails. But what you're looking at in that case is the *range* of the values. The range is very sensitive to the number of cases, and the tails of any distribution that's remotely similar to a normal curve have fewer observations than does its center. Ignore the visual evidence of the ranges. If you must worry about this assumption, worry about the variances, which (because they're averages) are insensitive to sample sizes.

Understanding LINEST( )'s F-ratio

When you take it down to its basics, the F-ratio is simply a ratio of one variance to another, given that the two variances are arrived at independently. That's pretty basic. However, the F-ratio underlies a broad array of inferential statistical analyses with single outcome variables. (Wilks' Lambda tends to replace F in multivariate analyses, those with multiple outcome variables.) Of particular interest here is the F-ratio's use in the analysis of variance, or *ANOVA*, where it helps you decide whether a meaningful result is also a reliable one.

NOTE

The *F* in *F-ratio* has nothing to do with its purpose; the term *F-ratio* was conferred by the statistician George Snedecor in honor of Sir Ronald Fisher, who developed the analysis of variance in the early twentieth century as a method of testing mean differences for one or more factors, as well as their interactions.

The Analysis of Variance and the F-ratio in Traditional Usage

You might have come across the F-ratio if you ever studied the analysis of variance in a college statistics course. In traditional ANOVA, you calculate a quantity called the *Sum of Squares Between Groups*, often shortened to *Sum of Squares Between* or abbreviated as SS_B. That quantity is directly analogous to regression's Sum of Squares Regression.

ANOVA also calculates a *Sum of Squares Within Groups*, or *Sum of Squares Within* or SS_W. It is directly analogous to regression's Sum of Squares Residual.

Used in this fashion, the analysis of variance and the F-ratio are meant to assess the likelihood that the observed differences in the means of sampled groups will stand up if subsequent samples are taken and tested. The best sort of experimentation obtains random samples from a population (of humans, or plants, or anything that can react to a treatment) and assigns them, again randomly, to two or more groups. Those groups nearly always include at least one group that receives an experimental treatment and a group that constitutes a control or comparison—one that receives no treatment or that receives a traditional treatment whose effects are already understood.

It simplifies discussion to act as though the treatments put the sampled groups into different populations: the population of subjects who receive Treatment A, the population of subjects who receive a placebo, and so on. In this sort of experimental design, it's usually feasible to infer a causal relationship between the treatments and their effects: The application of the treatment *causes* differences in their outcomes. Then, the variable that distinguishes the groups is often termed an *independent* variable.

After the treatments have concluded, a measure that's sensitive to the presumed treatment effects is administered and the mean on that variable is calculated for each group. Again with this sort of experimental design, that variable is often termed a *dependent* variable: Its values are thought to depend on the treatments—the independent variable—that are applied to the subjects.

The opportunity to infer causation, not simply a relationship, is a feature of the experiment's design (random selection, random assignment and so forth), not the statistical procedure that's employed. Therefore, many researchers and statisticians are careful about using terms such as *independent variable* and *dependent variable* when discussing research that does not use a rigorous experimental design. Many alternatives, such as *criterion variable* and *predicted variable* have been used, but it seems that there is little agreement in the literature.

When the emphasis is on the role that the variable plays, I tend to use the terms *predictor variable* and *predicted variable* because I find them both relatively descriptive and relatively innocuous. When the emphasis is on the way the variable is measured, many (myself included) prefer *factor* for a variable measured on a nominal scale and *covariate* for a variable measured on an interval scale.

The analysis of variance as traditionally applied to research into the reliability of mean differences measures the variance of the group means. A straightforward statistical calculation converts the variance of the means to an estimate of the total variance of all the subjects, usually termed the *mean square between*. Then, the variances within each group are calculated and averaged. This provides a separate, independent estimate of the variance of all the subjects, usually termed the *mean square within*.

Finally, the two calculated variance estimates are compared via the F-ratio, dividing the variance estimated from the differences in the group means by the average variance within the groups. If the differences between the means are large, due solely to differences in the treatments, the ratio's numerator will tend to be large relative to its denominator.

We know enough about the theoretical distribution of F-ratios to determine the likelihood of observing a given F-ratio in an experiment if the treatments have no reliable differential effect. So we can conclude, for example, that "The probability of observing this F-ratio is only 1% if the means of the populations of subjects are identical—and therefore have no variance. That seems to us extremely unlikely, and we prefer to infer that the population means in fact differ from one another."

The Analysis of Variance and the F-ratio in Regression

Most of the considerations discussed in the preceding section apply to regression every bit as much as they do to traditional analysis of variance. However, regression takes a different tack to reach the same conclusions. Instead of working solely with sums of squares, regression enriches the analysis by working with proportions of variance, in particular proportions of shared variance.

As you've seen, the regression approach is capable of using interval and ratio variables (such as Height in this chapter's example of predicting Weight) as predictor variables. As you'll see in the next section, regression is also capable of using a nominal variable, such as Type of Medication, Sex, or Ethnicity as a predictor—or, if the experimental design is appropriate for the term, as an independent variable.

So, the regression is more flexible than the traditional approach to the analysis of variance. Both approaches analyze variance, so you could refer to either as an ANOVA (regression applications typically include a table of results labeled "Analysis of Variance"). But the traditional approach does not have a feasible way to analyze a predictor variable that's measured on an interval scale, such as Height in the context of a Height-Weight investigation. It's true that traditional ANOVA can accommodate one or more covariates (interval variables) along with one or more factors (nominal variables) in an approach called analysis of covariance, or *ANCOVA*. This book discusses the regression approach to ANCOVA in Chapter 8, "The Analysis of Covariance." But traditional ANOVA becomes very clumsy if you want to investigate the relationship between a single predictor that's measured on an interval scale, and a predicted variable.

What about the opposite problem, the use of regression to analyze a nominal predictor? To use regression analysis with a factor such as Type of Medication rather than a covariate such as Height, it's necessary to code the factor with numeric codes. For example (one which I'll explore more fully in the next section), in a medical experiment you might code one level of the treatment factor, Statin, as a 1, and another level of the treatment factor, Diet, as a 0. By assigning numeric values to names such as Statin and Diet, you put any regression application—whether SAS, SPSS, Minitab, Stata, R, or Excel—in a position to calculate the regression coefficients, their standard errors, R^2, the standard error of estimate, the F-ratio, the sums of squares for the regression and for the residual, degrees of freedom, and related regression statistics such as R^2 shrinkage.

As you've seen earlier in this chapter, R^2 expresses the proportion of the total sum of squares that's associated with the regression of the predictor variable (or predictor variables) with the predicted variable. The quantity (1 – R^2) expresses the proportion of the total sum of squares associated with the residual, or errors of prediction.

Therefore, regression puts you in a position to work with percentages of the total sum of squares: intuitively meaningful figures such as 26.7% of the total sum of squares (or, as you'll see, 26.7% of the total variance), rather than frequently arbitrary values such as a Sum of Squares Regression (or Sum of Squares Between) such as 2919.471. See Figure 4.14.

Figure 4.14
In accounting for variance, it matters little whether you work with sums of squares or proportions of sums of squares.

G4 {=LINEST(B2:B21,A2:A21,,TRUE)}

	A	B	C	D	E	F	G	H	I	J
1	Height (inches)	Weight (pounds)		Predicted Weight	Residuals		=LINEST(B2:B21,A2:A21,,TRUE)			
2	72	131		147.0	-16.0		2.092	-3.591		
3	58	97		117.8	-20.8		0.818	54.216		
4	61	144		124.0	20.0		0.267	21.118		
5	60	120		121.9	-1.9		6.546	18		
6	63	100		128.2	-28.2		2919.471	8027.329		
7	68	150		138.7	11.3					
8	67	170		136.6	33.4					
9	58	130		117.8	12.2		Sum of Squares	DF	Mean Square	F-ratio
10	68	146		138.7	7.3		2919.471	1	2919.471	6.546
11	63	106		128.2	-22.2		8027.329	18	445.963	
12	65	105		132.4	-27.4					
13	58	150		117.8	32.2		% of Variance	DF	"Mean Square"	F-ratio
14	72	170		147.0	23.0		0.267	1	0.267	6.546
15	72	156		147.0	9.0		0.733	18	0.041	
16	63	127		128.2	-1.2					
17	75	172		153.3	18.7					
18	72	150		147.0	3.0					
19	76	121		155.4	-34.4					
20	60	106		121.9	-15.9					
21	70	141		142.9	-1.9					
22										
23	Sum of Squares	10946.800		2919.471	8027.329					

Figure 4.14 analyzes the Height and Weight data in three different ways.

Analysis via LINEST( )

The first of the three analyses in Figure 4.14 is shown as the results of the LINEST() function. In the range G2:H6 you can find the LINEST() results shown in several figures earlier in this chapter, most recently in Figure 4.13. Notice the sum of squares for the regression in cell G6 and for the residual in cell H6. They agree with the results of the DEVSQ() function in cells D23 and E23.

Analysis via Sums of Squares

The analysis in the range G10:J11 of Figure 4.14 represents the traditional approach to the analysis of variance, even though the predictor variable is a covariate rather than a factor. Were it a factor, the sum of squares in cell G10 would be termed the Sum of Squares Between, and the value in cell G11 would be termed the Sum of Squares Within, but with a covariate as a predictor, you would probably term them Sum of Squares Regression and Sum of Squares Residual. The values are linked directly to those calculated via DEVSQ() in D23 and E23.

The covariate is one variable only, and so the degrees of freedom for the regression equals 1. This is true in general. Whether you're using regression to analyze the association of covariates or that of factors with a predicted variable, the number of columns occupied by the predictors equals the degrees of freedom for the regression. That number reduces the degrees of freedom for the residual and increases the degrees of freedom for the regression.

Cells I10 and I11 contain the variance estimated by the regression and the residual variation. These variances, also termed *Mean Squares* in ANOVA jargon, are obtained by dividing each sum of squares by its degrees of freedom. Finally, the F-ratio is obtained by dividing the Mean Square Regression by the Mean Square Residual, in cell J10. Notice that its value is identical to the value returned by LINEST() in cell G5.

Analysis Using Proportions of Variance

The range G14:J15 in Figure 4.14 contains the same analysis as in G10:J11, but the sums of squares have been replaced by their proportions of the total variance. For example, in cell G14 the proportion of the total sum of squares attributable to the regression is 2919.471 / 10946.8 or 0.267 (see cell B23 for the total). That proportion is identical to the R^2 returned by LINEST() in cell G4. That's as it should be, given that R^2 is the proportion of the total sum of squares attributable to the regression.

The proportions of variance, regression and residual, are divided by the appropriate degrees of freedom and the results of the divisions appear in cells I14 and I15. They are not truly mean squares: Rather than the result of dividing a sum of squares by its degrees of freedom, they are the result of dividing a *proportion* of a sum of squares by the associated degrees of freedom. I have to report that I've never seen a generally accepted term for these ratios, so I've labeled them as "Mean Square," with the quotation marks, in Figure 4.14.

Lastly, notice that the ratio of the "Mean Squares" in I14 and I15 is precisely and unsurprisingly the same as the F-ratios in cells G5 and J10.

Partitioning the Sums of Squares in Regression

Just as do the Regression Sums of Squares and the Residual Sums of Squares in regression analysis, ANOVA's Sums of Squares Between and Within add up to the Total Sum of Squares, the sum of the squared deviations of the predicted or outcome variable, calculated without regard to its relationship to the predictor or grouping variable. That is:

$$SS_B + SS_W = SS_{Total}$$

$$SS_{Regression} + SS_{Residual} = SS_{Total}$$

However, differences exist between traditional analysis of variance and regression analysis. Those differences are due to the nature of the factor in traditional ANOVA, and to the rationale for the analysis.

4

In its simplest form, traditional analysis of variance uses two variables: a factor (a nominal variable used to distinguish the groups that individuals belong to), and an interval variable that measures the effect of belonging to a particular group. For example:

- In a medical experiment, patients might be randomly assigned to one of three groups: a group that takes statins, a group whose diet is restricted to vegetarian meals, and a group that receives a placebo. With respectable sample sizes and random assignment, you can be reasonably sure that the groups are equivalent at the time that treatments begin. After one month of treatment, the participants' cholesterol levels are checked to see whether there are any differences in the mean cholesterol of each group. Even this sketchy description sounds like a true experiment, in which you're able to infer causation. With random assignment you might term the grouping factor an *independent variable* and the cholesterol level a *dependent variable*.
- You might plant herb seeds in four identical garden plots and use a different fertilizer in each plot. After some period of time you harvest and weigh the herbs, to see whether there are differences in the fertilizers' effects on the size of the plants. Again, this sounds like a true experiment and you could plausibly infer that any differences in harvested weight are due to the fertilizer used.

In these two examples, there's a factor: a nominal variable (patient group and fertilizer used) with a limited number of values:

- Cholesterol Treatment Group: Statin, Diet, Placebo
- Fertilizer Used: Brand A, Brand B, Brand C, Brand D

The number of applicable values is limited with variables that are measured on a nominal scale. Furthermore, neither of the two factors is associated with an underlying continuum. Suppose that three groups consisted of makes of car: Ford, Toyota, and General Motors. The nominal values on that nominal scale do not imply a continuum of "car-ness," where Ford (say) has the most and General Motors has the least. There's no implication of distance between any two brands on the nominal scale. The nominal scale's values consist of names alone.

In contrast, the example that we've been considering throughout this chapter—the prediction of weight from knowledge of height—measures Height, the covariate or predictor variable, on an interval scale. Height is an interval scale: not only is 60" taller than 30", but 30" is as far from 0" as 60" is from 30".

So it's entirely reasonable to treat different heights as different values that vary quantitatively from one another. Suppose, though, that you decided to use the number 1 to identify a member of the Statin group, a 2 to identify a member of the Diet group, and a 3 to identify someone who takes a placebo.

The two situations (height influencing the weight outcome, treatment influencing the cholesterol outcome) are completely different. In one case, you're asking whether an increment in height is associated with an increment in weight. In the other case, you're trying to ask whether an increment in treatment group, going from 1 (Statin) to 2 (Diet) is associated with, or even causes, an increment in cholesterol level.

But the analysis does not, *cannot*, answer that question because the Treatment variable is measured on a nominal scale rather than an interval scale. The difference between 1 (Statin) to 2 (Diet) is just a qualitative difference, not a quantitative increment. It's arbitrary whether Statin is 1 and Diet is 2, or vice versa. Going from 70 inches in height to 71 inches is not arbitrary: It's objective and meaningful, and the meaning is intrinsic to the nature of the scale. Seventy-one inches is more than 70 inches.

As it turns out, it's necessary to represent the nominal variable numerically, but in a different way. Split its information between two different variables: Let Variable 1 equal 1 if a person is in the Statin group, and let it equal 0 otherwise. Let Variable 2 equal 1 if a person is in the Diet group, and let it equal 0 otherwise. All members of the Placebo group have a 0 on both variables. Figure 4.15 shows how this layout looks on a worksheet.

Figure 4.15
This method of representing a nominal variable with numerals is called *dummy coding*.

	A	B	C	D
1	Treatment	Variable 1	Variable 2	LDL
2	Statin	1	0	163
3	Statin	1	0	182
4	Statin	1	0	179
5	Statin	1	0	170
6	Diet	0	1	175
7	Diet	0	1	196
8	Diet	0	1	173
9	Diet	0	1	193
10	Placebo	0	0	161
11	Placebo	0	0	173
12	Placebo	0	0	178
13	Placebo	0	0	187

The coding scheme converts a nominal variable with three values to two nominal variables that use numbers to distinguish the nominal values from one another—and that's all that Variable 1 is doing here: distinguishing the Statin treatment from the others, just as Variable 2 distinguishes the Diet treatment from the others. Together, Variable 1 and Variable 2 distinguish the Placebo treatment from the others—that is, anyone who has a 0 on both Variable 1 and Variable 2 gets a placebo.

The coded variables do not imply, with their specific values, more or less of anything, the way that the Height variable (60", 61", 62" and so on) does in earlier examples in this chapter. All the 1s and 0s are doing is distinguishing one category of the original, nominal variable from other categories.

This is true in general: When you use a coding system such as this one (which is called *dummy coding*) to convert a nominal variable, you wind up with one new variable fewer than original categories. If your original variable has five categories (such as Ford, Toyota, GM, Nissan, and Volvo), you wind up with four new, coded variables, one fewer than the original number of categories.

I'll get into these issues more fully in Chapter 7, "Using Regression to Test Differences Between Group Means." In the meantime I want to clarify where the number of degrees of freedom is coming from and going to when you use regression with a coded variable. When you add a new predictor variable to the analysis, you lose an additional degree of freedom from the Sum of Squares Residual—and the Sum of Squares Regression gets it right back.

The F-ratio in the Analysis of Variance

Traditionally, the route to calculating an F-ratio begins with the sums of squares, between groups and within groups. The sums of squares are converted to independent estimates of variance, by dividing by the degrees of freedom. Then, the variances are compared in a ratio, which is termed the *F-ratio*. If the ratio is large enough, you can conclude that the groups did not all come from the same population, but from different populations with different means.

Figure 4.16 provides an example of an ANOVA using a sample of data that comes from one population. Suppose that you randomly select 20 people standing in a grocery checkout line, randomly assign them to one of two groups, and measure their heights. You have no reason to suppose that the average heights of the two groups will differ, but you run the numbers anyway. See Figure 4.16.

Figure 4.16 A traditional ANOVA, evaluating the reliability of the difference in heights between two groups of randomly selected and assigned people.

	A	B	C	D	E	F	G	H	I	J
1	Group A	Group B		Anova: Single Factor						
2	63.7	70.3								
3	64.7	67.3		SUMMARY						
4	62.7	63.3		*Groups*	*Count*	*Sum*	*Average*	*Variance*		
5	71.7	68.3		Group A	10	677	67.7	18.89		
6	63.7	67.3		Group B	10	658	65.8	17.39		
7	73.7	57.3								
8	63.7	61.3								
9	71.7	64.3		ANOVA						
10	71.7	69.3		*Source of Variation*	*SS*	*df*	*MS*	*F*	*P-value*	*F crit*
11	69.7	69.3		Between Groups	18.05	1	18.05	0.995	0.332	4.414
12				Within Groups	326.5	18	18.14			
13										
14				Total	344.55	19				
15										
16										
17	Between	18.05	=DEVSQ(G5:G6)*10							
18	Within	326.5	=DEVSQ(A2:A11)+DEVSQ(B2:B11)							

Traditionally, ANOVA is used to gauge whether subsequent replications of the same experiment, using different subjects, can be expected to have the same results. Also traditionally, this has meant that if no "significant" difference between group means was found in the actual experiment, no differences in future hypothetical replications of the experiment are expected.

ANOVA reaches its conclusions about group means via sums of squares and variances. It begins with a rearrangement of the standard error of the mean. The formula for that statistic is:

$$S_{\bar{X}} = S/\sqrt{n}$$

In the formula:

- $S_{\overline{X}}$ is the standard error of the mean.
- S is the sample standard deviation.
- n is the sample size.

The standard error of the mean is just another standard deviation. Suppose you took, say, 100 samples from a population and calculated the mean of each sample. If you then found the mean of those 100 means, that mean would be very, very close to the (unobservable) mean of the population from which you took the samples.

Furthermore, the standard deviation of all those sample means—known as the standard error of the mean—is expected to be close to the result of the formula just given:

$$S_{\overline{X}} = S/\sqrt{n}$$

In other words, you can get a good estimate of the standard error of the mean from a single sample, by dividing that sample's standard deviation by the square root of the sample size. (Of course, the accuracy of the estimate becomes greater as the sample size increases.) I won't give the proof of the formula here, but if you want you can find it in Chapter 8 of *Statistical Analysis: Microsoft Excel 2013*.

Now, rearrange the formula to get this:

$$S_{\overline{X}}\sqrt{n} = S$$

and square both sides of the equation:

$$S_{\overline{X}}^2 n = S^2$$

In words, the variance S^2 is estimated by multiplying the variance error of the mean $S_{\overline{X}}^2$ by n.

In an analysis of variance, you can get the standard error of the mean from your sample. You could simply take the group means, two of them in this example, and calculate their variance. For example, in Figure 4.16, you could use this formula:

=VAR.S(G5:G6)

which returns 1.805.

Another method, which I use here because it's the way that traditional analysis of variance manages things, is to get the sum of the squared deviations of the group means from the grand mean, and total them. Recall that's exactly what Excel's DEVSQ() function does, and Figure 14.16 uses it in that way in cell B17:

=DEVSQ(G5:G6)

If you then multiply by the number of subjects per group (here, that's 10), you get the Sum of Squares Between. So the full formula in cell B17 is:

=DEVSQ(G5:G6)*10

or 18.05. That's the value of the Mean Square Between, which itself is not the variance error of the mean, but is the estimate of the total variance *based on* the variance error of the mean.

NOTE

The rearranged formula given earlier:

$$S_{\bar{X}}^2 n = S^2$$

states that the total variance S^2 is estimated by multiplying the variance error of the mean $S_{\bar{X}}^2$ by the number of observations per group. In this case, the variance error of the mean equals the sum of the squared deviations. The reason is that when you calculate the estimate of the population variance from the sample variance, you divide the sum of the squared deviations by (n – 1). Here we have the variance of two group means, so (n – 1) equals (2 – 1), you divide the sum of squares by the DF of 1, and the sum of the squared deviations equals the variance.

Figure 4.16 also provides the results of running the Data Analysis add-in's Single Factor ANOVA tool on the data in A2:B11. Notice that its Sum of Squares Between, in cell E11, is identical to the Sum of Squares Between calculated in cell B17. *Bear in mind* that cell B17 obtains its figure for the Sum of Squares Between using only the group means and the number of observations per group, 10. The sum of squares between is independent of the variation among individuals around either the group means or around the grand mean.

You can get another estimate of the total variance quite simply, by starting with the sum of the squared deviations within each group. Cell B18 contains this formula:

```
=DEVSQ(A2:A11)+DEVSQ(B2:B11)
```

Notice that this formula returns the sum of the squared deviations around each group mean. It makes no difference how high or low those two means are. What matters is the distance of each observation from the mean of its own group.

The result, 326.5, is the sum of squares within groups and is identical to the value returned by the Data Analysis add-in in cell E12. When the two sums of squares are divided by their respective degrees of freedom in F11 and F12, you get two estimates of the total variance in G11:G12.

It's important here to note that these are two *independent* estimates of the total variance. The Mean Square Between is based on the variance of the group means, and pays no attention to the variability of the individual observations within each group. The Mean Square Within uses only information about the variability of the individual observations around their respective group means, and pays no attention to the size of either mean relative to the size of the other.

So, if the Mean Square Between is large relative to the Mean Square Within—that is, if the F-ratio itself is relatively large—there is evidence that the group means differ. A difference in the group means tends to increase their variance and, therefore, the numerator's estimate of the total variance.

In the case of the data shown in Figure 4.16, the F-ratio is small, because the group means differ very slightly (67.6 versus 65.8, in the context of a standard deviation of 4.25). Excel offers several worksheet functions that tell you the probability of calculating a given F-ratio under the assumption that the F-ratio in the population is 1.0—that is, that the two variance estimates, between and within, are estimating the same population parameter. In this case, the probability is 0.332 (see cell I11) of getting an F-ratio of 0.995 in a sample from a population whose true F-ratio is 1.0. So if you repeated the experiment many times, you would observe an F-ratio this large one time in three when the group means in the population are equal. That's generally considered far too likely an outcome to reject the notion of equal group means.

Figure 4.17 provides another example of the use of the F-ratio in ANOVA, this time with a difference in group means that most would consider significant, and I'll follow it up with an example of regression analysis of the same data.

Figure 4.17
Height in inches of 10 randomly selected men and women, analyzed using ANOVA.

	A	B	C	D	E	F	G	H	I	J
1	Males	Females		Anova: Single Factor						
2	71	61								
3	64	66		SUMMARY						
4	62	62		*Groups*	*Count*	*Sum*	*Average*	*Variance*		
5	71	64		Males	10	679	67.9	18.99		
6	63	64		Females	10	640	64.0	7.33		
7	73	64								
8	63	60								
9	71	63		ANOVA						
10	72	68		*Source of Variation*	*SS*	*df*	*MS*	*F*	*P-value*	*F crit*
11	69	68		Between Groups	76.05	1	76.05	5.778	0.027	4.414
12				Within Groups	236.90	18	13.16			
13										
14				Total	312.95	19				
15										
16				**SS Between**	76.05	**=DEVSQ(G5:G6)*10**				
17				**SS Within**	236.90	**=DEVSQ(A2:A11)+DEVSQ(B2:B11)**				

Figure 4.17 shows the same analysis used in Figure 4.16, but with different data. Figure 4.17 assesses the difference in the mean heights of ten men and ten women, all randomly selected. Notice the calculation of the Sum of Squares Between and the Sum of Squares Within, in cells E16 and E17, using Excel's DEVSQ() function and the number of cases per group. The values returned are identical to those calculated by the Data Analysis add-in, in cells E11 and E12.

The route to the F-ratio is the same as in Figure 4.16: divide the sums of squares by their respective degrees of freedom to get estimates of the total variance. The estimate labeled Mean Square Between is based on the standard error of the mean (and thus ignores individual variation within groups). The estimate labeled Mean Square Within is based on the within-group variation (and thus ignores differences between the group means).

In this case, though, the group means are far enough apart that the estimate of total variance in the Mean Square Between is large relative to the estimate of total variance based on variability within the groups. An F-ratio of almost 6 is quite large. Excel's Data Analysis add-in shows that it would occur by chance a little less than 3% of the time (0.027 in cell I11) with 1 and 18 degrees of freedom (cells F11 and F12) under the assumption that, in the population, mean heights were equal for men and women. (In Figure 4.18 I'll show how the F.DIST.RT() function confirms that probability.) On the basis of the probability estimate, less than 3% by chance, most reasonable people would reject the assumption of equal heights in the population (also termed the *null hypothesis*).

The F-ratio in Regression Analysis

Figure 4.18 shows the same analysis using regression instead of ANOVA.

Figure 4.18
The regression analysis is identical to ANOVA where it needs to be, but it returns a richer set of information.

L2 =F.DIST.RT(K2,I2,I3)

	A	B	C	D	E	F	G	H	I	J	K	L
1	Sex	Sex (Coded)	Height	Predicted Height	Residual Height		Source	Prop. Of Variance	DF	"Mean Square"	F	Prob. Of F
2	Male	0	71	67.9	3.1		Regression	0.243	1	0.243	5.778	0.027
3	Male	0	64	67.9	-3.9		Residual	0.757	18	0.042		
4	Male	0	62	67.9	-5.9							
5	Male	0	71	67.9	3.1							
6	Male	0	63	67.9	-4.9							
7	Male	0	73	67.9	5.1		=LINEST(C2:C21,B2:B21,,TRUE)					
8	Male	0	63	67.9	-4.9		-3.9	67.9				
9	Male	0	71	67.9	3.1		1.622	1.147				
10	Male	0	72	67.9	4.1		0.243	3.628				
11	Male	0	69	67.9	1.1		5.778	18				
12	Female	1	61	64.0	-3		76.05	236.90				
13	Female	1	66	64.0	2							
14	Female	1	62	64.0	-2							
15	Female	1	64	64.0	0							
16	Female	1	64	64.0	0							
17	Female	1	64	64.0	0							
18	Female	1	60	64.0	-4							
19	Female	1	63	64.0	-1							
20	Female	1	68	64.0	4							
21	Female	1	68	64.0	4							
22												
23		Sum of Squares		76.05	236.9							

Compare Figure 4.18 to Figure 4.17. Figure 4.17 provides the raw data in two columns, column A and column B, to conform to the layout requirements of the Single Factor ANOVA tool in the Data Analysis add-in. In contrast, Figure 4.18 lists the subjects' heights in a single column, column C, and distinguishes the men from the women by different values, 0 and 1, in column B—*not* by placing their heights in different columns, as in Figure 4.17.

Figure 4.18 also calculates the predicted heights explicitly on the worksheet. This step is not necessary for the regression analysis but it does make a little more clear what's going on in the calculation of the Sum of Squares Regression in cell D23 (which is numerically equal to the Sum of Squares Between in cell E11 of Figure 4.17).

Notice that the predicted value is identical for each member of each of the two groups. That's not unexpected: All the males have an observed (but coded) value of 0 on the Sex variable, and all the women have a value of 1.

It's also worth noting that the predicted value for each male is 67.9, and for females it is 64.0. Those values are the mean heights for each group: Compare them with the group means shown in cells G5 and G6 in Figure 4.17. Here's the reason that the predicted heights equal the group means:

The mean and the standard deviation (equivalently, the sum of squared deviations from the mean) have complementary properties. One is that the mean minimizes the standard deviation. Take these steps if you want to demonstrate the property:

1. Take a set of numbers.
2. Pick some other, random number—call it the *stand-in*.
3. Calculate the deviation of each number of your set from the stand-in, and square each deviation.
4. Find the sum of the squared deviations.

Now get the mean of the set of numbers you selected in step 1, and in step 3 use it instead of the stand-in. The result of step 4 will be smaller than if you use any number other than the mean to calculate the deviations in step 3.

The comparison as I've outlined this exercise is the sum of the squared deviations, but the result is the same if you divide that sum by the number of observations (to get the variance) or take the square root of that variance (to get the standard deviation).

So what? Well, one primary task assigned to regression analysis is to minimize the squared deviations between the actually observed values and the predicted values. In the present example, all the members of a particular group have the same observed value on the predictor variable—that's the code value 0 for men and 1 for women. Therefore, all members of a particular group must have the same predicted value.

In terms of the variables used in Figure 4.18:

- No value returns a smaller sum of the squared deviations of individual values than does the mean of those values.
- Regression finds the predicted heights that minimize the squared deviations from the actual heights.
- The Sex values are constant: 0 for men and 1 for women.
- Therefore the predicted heights must also be constant for a given group, and *must* equal the mean height of that group.

That constant-predicted height clarifies the reason for multiplying the sum of the squared deviations in Figure 4.17 by the number of observations per group. That figure calculates the Sum of Squares Between, in cell E16, via this formula:

=DEVSQ(G5:G6)*10

where cells G5 and G6 contain the mean heights for men and for women. The result returned by the DEVSQ() function is 7.605. However, it's not just two values, but ten pairs of values that contribute to the *sum* of the squared deviations from the grand mean. Therefore, to get the true value of the sum of squares based on the group mean differences, we multiply the result of the DEVSQ() function by the number in the group. (The calculation is a little more complicated, but only a little, when the groups have different numbers of observations.)

Figure 4.18 demonstrates this issue more directly, and I believe it's yet another reason to prefer the regression approach to General Linear Model problems. In Figure 4.18 you can see the predicted values, the mean of each group, in the range D2:D21. Pointing the DEVSQ() function at that range totals the sum of the squared deviations of each group mean from the grand mean—not just for two group means, as in Figure 4.17, but once for each observation. Therefore the multiplication by group size that is needed in Figure 4.17 is unnecessary in Figure 4.18, which adds 10 squared deviations in each group instead of multiplying a constant squared deviation by 10.

Figure 4.18 also uses a different approach to calculating the Sum of Squares Residual than does traditional ANOVA (where it's termed the Sum of Squares Within). The regression approach calculates the Sum of Squares Regression first by subtracting the predicted values from the actual values—in the present example, that entails subtracting the predicted heights from the actual heights.

That leaves a residual value for each observation: In fact, it's the part of each person's weight that's not predictable from knowledge of the person's sex. By pointing the DEVSQ() function at those residual values, as in cell E23 of Figure 4.18, you get the sum of the squared deviations for the residual.

In Figure 4.17, you got the Sum of Squares Within by subtracting each person's height from the mean of his or her group, squaring the result, and adding them up. In Figure 4.18, you obtain the residual values by subtracting the predicted values from the original values—*but the predicted values are the group means*. Therefore, the two quantities, Sum of Squares Within for traditional ANOVA and Sum of Squares Residual for regression analysis, must equal one another. Really, they just take slightly different routes to reach the same conclusion.

By the way, the coding scheme you use makes no difference to any of what I've called LINEST()'s ancillary statistics, those in its final three rows. Those of you playing along at home, using the workbook downloaded from the publisher's website, could use 3.1416 and –17 instead of 0 and 1 to identify men and women, and the R^2, the standard error of estimate, the F-ratio, the residual degrees of freedom, and the two sums of squares would remain the same. But the regression coefficients and the intercept would change, along with their standard errors.

So why use 0 and 1 to encode Male and Female? The reason is with that approach, the regression coefficients and the intercept take on particularly useful values. The example in Figure 4.18 uses dummy coding, one of the three main methods discussed in Chapter 7, "Using Regression to Test Differences Between Group Means." A more extensive discussion will have to wait until then, but for the moment these two points are worth noting:

- The regression coefficient in cell G8 is the difference between the mean height of the sampled men and the mean height of the sampled women.
- The intercept in cell H8 is the average height of the sampled men, who are coded as 0.

These features of the coefficient and the intercept are useful in the more complex designs discussed in later chapters. And the use of 1s and 0s in dummy coding turns out to be particularly useful in some types of logistic regression, a maximum likelihood technique that uses predicted variables measured on nominal scales.

To recapitulate, the LINEST() results in Figure 4.18 provide all the figures necessary to reconstruct a traditional analysis of variance, should you want to. Figure 4.18 shows that traditional table in the range G1:L3. Here's where the values come from:

- Proportion of Variance, Regression in cell H2. In this simple design, that's the same as R^2 returned by LINEST() in cell G10.
- Proportion of Variance, Residual in cell H3. 1.0 minus the proportion in cell H2. Note that if you multiply the two proportions in H2 and H3 by the total sum of squares from Figure 4.17, you get the Sum of Squares Between and Sum of Squares Within, also from Figure 4.17.
- DF Regression, cell I2. The number of predictor variables presented to LINEST(). Here, that's 1: the variable in the range B2:B21.
- DF Residual, cell I3. The number of observations less the number of predictors less 1, or n – k – 1. Part of the LINEST() results and found in cell H11.
- "Mean Square" Regression, cell J2. Again, this is not truly a mean square (another and not entirely accurate term for *variance*). It is the proportion of variance for the regression in H2 divided by the degrees of freedom for the regression in I2.
- "Mean Square" Residual, cell J3. The proportion of residual variance divided by the degrees of freedom for the residual in I3.
- F-ratio, cell K2. The "Mean Square" Regression in J2 divided by the "Mean Square" Residual in J3. Note that it is identical to the F-ratio returned in cell H11 of Figure 4.17.

Figure 4.18's ANOVA table has one additional value, found in cell L2 and labeled Prob. Of F, which is not available directly from the LINEST() results. In Figure 4.18, the formula that returns the value 0.027 is as follows:

```
=F.DIST.RT(K2,I2,I3)
```

4

The function F.DIST.RT() returns the probability of obtaining an F-ratio as large as the one found (in this example) at cell K2, from a central F distribution with the number of degrees of freedom found at cells I2 and I3.

When, in the population that is too large to be observed directly, two variances are equal, F-ratios sampled from that distribution follow what's termed a *central F distribution*. If you chart the frequency of occurrence of different F-ratios, sampled representatively from a central F distribution, you get a chart that looks something like the one shown in Figure 4.19.

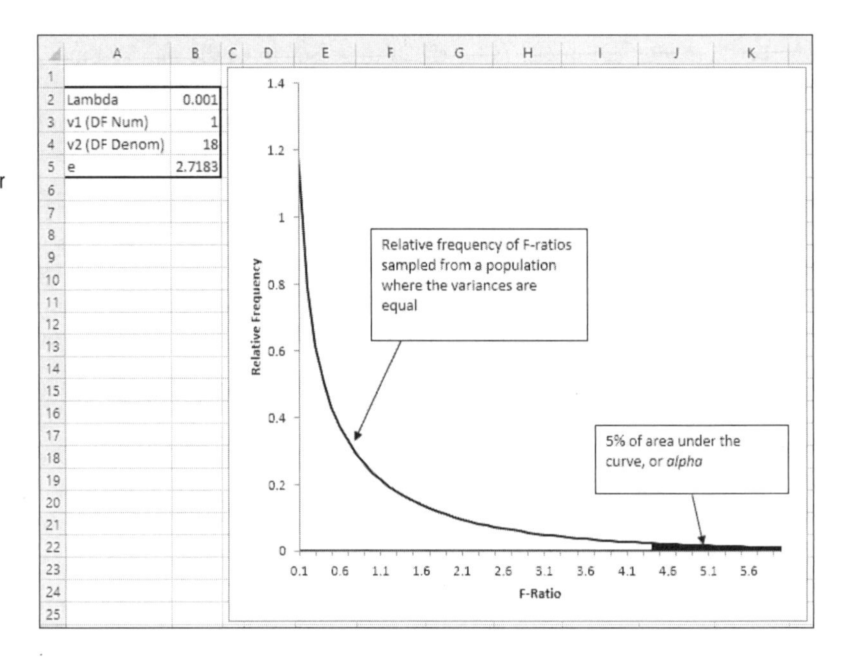

Figure 4.19
The mean of the F distribution is 1.0 only when the degrees of freedom for the numerator equals that for the denominator.

In contrast to the central F distribution that Figure 4.19 depicts, there are *non-central* F distributions that look very different from central F distributions. Several appear in Figure 4.20.

Figure 4.20 shows five different F distributions, each characterized by a different value for *lambda*. That value is often termed the *non-centrality parameter*. (The field of statistics itself is characterized by a tendency to make the simplest concepts sound esoteric.)

The non-centrality parameter is just a way of stretching out the F distribution. Notice in Figure 4.20 that as lambda gets larger, the associated F distribution becomes less skewed and flatter. This happens when the numerators of the F-ratio become large relative to the denominators. Then, the relative frequency of the larger ratios tends to become greater and the distribution gets stretched out to the right.

While I'm at it, I might as well point out that lambda itself is just the ratio of the Sum of Squares Between to the mean square residual. So, as group means become further apart (ANOVA) or as the relationship between predictor and predicted variables becomes stronger (regression), lambda becomes larger and the F distribution stretches out to the right. In such a distribution, the larger F-ratios become less and less rare.

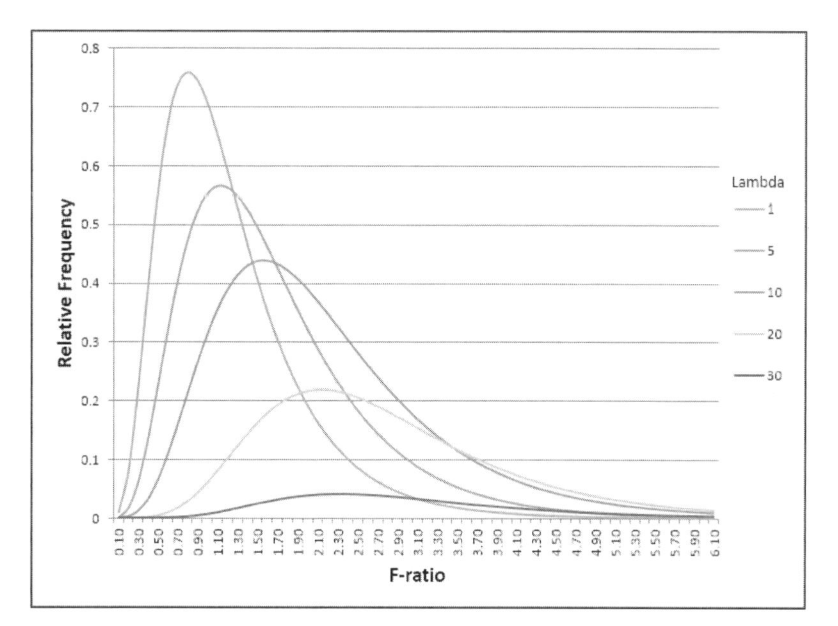

Figure 4.20
The shape of the non-central F distribution depends on the degrees of freedom *and* on the non-centrality parameter.

Compare Figure 4.19 (which depicts a central F distribution and therefore a lambda which equals 0) with Figure 4.20. The F distributions depicted in Figure 4.20 have much more room for the larger F-ratios in the right tail of the distributions than is the case with the central F-ratio in Figure 4.19. In that particular central F distribution, an F-ratio of 4.4 cuts off only 5% of the area under the curve. Much more of the area is cut off under the non-central F distributions is cut off by an F-ratio of 4.4.

That's as it should be. In a central F distribution, where the variance based on group means is the same as the variance based on within-groups variability, you regard a relatively large F as evidence that the central F distribution doesn't fit the observed data—and you are more willing to believe that the group means are different in the population. That leads to a larger non-centrality parameter, a stretched-out F distribution, and a greater likelihood of observing a large F-ratio.

(The same line of thought applies to a relatively strong R^2 between the predictor variable and the predicted variable, which produces a relatively large regression sum of squares, a relatively small residual sum of squares, and therefore a lambda that's larger than 0.)

It's when you get into probability inferences such as those discussed in this section that the assumptions about your raw data, and how you collected it, become critical. This book takes a longer look at those assumptions in Chapter 6, "Assumptions and Cautions Regarding Regression Analysis."

The F-ratio Compared to R^2

It's helpful at this point to compare how two important ratios are calculated. Recall that the R^2 statistic is a ratio: To determine the value of R^2, you divide the sum of squares for the regression by the total sum of squares:

R^2 = SS Regression / (SS Regression + SS Residual)

The R^2 ratio expresses the proportion of the total sum of squares attributable to the regression. Another way of viewing the R^2 is to note that it is the percentage of variability in the predictor variable that is shared with the predicted variable.

In contrast, it's generally less helpful to view the F-ratio as a proportion. Here's the basic definition of the F-ratio:

$$F = \frac{SS\ Between / DF\ Between}{SS\ Within / DF\ Within}$$

where DF Between is the degrees of freedom for the sum of squares between groups, and the DF Within is the degrees of freedom for the sum of squares within groups. So the F-ratio depends in large measure on how many times greater the sum of squares between groups is vis-à-vis the sum of squares within groups. Of course, the size of that ratio changes as a function of the ratio of the DF Between to the DF Within. In practice, and as the F-ratio is used in the analysis of variance, DF Between is much smaller than DF Within, and that relationship generally increases the size of the ratio of the Mean Square Between to the Mean Square Within.

At any rate, the point to remember here is that R^2 is the ratio of the sum of squares due to regression to the total sum of squares. That ratio represents a proportion, a subset divided by its total set.

In contrast, the F-ratio divides one estimate of the total variance (due to differences between groups) by another estimate of the total variance (due to differences within groups). The numerator and the denominator each estimate the same quantity. Therefore you can think of the F-ratio as a multiple that expresses the degree to which the differences between means contribute to the estimate of the total variance, over and above the contribution of individual differences to the estimate of total variance.

The General Linear Model, ANOVA, and Regression Analysis

You can understand the differences and similarities between traditional analysis of variance and regression analysis perfectly well without resorting to this section. However, if you're curious about how the two approaches take different routes to identical conclusions, you might want to look this material over.

A model termed the *General Linear Model* underlies both the ANOVA and the regression methods. Per that model, each individual score in a study can be expressed as three components:

- The grand mean (symbolized here as $\bar{X}_{..}$)
- The deviation of the group mean (symbolized here, for group j, as $\bar{X}_{.j}$) from the grand mean
- The deviation of the individual score (symbolized here for the ith individual in the jth group, as X_{ij}) from the mean of the jth group

Then:

$$X_{ij} = \bar{X}_{..} + (\bar{X}_{.j} - \bar{X}_{..}) + (X_{ij} - \bar{X}_{.j})$$

That is, the ith observation in the jth group is the sum of the grand mean, plus the difference between the grand mean and the jth group mean, plus the difference between the ith observation in the jth group and the mean of the jth group.

Now subtract the grand mean from each side of the equation:

$$X_{ij} - \bar{X}_{..} = (\bar{X}_{.j} - \bar{X}_{..}) + (X_{ij} - \bar{X}_{.j})$$

and square both sides, and sum across the i observations in the j groups. Suppose that the number of groups is 3, and the number of subjects per group is 5. So:

$$\sum_{j=1}^{3}\sum_{i=1}^{5}(X_{ij} - \bar{X}_{..})^2 = \sum_{j=1}^{3}\sum_{i=1}^{5}((\bar{X}_{.j} - \bar{X}_{..}) + (X_{ij} - \bar{X}_{.j}))^2$$

Notice that the expression now gives the total sum of squares: the sum of the squared differences of all the observations from $\bar{X}_{..}$, the grand mean.

Recall from your algebra classes that:

$$(x + y)^2 = x^2 + 2xy + y^2$$

Applying the same expansion to the sum of squares, we get:

$$\sum_{j=1}^{3}\sum_{i=1}^{5}(X_{ij} - \bar{X}_{..})^2 = \sum_{j=1}^{3}\sum_{i=1}^{5}((X_{.j} - \bar{X}_{..})^2 + 2(X_{.j} - \bar{X}_{..})(X_{ij} - \bar{X}_{.j}) + (X_{ij} - \bar{X}_{.j})^2)$$

Distribute the summation operators through the right half of the equation:

$$\sum_{j=1}^{3}\sum_{i=1}^{5}((X_{.j} - \bar{X}_{..})^2 + 2\sum_{j=1}^{3}\sum_{i=1}^{5}(X_{.j} - \bar{X}_{..})(X_{ij} - \bar{X}_{.j}) + \sum_{j=1}^{3}\sum_{i=1}^{5}(X_{ij} - \bar{X}_{.j})^2$$

The second term in the prior expression is:

$$2\sum_{j=1}^{3}\sum_{i=1}^{5}(X_{.j} - \bar{X}_{..})(X_{ij} - \bar{X}_{.j})$$

4

Notice that the second factor in that expression takes the sum of the raw deviation scores in the jth group. The sum of a set of deviation scores—the sum of the differences between each score in a group and the group's mean—is always zero. Because that sum is always zero, we have for the prior expression:

$$2\sum_{j=1}^{3}\sum_{i=1}^{5}(X_{.j} - \overline{X}_{..})(0)$$

which results in zero, for any number of groups and for any set of values within those groups. That means that the middle term of the expansion drops out, because it always equals zero, and we're left with:

$$\sum_{j=1}^{3}\sum_{i=1}^{5}(X_{ij} - \overline{X}_{..})^2 = \sum_{j=1}^{3}\sum_{i=1}^{5}(X_{.j} - \overline{X}_{..})^2 + \sum_{j=1}^{3}\sum_{i=1}^{5}(X_{ij} - \overline{X}_{.j})^2$$

The first term on the right side of the equal sign calculates the deviation of the jth group's mean from the grand mean, and squares the difference:

$$(X_{.j} - \overline{X}_{..})^2$$

This expression is a constant for each of the 5 observations in a particular group; therefore, we can simply convert the summation symbol to a multiplication:

$$\sum_{j=1}^{3}\sum_{i=1}^{5}(X_{.j} - \overline{X}_{..})^2$$

resulting in this expression:

$$5\sum_{j=1}^{3}(X_{.j} - \overline{X}_{..})^2$$

So here is the total of the squared deviations of each observation from the grand mean:

$$\sum_{j=1}^{3}\sum_{i=1}^{5}(X_{ij} - \overline{X}_{..})^2 = 5\sum_{j=1}^{3}(X_{.j} - \overline{X}_{..})^2 + \sum_{j=1}^{3}\sum_{i=1}^{5}(X_{ij} - \overline{X}_{.j})^2$$

This results in the separation of the total variation in the data set (computed on the left side of the equal sign) into the two components on the right side: the sum of squares between groups plus the sum of squares within groups.

Traditional ANOVA calculates the Sum of Squares Between and the Sum of Squares Within directly from the two terms on the right side of the prior equation's equal sign.

Regression analysis picks up the Sum of Squares due to Regression from the predicted values, and the Residual Sum of Squares from the actual minus the predicted values, as demonstrated earlier in this chapter, in the section titled "The F-ratio in Regression Analysis."

Other Ancillary Statistics from LINEST( )

The remaining statistics returned by LINEST() are the degrees of freedom for the residual (fourth row, second column of the LINEST() results), the Sum of Squares for the Regression (fifth row, first column) and the Sum of Squares for the Residual (fifth row, second column).

We have already covered those statistics at some length in this chapter. It's necessary to do so in order to provide meaningful explanations of other statistics such as R^2, the F-ratio, and the standard error of estimate. There is more to say about them as we move into more complicated designs, with both factors *and* covariates, and with more than just one factor such as sex and medical treatment.

For the time being, though, there is little more to say regarding these three statistics, other than to note that LINEST() returns them in the cells as just noted, and that changes to the experimental design that underlies the acquisition of the raw data can have profound effects on the values calculated and returned by LINEST().

Multiple Regression

5

When you move from simple, single-predictor regression to multiple regression, you add one or more predictor variables to the mix. Doing so is very easy to accomplish in the Excel worksheet structure and the function syntax. But it has a major impact on the nature of the analysis, on how you interpret the results, and on the statistical power of the regression.

As to using LINEST() and TREND(), Excel's two principal regression functions, there are just three items to keep in mind when you go from one to multiple predictor variables:

- Store the values of your predictor values in contiguous columns, such as columns A and B for two predictors, or columns A, B, and C for three predictors. (You could instead store them in contiguous rows, but that would be an unusual choice.)
- Before actually entering the formula with the LINEST() function, select five rows and as many columns as the number of predictors, plus one for the intercept. There's no change to the range you select for a formula that uses the TREND() function. Regardless of the number of predictors, the TREND() function takes as many rows as you have for the predicted variable, and one column.
- Specify the full range of the *known x's* for both LINEST() and TREND(). You might use this structure in a two-predictor regression:

  =LINEST(A2:A41,B2:C41,,TRUE)

  Or this structure in a three-predictor regression:

  =LINEST(A2:A61,B2:D61,,TRUE)

IN THIS CHAPTER

A Composite Predictor Variable 152

Understanding the Trendline 160

Mapping LINEST()'s Results to the Worksheet 163

Building a Multiple Regression Analysis from the Ground Up 166

Using the Standard Error of the Regression Coefficient 181

Using the Models Comparison Approach to Evaluating Predictors 192

Estimating Shrinkage in R^2 197

Regardless of the number of predictors in use, you still need to array-enter the formula that contains the LINEST() or TREND() function, by means of Ctrl+Shift+Enter.

You'll see many examples of adapting the LINEST() and TREND() syntax to different numbers and types of variables through the remainder of this book. Let's move from the pure mechanics of entering the formulas to problems of interpretation.

A Composite Predictor Variable

Suppose that you are interested in investigating the relationship of age and diet to levels of LDL, the "low density" cholesterol thought responsible for the buildup of plaque in arteries and, therefore, coronary disease. You obtain a sample of 20 adults and administer blood tests that result in an estimate of the LDL level in each of your subjects.

You also record each subject's age in years and obtain self-reports of the number of times each week that your subjects eat red meat.

The results of your research appear in Figure 5.1.

Figure 5.1
It's best to keep the predictor variables in adjacent columns when you're preparing to use LINEST() or TREND().

E2 | {=LINEST(C2:C21,A2:A21,,TRUE)}

	A	B	C	D	E	F	G	H	I
1	Diet	Age	LDL		LDL Regressed on Diet			LDL Regressed on Age	
2	1	41	81		11.338	89.878		1.019	90.554
3	2	20	82		4.797	23.865		0.374	20.765
4	5	87	198		0.237	37.938		0.292	36.536
5	8	77	160		5.587	18		7.432	18
6	2	52	115		8041.001	25907.799		9921.076	24027.724
7	4	30	77						
8	5	38	117						
9	4	83	174						
10	4	70	103						
11	4	40	138						
12	5	45	170						
13	4	61	196						
14	7	76	174						
15	7	78	168						
16	3	39	133						
17	7	28	125						
18	6	71	166						
19	5	20	86						
20	5	22	190						
21	5	44	199						

The LDL values, or the *known y's*, appear in the range C2:C21 in Figure 5.1. The *known x's*, which are the subjects' Diet and Age, appear in A2:A21 and B2:B21. (The weekly frequency with which each subject eats red meat is labeled *Diet*.)

Figure 5.1 also shows two instances of LINEST(). The statistics that describe the relationship between LDL and Diet appear in E2:F6. Those that apply to the relationship between LDL and Age appear in H2:I6.

The values of R^2 for the two analyses show up in cells E4 and H4. (For more information about the meaning of the R^2 statistic, see Chapter 3, "Simple Regression" in the section "R^2 in Simple Linear Regression.") Variability in Diet is associated with 23.7% of the variability in LDL, and the subjects' variability in Age is associated with 29.2% of the variability in LDL.

Although this is useful information, it doesn't go anywhere near far enough. It does not address the *combined* effects of diet and age. Given that the two variables Diet and Age do exert a combined effect on LDL, the simple regression analysis of LDL on Diet does not account for the presence of the Age variable, and LDL on Age does not account for the presence of the Diet variable.

I'll discuss these issues in detail later in this chapter, but first it's useful to have a look at how to move from simple, one-predictor regression to multiple-predictor regression.

Generalizing from the Single to the Multiple Predictor

One way of thinking about simple regression is that you multiply a predictor variable by a coefficient and then add a constant called the intercept. If you do that for every observation of the predictor variable, the result is variously termed the *regression line*, the *predictions*, the *trendline*, and (more rarely) a *linear transformation of the predictor*. You'll find the regression line in Figure 5.2, which charts the simple regression between Age and LDL in your fictitious study.

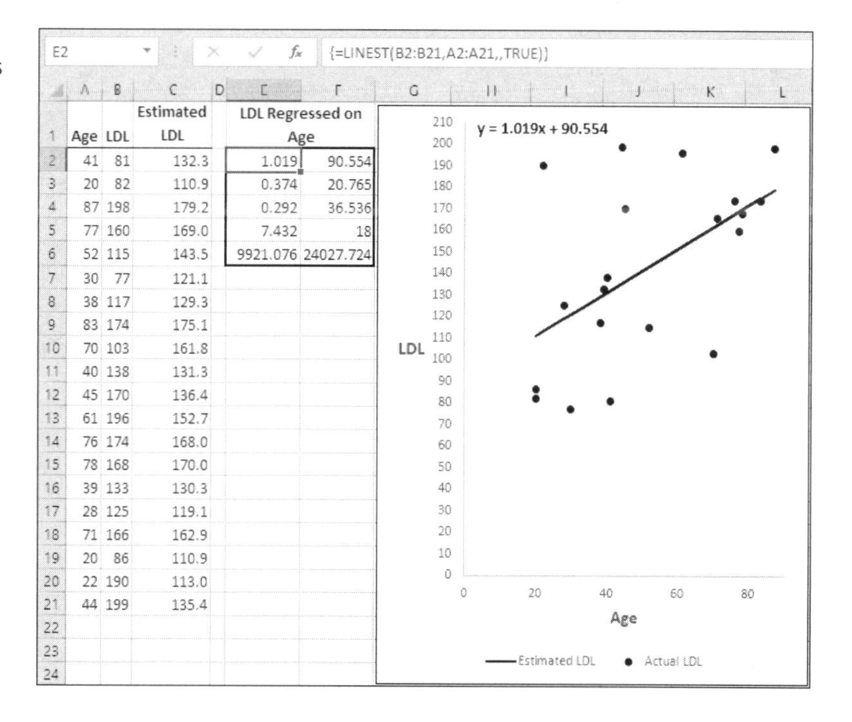

E2 {=LINEST(B2:B21,A2:A21,,TRUE)}

	A	B	C	D	E	F
1	Age	LDL	Estimated LDL		LDL Regressed on Age	
2	41	81	132.3		1.019	90.554
3	20	82	110.9		0.374	20.765
4	87	198	179.2		0.292	36.536
5	77	160	169.0		7.432	18
6	52	115	143.5		9921.076	24027.724
7	30	77	121.1			
8	38	117	129.3			
9	83	174	175.1			
10	70	103	161.8			
11	40	138	131.3			
12	45	170	136.4			
13	61	196	152.7			
14	76	174	168.0			
15	78	168	170.0			
16	39	133	130.3			
17	28	125	119.1			
18	71	166	162.9			
19	20	86	110.9			
20	22	190	113.0			
21	44	199	135.4			

Figure 5.2
The estimated LDL values are obtained using the TREND( ) function.

To review some material from Chapter 3:

- The regression line is the solid diagonal line in Figure 5.2.
- The formula to create the regression line appears in the chart. You could choose a new value for x, multiply it by 1.019, and add 90.554 to get the estimated y value for the new x value. If you plotted the new x value and the resulting estimated y value on the chart, the new point would lie directly on the regression line.
- The value 1.019 is termed the *regression coefficient* and the value 90.554 is termed the *intercept* or *constant*. Note that the two values shown in the chart's formula are identical to those returned in E2:F2 by the LINEST() function. The TREND() function calculates the coefficient and the intercept behind the scenes, and then uses them to derive the values for the regression line in the range C2:C21.
- The actual observed values of Age and LDL, shown in A2:B21, are charted as the individual points on the chart. The vertical distance between each data point and the regression line represents the amount and direction of the estimation error for that data point. The amount of error in the estimates is closely related to regression statistics such as R^2, the standard error of estimate, the F ratio, and the sum of squares for the regression—each of which, by the way, is calculated and returned to the worksheet by the LINEST() function in E2:F6.

Each of the preceding four bulleted points is, of course, also true of the regression of LDL onto Diet. See the chart and the LINEST() results in Figure 5.3.

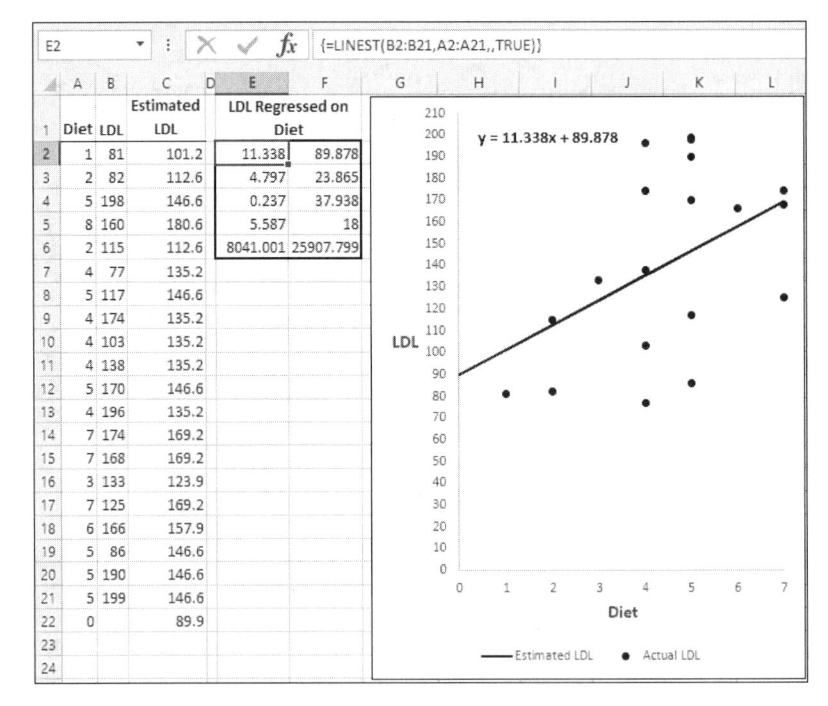

	A	B	C	D	E	F
1	Diet	LDL	Estimated LDL		LDL Regressed on Diet	
2	1	81	101.2		11.338	89.878
3	2	82	112.6		4.797	23.865
4	5	198	146.6		0.237	37.938
5	8	160	180.6		5.587	18
6	2	115	112.6		8041.001	25907.799
7	4	77	135.2			
8	5	117	146.6			
9	4	174	135.2			
10	4	103	135.2			
11	4	138	135.2			
12	5	170	146.6			
13	4	196	135.2			
14	7	174	169.2			
15	7	168	169.2			
16	3	133	123.9			
17	7	125	169.2			
18	6	166	157.9			
19	5	86	146.6			
20	5	190	146.6			
21	5	199	146.6			
22	0		89.9			
23						
24						

Figure 5.3
Note that the individual errors of estimate are somewhat different than in Figure 5.2.

Notice again that the R^2 for LDL on Age is 0.292 and 0.237 on Diet. (These values appear in cell E4 in Figures 5.2 and 5.3.) Between 20% and 30% of the variability in LDL can be estimated given knowledge of a person's age or the frequency of consuming red meat—given this specific set of observations as our source of knowledge.

We can nearly always do better than either of two or more R^2 values by themselves. We do so by *combining the predictor variables (here, Age and Diet) into a single composite variable.* Figure 5.4 shows how multiple regression manages that.

Figure 5.4
The LINEST( ) results in the range F11:H15 occupy an additional column to accommodate the additional predictor variable.

F11 | {=LINEST(C2:C21,A2:B21,,TRUE)}

	A	B	C	D	E	F	G	H	I
1	Diet	Age	LDL		LDL Regressed on Diet			LDL Regressed on Age	
2	1	41	81		11.338	89.878		1.019	90.554
3	2	20	82		4.797	23.865		0.374	20.765
4	5	87	198		0.237	37.938		0.292	36.536
5	8	77	160		5.587	18		7.432	18
6	2	52	115		8041.001	25907.799		9921.076	24027.724
7	4	30	77						
8	5	38	117						
9	4	83	174						
10	4	70	103		Multiple Regression of LDL on Diet and Age				
11	4	40	138			0.789	7.784	66.076	
12	5	45	170			0.384	4.742	24.830	
13	4	61	196			0.389	34.929	#N/A	
14	7	76	174			5.413	17	#N/A	
15	7	78	168			13208.033	20740.767	#N/A	
16	3	39	133						
17	7	28	125						
18	6	71	166						
19	5	20	86						
20	5	22	190						
21	5	44	199						

From the standpoint of understanding what multiple regression is all about, perhaps the most interesting aspect of Figure 5.4 is the different values of R^2 in cells E4, H4, and F13. The R^2 for the multiple regression in cell F13, 0.389, which combines the information about two predictor variables instead of using just one, is larger than either of the single-predictor analyses.

It's called *multiple* regression, in contrast to simple regression, because you use more than one predictor variable simultaneously. When you add a predictor to a regression analysis, you nearly (but not quite) always find that the R^2 value increases. Here, we have added Age to the regression of LDL on Diet (equivalently, we have added Diet to the regression of LDL on Age). One result is that the R^2 has increased to 38.9% in cell F13. That means the inclusion of both Age and Diet in the same regression analysis increases to the amount of explained variability in LDL by 15.2% over Diet alone (from 23.7% to 38.9%) and by 9.7% over Age alone (from 29.2% to 38.9%).

From an intuitive perspective, you could simply note that Age explains some variability in LDL that Diet doesn't: variability that's missing from the regression of LDL on Diet. Similarly, Diet explains some variability in LDL that Age doesn't, and that's missing from the regression of LDL on Age.

Only when you combine the two predictors via multiple regression does all the variance in LDL that's attributable to either predictor enter the analysis. The next section begins to discuss a little more formally how that comes about.

Minimizing the Sum of the Squared Errors

Notice in Figure 5.4 that the regression coefficients for Diet as the sole predictor (cell E2) and for Age as the sole predictor (cell H2) differ both from one another and from the regression coefficients for Diet and Age as simultaneous predictors (respectively, cells G11 and F11). One fundamental characteristic of regression analysis is that it returns the coefficients and the intercept that minimize the sum of the squared errors of estimate for the data in question. Put another way: When you apply the regression coefficients to the observed values and total the results along with the intercept, the total of the squared differences between the predicted values and the observed values is smaller than using any other coefficients.

However, that's true only for a particular set of observations. The data presented to LINEST() in E2:F6 is missing the Age variable, and the data presented to LINEST() in H2:I6 is missing the Diet variable. Both predictors are present in the LINEST() results shown in F11:H15, so it's not surprising that the coefficients and intercept in F11:H11 differ from those in E2:F2 and H2:I2. The multiple regression analysis in F11:H15 is working with *all* the variance shared by Age, Diet, and LDL. Using all the available variance means that the sum of the squared errors of estimate can be reduced even further than in either single-predictor regression equation.

Suppose you calculated by brute force the squared errors of estimate from Diet alone, as is done in Figure 5.5.

The calculations in Figure 5.5 are as follows:

1. The LINEST() function is array-entered in the range D2:E6 with this formula: =LINEST(B2:B21,A2:A21,,TRUE).
2. The predicted values are calculated using this formula in cell G2: =A2*$D$2+$E$2. That formula is copied and pasted into G3:G21.
3. The errors of estimate, or *residuals*, are calculated by subtracting the estimated values in column G from the associated, observed values in column B. The differences between the observed and estimated values—the errors of estimate—are in column H.
4. The squares of the errors of estimate appear in column I. The formula in cell I2, for example, is =H2^2.
5. Cell I23 returns the total of the squared errors of estimate with this formula: =SUM(I2:I21).

Figure 5.5
This is where the term *least squares* comes from.

H2 =B2-G2

	A	B	C	D	E	F	G	H	I
1	Diet	LDL		LDL Regressed on Diet			LDL Estimated from Diet, Using LINEST() Stats	Errors of Estimate	Squared Errors of Estimate
2	1	81		11.338	89.878		101.2	-20.2	408.7
3	2	82		4.797	23.865		112.6	-30.6	933.5
4	5	198		0.237	37.938		146.6	51.4	2645.2
5	8	160		5.587	18		180.6	-20.6	423.6
6	2	115		8041.001	25907.799		112.6	2.4	6.0
7	4	77					135.2	-58.2	3390.8
8	5	117					146.6	-29.6	874.3
9	4	174					135.2	38.8	1503.1
10	4	103					135.2	-32.2	1038.8
11	4	138					135.2	2.8	7.7
12	5	170					146.6	23.4	549.0
13	4	196					135.2	60.8	3693.0
14	7	174					169.2	4.8	22.6
15	7	168					169.2	-1.2	1.5
16	3	133					123.9	9.1	83.0
17	7	125					169.2	-44.2	1957.6
18	6	166					157.9	8.1	65.5
19	5	86					146.6	-60.6	3668.5
20	5	190					146.6	43.4	1886.3
21	5	199					146.6	52.4	2749.1
22									
23								Sum	25907.8
24							Standard Error		37.938

The principle of least squares takes effect on the example in Figure 5.5 in this way: The sum of the squared errors in cell I23 is smaller than it would be if *any other value were used for the regression coefficient or the intercept.*

In other words, regression minimizes the sum of the squared errors. The values it calculates for the regression coefficient and the intercept bring about the most accurate set of estimates possible, given the data that you hand off to LINEST() and given that the sum of the squared errors is the chosen method of assessing the accuracy of the estimates.

NOTE

The final assumption in the prior sentence is not always apt. Many legitimate questions, addressed with legitimate data, require some method of analysis other than multiple regression. A common example is an outcome variable with two nominal values, such as "Buys" and "Doesn't buy." Circumstances alter cases, of course, but many statisticians would prefer to analyze such a data set by means of logistic regression (using odds ratios to form the criterion) than by means of the more traditional approach, discriminant function analysis (using squared deviations to form the criterion).

Binomial outcomes notwithstanding, though, you'll often find that regression is an appropriate method for analyzing your data. The least squares criterion is generally accepted as the best measure for evaluating the accuracy of regression analyses (as well as most other approaches that share its theoretical basis, the General Linear Model). Other methods to assess accuracy exist, but they don't have the useful properties of squared deviations. For example, the errors of estimate usually have both positive and negative values, and in fact cancel one another out and sum to 0.0. Squaring the errors eliminates that problem. (It is possible to use the absolute values of the errors instead, leading to a *median regression line*. This creates other problems with the interpretation of the regression, many of them quite difficult to deal with.)

You can demonstrate the least squares property of regression analysis for yourself, using data set up as in Figure 5.6 in conjunction with Excel's Solver.

Figure 5.6
This setup alters the regression coefficient and intercept used for the estimates to two randomly chosen numbers.

G2 =A2*$D$8+$E$8

	A	B	C	D	E	F	G	H	I
1	Diet	LDL		LDL Regressed on Diet			LDL Estimated from Diet, Using LINEST() Stats	Errors of Estimate	Squared Errors of Estimate
2	1	81		11.338	89.878		100.0	-19.0	361.0
3	2	82		4.797	23.865		120.0	-38.0	1444.0
4	5	198		0.237	37.938		180.0	18.0	324.0
5	8	160		5.587	18		240.0	-80.0	6400.0
6	2	115		8041.001	25907.799		120.0	-5.0	25.0
7	4	77					160.0	-83.0	6889.0
8	5	117		20	80		180.0	-63.0	3969.0
9	4	174					160.0	14.0	196.0
10	4	103					160.0	-57.0	3249.0
11	4	138					160.0	-22.0	484.0
12	5	170					180.0	-10.0	100.0
13	4	196					160.0	36.0	1296.0
14	7	174					220.0	-46.0	2116.0
15	7	168					220.0	-52.0	2704.0
16	3	133					140.0	-7.0	49.0
17	7	125					220.0	-95.0	9025.0
18	6	166					200.0	-34.0	1156.0
19	5	86					180.0	-94.0	8836.0
20	5	190					180.0	10.0	100.0
21	5	199					180.0	19.0	361.0
22									
23								Sum	49084.0
24								Standard Error	52.220

Two crucial differences distinguish Figure 5.5 from Figure 5.6. In Figure 5.6, I entered two numbers, which I pulled out of my hat, in cells D8 and E8. I then treated those random numbers as the regression coefficient and the intercept by changing the formula in cell G2 from this:

=A2*$D$2+$E$2

to this:

=A2*$D$8+$E$8

and copied that formula down through cell G21. The eventual result is a total of the squared deviations in cell I23, or 49084.0, that is larger than the corresponding value in Figure 5.5, or 25907.8.

Over to you. Activate the downloaded version of Figure 5.6, or any similarly structured worksheet. Take these steps:

1. Select cell I23, or whatever cell you have chosen to contain the sum of the squared errors.
2. Click the Ribbon's Data tab, and then click Solver in the Analysis group. (If you don't see the Solver link, you'll need to make its add-in accessible to Excel via the Options link on Excel's File navigation bar. You might first have to install Solver from whatever source you used to get Excel onto your computer.)
3. The Solver dialog box appears. See Figure 5.7. Make sure that cell I23 is specified in the Set Objective edit box.
4. Select the Min radio button. So far you have told Solver to minimize the sum of the squared errors in cell I23.
5. In the By Changing Variable Cells edit box, enter the addresses D8 and E8. You have now told Solver to minimize the value in cell I23 by changing the value in cell D8 that's used (by the formulas in the range G2:G21) as the regression coefficient, and the value in cell E8 that's used as the intercept.
6. Make sure that GRG Nonlinear is selected in the Select a Solving Method dropdown, and click Solve.

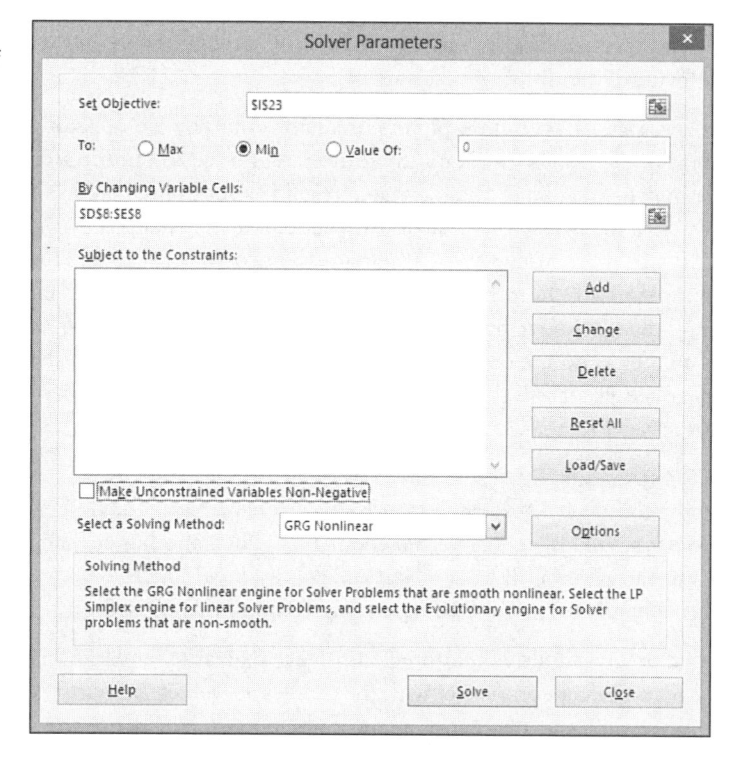

Figure 5.7
If you're using a version of Excel that predates 2013, the Solver dialog box might appear somewhat different than shown in this figure.

5

If you have entered the data as it exists in Figure 5.6, and if you have followed steps 1 through 6, Solver should—within a few seconds at most—return the values that it finds to minimize the sum of the squared errors of estimate. Those values will be the same as those returned by LINEST() in cells D2 and E2, within the tolerances you can specify in Solver's options. Solver reaches the same conclusions using maximum likelihood techniques as LINEST() reaches using differential calculus (or, since 2003, using QR decomposition).

By the way, before we leave Figures 5.5 and 5.6, I should point out that the sum of the squared errors of estimate in cell I23 of Figure 5.5 is exactly identical to the value in cell E6, which LINEST() documentation correctly terms the "sum of squares residual." The same is true of Figure 5.6 after Solver has minimized the sum of the squared errors of estimate.

Furthermore, the value in cell I24 is the square root of the ratio of value in I23 to its degrees of freedom. The result is precisely equal to the value in cell E4, which LINEST() documentation correctly terms the "standard error of estimate." I'll have much more to say about those statistics in subsequent chapters, but you can see how closely related they are to the sum of the squared errors of estimate—as minimized by multiple regression.

Understanding the Trendline

In simple, single-predictor regression, a perfect relationship exists between the predictor variable and the resulting, estimated values of the predicted variable. Things aren't quite so straightforward in multiple regression.

In simple regression, you multiply the predictor values by a regression coefficient and add a constant to the results of the multiplication. As I discussed in the section titled "Partial Correlation" in Chapter 3, the inevitable result is a set of predicted values that correlate perfectly with the predictor variable. All you're doing is taking the original predictor, multiplying it by a constant value (the regression coefficient) and adding the intercept (also termed the *constant*). In those circumstances—multiplying by a constant and adding another constant—the arithmetic produces a set of predicted values that bear the same relationship to one another as do the original predictor values. Let's have a look at a specific example of that effect because it will then be easier to contrast with the analogous situation in multiple regression. See Figure 5.8.

5

Figure 5.8 starts out with a set of random values in the range A2:A12. It multiplies those values by a constant, 6.5, in the range B2:B12. That action is analogous to multiplying a predictor variable by its regression coefficient. Then another constant, 2.3, is added in C2:C12 to the values in B2:B12. That's analogous to adding the intercept in a regression equation. The results in C2:C12 have a correlation of 1.000 with those in A2:A12.

You'll find in any simple, single-predictor linear regression equation that the predicted values have a perfect correlation with the predictor's values, because the equation multiplies the predictor values by one constant and adds another constant to the result. This effect is not particularly interesting or useful in and of itself. But note that when you subtract the predicted values from the observed outcomes, you get a set of residuals that has a correlation of 0.000 with the predictors.

Figure 5.8
Changing one variable with constants results in the same distribution of values.

C2 =B2+2.3

	A	B	C	D
1	Original Data	Multiply by 6.5	Add 2.3	
2	16	104	106.3	
3	17	110.5	112.8	
4	18	117	119.3	
5	15	97.5	99.8	
6	17	110.5	112.8	
7	3	19.5	21.8	
8	30	195	197.3	
9	1	6.5	8.8	
10	15	97.5	99.8	
11	2	13	15.3	
12	19	123.5	125.8	
13				
14			1.0000	=CORREL(A2:A12,C2:C12)

The effect that Figure 5.8 demonstrates—that multiplying Variable A by a constant and then adding another constant results in a set of values that correlate perfectly with Variable A—does not hold up in multiple regression. The reason is that although a multiple regression equation has just one intercept, it has as many regression coefficients as it has predictors. In all but the most trivial, contrived situations, it's overwhelmingly likely that the regression coefficients will differ from one another. Therefore, no single predictor will correlate perfectly with the predicted values. Figure 5.9 shows an example.

Figure 5.9
Correlations between individual predictors and the multiple regression's predicted values are almost sure to be imperfect.

G2 {=TREND(C2:C21,A2:B21)}

	A	B	C	D	E	F	G	H	I	J
1	Diet	Age	LDL		LDL Predicted by Diet Only	LDL Predicted by Age Only	LDL Predicted by Diet and Age		Correlations of Predicted Values with Predictors	
2	1	41	81		101.2	132.3	106.2		1.00	=CORREL(A2:A21,E2:E21)
3	2	20	82		112.6	110.9	97.4		1.00	=CORREL(B2:B21,F2:F21)
4	5	87	198		146.6	179.2	173.7			
5	8	77	160		180.6	169.0	189.1		0.78	=CORREL(A2:A21,G2:G21)
6	2	52	115		112.6	143.5	122.7		0.87	=CORREL(B2:B21,G2:G21)
7	4	30	77		135.2	121.1	120.9			
8	5	38	117		146.6	129.3	135.0			
9	4	83	174		135.2	175.1	162.7		Correlations of Actual with	
10	4	70	103		135.2	161.8	152.5		Predicted LDL	
11	4	40	138		135.2	131.3	128.8		0.49	=CORREL(C2:C21,E2:E21)
12	5	45	170		146.6	136.4	140.5		0.54	=CORREL(C2:C21,F2:F21)
13	4	61	196		135.2	152.7	145.4		0.62	=CORREL(C2:C21,G2:G21)
14	7	76	174		169.2	168.0	180.5			
15	7	78	168		169.2	170.0	182.1			
16	3	39	133		123.9	130.3	120.2			
17	7	28	125		169.2	119.1	142.7			
18	6	71	166		157.9	162.9	168.8			
19	5	20	86		146.6	110.9	120.8			
20	5	22	190		146.6	113.0	122.4			
21	5	44	199		146.6	135.4	139.7			

5

Figure 5.9 draws together a number of the points discussed in this chapter. The context is a multiple regression analysis of Diet and Age as predictor variables with LDL as the predicted variable. The raw data is in columns A, B, and C.

Excel's TREND() function predicts the LDL values in columns E, F, and G. Recall that the TREND() function calculates the regression coefficients and the intercept based on the data you give it, and reports back the predicted values. So:

- The range E2:E21 uses this array formula to predict LDL levels on the basis of the Diet variable only:

  =TREND(C2:C21,A2:A21)

- The range F2:F21 uses this array formula to predict LDL levels on the basis of the Age variable only:

  =TREND(C2:C21,B2:B21)

- The range G2:G21 uses this array formula to predict LDL levels on the basis of *both* Diet and Age:

  =TREND(C2:C21,A2:B21)

So, columns E and F contain the predicted values from two different simple, single-predictor regression analyses, and column G contains the predicted values from a multiple regression analysis.

Column I contains several pertinent correlations between raw data and predicted values, and column J shows the formulas used to calculate each correlation in column I.

- Cell I2 shows that there is a perfect, 1.0 correlation between the Diet values in column A and the LDL values predicted from Diet in column E.
- Cell I3 shows another perfect, 1.0 correlation between the Age values in column B and the LDL values predicted from Age in Column F.

These perfect correlations are exactly what we expect to find. The predicted values are the results of simple, single-predictor regression analyses and therefore are based solely on multiplying the predictor values by one constant and then adding another. The inevitable result is a perfect correlation between the predictor variable and the predicted variable.

- Cell I5 shows a strong but imperfect correlation between Diet and LDL as predicted simultaneously by Diet *and* Age.
- Cell I6 also shows a strong but imperfect correlation between Age and LDL as predicted by Diet and Age. The value in I6 is different from the value in cell I5.

We can anticipate the outcomes in cells I5 and I6, although not the actual correlations returned. The predicted values in column G are based on the combination of Diet and

Age, so there's no particular reason to expect a perfect correlation between just one of the predictors and the predicted values. The relationship between the predictor and the predicted values in simple regression does not apply in multiple regression, because it's, well, multiple.

So what does multiple regression get us, compared to simple regression? The statistics that really count in this case are in cells I11:I13.

- Cell I11 contains the correlation between the observed LDL values and the LDL values predicted by Diet alone.
- Cell I12 contains the correlation between the observed LDL values and the LDL values predicted by Age alone.
- Cell I13 contains the correlation between the observed LDL values and the LDL values predicted by the combination of Diet and Age.

Note that the predictions made by multiple regression in column G correlate more strongly (cell I13) with the actual observations than do the predictions made by either simple regression analysis (cells I11 and I12).

So although the correlations between the predictor variables and the predicted values are interesting and useful—and will become more interesting yet when we start to examine the effect of adding predictors one-by-one—the really important matter is the accuracy of the predictions. That's the issue addressed in cells I11:I13 of Figure 5.9. It's clear that multiple regression outperforms simple regression.

And while that improved performance is also the typical result, interpreting it requires some care. You might find yourself capitalizing on chance. I'll return to that topic in Chapter 6, "Assumptions and Cautions Regarding Regression Analysis."

Mapping LINEST( )'s Results to the Worksheet

I've had a lot of good things to say about Excel's regression functions, and other worksheet functions that you can use in support of the regression functions. I'll have more along those lines in subsequent chapters.

But there's one truly egregious problem with LINEST() that Microsoft has never properly addressed, and that it probably never will. I've waited until this point to bring it up because the problem doesn't arise in single-predictor situations—only in multiple regression.

It's not a bug in the conventional sense. It's more of an annoying oversight. Figure 5.10 illustrates it.

I've added a third predictor variable, HDL, to Age and Diet, so as to make it a little clearer what's going on with the order of LINEST()'s results. (I also added a blank column between the predictors in columns A through C and the predicted variable LDL in column E.)

Figure 5.10
The coefficients and their standard errors appear in the reverse of worksheet order.

H10 =G2*C2+H2*B2+I2*A2+J2

	A	B	C	D	E	F	G	H	I	J
1	Diet	Age	HDL		LDL		Coefficient for HDL	Coefficient for Age	Coefficient for Diet	Intercept
2	1	41	65		81		-1.823	0.437	12.274	167.287458
3	2	20	55		82		1.016	0.410	5.113	61.011616
4	5	87	40		198		0.492	32.846	#N/A	#N/A
5	8	77	70		160		5.155	16	#N/A	#N/A
6	2	52	45		115		16686.539	17262.261	#N/A	#N/A
7	4	30	64		77					
8	5	38	61		117					
9	4	83	52		174					
10	4	70	45		103		Correct	78.98	=G2*C2+H2*B2+I2*A2+J2	
11	4	40	60		138					
12	5	45	51		170		Incorrect	981.23	=G2*A2+H2*B2+I2*C2+J2	
13	4	61	54		196					
14	7	76	67		174					
15	7	78	56		168					
16	3	39	60		133		TREND()	78.98	=TREND(E2:E21,A2:C21)	
17	7	28	69		125					
18	6	71	61		166					
19	5	20	62		86					
20	5	22	63		190					
21	5	44	42		199					

NOTE

You can put as many columns as you want, including none, between the columns that contain the raw data for the predictors and the column that contains the predicted variable. But you can't put columns between the predictor variables themselves. Those must be in contiguous columns, as is the data in columns A through C in Figure 5.10.

The problem is as follows. The order of the variables used as arguments to LINEST(), in the range A2:C21 and reading left to right, is Diet, then Age, then HDL. But the LINEST() results, in G2:J6, report the regression coefficients in the reverse of that order: HDL, then Age, then Diet. The intercept is always the rightmost value in the first row of the LINEST() results, so you need not worry about its position.

Also, LINEST()'s ancillary statistics in the third through the fifth row of the results are unaffected. You always get the R^2, the standard error of estimate, the F ratio, the degrees of freedom for the residual, and the two sums of squares if you set LINEST()'s fourth argument to TRUE. The position of those statistics in the LINEST() results remains the same, regardless of the number of predictor variables you call for or where they're found on the worksheet.

At least the standard errors of the regression coefficients are properly aligned with the coefficients themselves. For example, in Figure 5.10, the regression coefficient for Diet is in cell I2, and the standard error for the Diet coefficient is in cell I3.

Why is this reversal of the worksheet order so bad? First, there's no good argument in favor of reversing the order of the regression coefficients from that in which the underlying data is presented. So doing is not necessitated by the underlying programming used to code LINEST(). And it helps nothing that you subsequently do with the results for LINEST() to reverse the order in which the underlying values are found on the worksheet.

Furthermore, it can easily lead to errors in the use of the results. One way to obtain the predicted values from a regression analysis is to multiply the observed values of the predictor variables by the associated regression coefficients, and then total the resulting products and add the intercept.

Cell H10 in Figure 5.10 shows an example of this process. The formula used in cell H10 is repeated as text in cell I10. The formula returns the correct value: the predicted LDL for the first observation in the data set. Notice that the value in cell H16, returned by the TREND() function, confirms the value in cell H10.

Take a close look at the formula used in cell H10:

```
=G2*C2+H2*B2+I2*A2+J2
```

It tells Excel to multiply the values in columns C, B, and A by the values in columns G, H, and I. As you move left to right from column G to column I, you move right to left from column C to column A. (And at the end you add the intercept in cell J2.)

This sort of bidirectional movement is utterly contrary to the way that things are done in Excel. Take Excel's lists and tables. In, say, a two column list that occupies columns A and B, the value in cell A3 is associated with the value in cell B3, A4 with B4, A5 and B5, and so on. You don't set things up to associate A3 with B10, A4 with B9, A5 with B8, and so on, but that's exactly what LINEST() is asking you to accept with its ill-considered design.

We've all been led to expect that the formula to get the predicted value from the regression equation *should* look like this:

```
=G2*A2+H2*B2+I2*C2+J2
```

so that as you move from G to H to I, you also move from A to B to C. But with LINEST(), that's wrong and it gets you the erroneous value in Figure 5.10, cell H12.

So why use LINEST() at all? Why not rely instead on TREND()? Most of the answer is that you need to know information from LINEST() that TREND() doesn't supply: R^2, the standard error of estimate, the F ratio, and so on. Furthermore, you often want to test the significance of the regression coefficients by means of their standard errors, so you need the results of both functions. There are plenty of other reasons you need to use LINEST() to do regression analysis in Excel and to do so accurately you'll just have to learn to allow for and deal with this idiosyncrasy.

Microsoft has had plenty of opportunities to fix it but it has declined each of them, and there's little reason to expect that this will get fixed in the future. A lot of worksheets out there are designed to accommodate this problem. If some new release of Excel suddenly reversed the order in which the regression coefficients appear, that release would cause those worksheets to interpret LINEST()'s results incorrectly.

Building a Multiple Regression Analysis from the Ground Up

It's not difficult to state in a few sentences how multiple regression differs from simple, single-predictor regression. However, I've found that just citing the differences doesn't help much in building an understanding of how the addition of a predictor affects the outcome.

I want this section of Chapter 5 to demonstrate for you how the principles and procedures used in single-predictor regression are easily adapted to account for multiple predictors. As it turns out, much of that adaptation depends on removing the effects that the predictors have *on one another*. So we begin with a brief review of material on partial and semipartial correlation.

Holding Variables Constant

Chapter 3 of this book, in the section "Partial and Semipartial Correlations," discussed how you can go about removing the effect of Variable B from both Variable A and Variable C. You then have in hand the *residuals* of Variable A and Variable C. If you now calculate the correlation between the two sets of residuals, you have what's termed the *partial correlation* between Variable A and Variable C with Variable B "held constant."

To put some meat on those bones, suppose that you have data on two sets of 100 hospitals each. One set is located in suburban areas with relatively high per capita income levels, and the other set is located in relatively poorer suburban neighborhoods. You're interested in whether a hospital's overall mortality rate varies according to the socio-economic status of a hospital's location.

Because mortality rates might be related to hospitals' staffing rates, often measured by nurse-to-patient ratios, you decide to "control for" nurse-to-patient ratio by removing its potential relationship to both an area's per capita income and the hospital's mortality rate. You do so by regressing each hospital's annual mortality rate and its area's income measure on its nurse-to-patient ratio. You then subtract the predicted mortality rate from the actual mortality rate, and the predicted income from the actual income.

The result is two sets of residuals, each of which is in theory free of the effect of nurse-to-patient ratios. Thus you have a way to correlate an area's income level with its hospital's mortality rate, unaffected by the hospital's policies regarding staffing. You'll also see this sort of analysis described as "controlling for" staffing levels.

The general approach, partial correlation, is closely related to an approach termed *semipartial* correlation (sometimes, unfortunately, referred to as *part* correlation). Partial correlation removes the effect of a third variable from both of two others. Semipartial correlation removes the effect of a third variable from just one of two other variables. As you'll see, when semipartial correlation removes the effect of a third variable from just one of the remaining two, that procedure turns out to be useful in deciding whether to add (or retain) a predictor variable in a multiple regression equation.

> **NOTE** Higher-order partial and semipartial correlations are both feasible and useful. In a higher-order partial or semipartial correlation, the influence of more than just one variable is removed from both, or just one, of two other variables.

Semipartial Correlation in a Two-Predictor Regression

Figure 5.11 shows how semipartial correlation works out in the simplest possible multiple regression analysis, one with just two predictors.

Figure 5.11 uses data that you've seen before, most recently in Figure 5.9. In Figure 5.11 the approach is to use the idea behind semipartial correlations to start building up LINEST()'s results. Of course it's wildly unlikely that you would ever go through this exercise as a practical matter—you would just deploy LINEST() instead—but it helps lay the conceptual groundwork for what goes on in multiple regression.

Figure 5.11
To make an easier comparison, I have retained the order in which LINEST() presents the regression coefficients.

E3 =SLOPE(C2:C21,L2:L21)

	A	B	C	D	E	F	G	H	I	J	K	L	M
1	Diet	Age	LDL		Multiple Regression of LDL on Diet and Age				Age Predicted by Diet	Diet Predicted by Age		Residual Age on Diet	Residual Diet on Age
2	1	41	81						34.66	4.35		6.34	-3.35
3	2	20	82		0.789	7.784	66.076		39.17	3.73		-19.17	-1.73
4	5	87	198						52.68	5.71		34.32	-0.71
5	8	77	160	=LINEST(C2:C21,A2:B21,,TRUE)					66.19	5.41		10.81	2.59
6	2	52	115		0.789	7.784	66.076		39.17	4.68		12.83	-2.68
7	4	30	77		0.384	4.742	24.830		48.17	4.03		-18.17	-0.03
8	5	38	117		0.389	34.929	#N/A		52.68	4.26		-14.68	0.74
9	4	83	174		5.413	17	#N/A		48.17	5.59		34.83	-1.59
10	4	70	103		13208	20741	#N/A		48.17	5.21		21.83	-1.21
11	4	40	138						48.17	4.32		-8.17	-0.32
12	5	45	170						52.68	4.47		-7.68	0.53
13	4	61	196						48.17	4.94		12.83	-0.94
14	7	76	174						61.68	5.38		14.32	1.62
15	7	78	168						61.68	5.44		16.32	1.56
16	3	39	133						43.67	4.29		-4.67	-1.29
17	7	28	125						61.68	3.97		-33.68	3.03
18	6	71	166						57.18	5.24		13.82	0.76
19	5	20	86						52.68	3.73		-32.68	1.27
20	5	22	190						52.68	3.79		-30.68	1.21
21	5	44	199						52.68	4.44		-8.68	0.56
22													
23	4.7	51.1	142.6		Variable Means								

We begin by ignoring the variable to be predicted, LDL, and focus on the two predictor variables, Age and Diet. The range I2:I21 uses the TREND() function to predict Age from Diet, using this array formula:

```
=TREND(B2:B21,A2:A21)
```

5

Then the process is reversed in the range J2:J21, where TREND() predicts Diet from Age, using this array formula:

=TREND(A2:A21,B2:B21)

The results in I2:I21 tell us what values of Age to expect, given the relationship in our sample between Diet and Age. Similarly, the results in J2:J21 tell us what Diet values to expect, given the same information.

Then to get the residuals, the range L2:L21 subtracts the predicted ages in I2:I21 from the actual ages in B2:B21. The result in column L is Age with the effect of Diet removed. In M2:M21, we subtract the predicted Diet values in J2:J21 from the actual Diet values in column A, to get Diet residuals with the effect of Age removed.

At this point we could use Excel's CORREL() function to get the semipartial correlation between, say, the residual Age values and the LDL observations—with Diet partialled out of Age. However, at present our interest centers on the regression coefficient for Age *in the context of multiple regression*, and we can get that using this formula:

=SLOPE(C2:C21,L2:L21)

That formula is entered in cell E3 of Figure 5.11, and the value 0.789 it returns is precisely equal to the value returned by LINEST() in cell E6. The SLOPE() function is not designed for use in a multiple regression context, but it works here because we use it with the *residual* values for Age, having removed the variability that Age and Diet share. What variability is left in the Age residuals belongs to Age, some of which is associated with LDL, and none with Diet.

NOTE

We can also return to an approach discussed in Chapter 3. In discussing Figure 3.5 I noted that one way to calculate the slope in a single-predictor regression analysis is to use this equation:

Slope of Y regressed on X = $r_{xy} (s_y / s_x)$

However, the present example takes place in a multiple regression context, so we have to be concerned about variance shared by the predictors. In Figure 5.11, we have in L2:L21 the residuals for Age after its shared variance with Diet has been removed, so we can use this Excel formula:

=CORREL(C2:C21,L2:L21)*(STDEV.S(C2:C21)/STDEV.S(L2:L21))

to return the regression coefficient for Age. The latter formula takes longer to type, but it's a better description of what's going on with the derivation of regression coefficients.

Similarly, this formula:

=SLOPE(C2:C21,M2:M21)

in cell F3 returns the regression coefficient for Diet. Its value is identical to the coefficient that LINEST() returns in cell F6. The effect of Age has been removed from the original Diet observations and the SLOPE() function takes as its arguments the residual Diet observations and the original LDL values.

It's easy enough to calculate the regression equation's intercept using (mostly) the work we've done so far. Figure 5.11 shows the mean of each of the three original variables in row 23. Using those means along with the regression coefficients in E3 and F3 we can get the intercept using this formula:

=C23 – E3*B23 – F3*A23

That is, find the product of each predictor's mean and its regression coefficient. Subtract those products from the mean of the predicted variable to get the value of the intercept. In Figure 5.11, cell G3 contains that formula.

In sum, by stripping from Age the variability it shares with Diet, and by stripping from Diet the variability it shares with Age, we have in effect converted a single problem in multiple regression to two problems in simple regression.

There's more work to be done, but the process of getting the regression coefficients from the residual values lays the groundwork.

Finding the Sums of Squares

The next step in pulling together a multiple regression analysis is to calculate the sum of squares for the regression and the residual sum of squares. See Figure 5.12.

Figure 5.12
The next step is to make the predictions and figure the residuals.

E7 =DEVSQ(I2:I21)

	A	B	C	D	E	F	G	H	I	J
1	Diet	Age	LDL		Multiple Regression of LDL on Diet and Age				LDL Predicted by Age and Diet	Residual LDL
2	1	41	81						106.22	-25.22
3	2	20	82		0.789	7.784	66.075599		97.43	-15.43
4	5	87	198						173.66	24.34
5	8	77	160						189.12	-29.12
6	2	52	115						122.68	-7.68
7	4	30	77		13208.03	20740.77			120.89	-43.89
8	5	38	117						134.99	-17.99
9	4	83	174						162.72	11.28
10	4	70	103		=LINEST(C2:C21,A2:B21,,TRUE)				152.46	-49.46
11	4	40	138		0.789	7.784	66.076		128.78	9.22
12	5	45	170		0.384	4.742	24.830		140.51	29.49
13	4	61	196		0.389	34.929	#N/A		145.35	50.65
14	7	76	174		5.413	17	#N/A		180.54	-6.54
15	7	78	168		13208.03	20740.77	#N/A		182.12	-14.12
16	3	39	133						120.21	12.79
17	7	28	125						142.66	-17.66
18	6	71	166						168.81	-2.81
19	5	20	86						120.78	-34.78
20	5	22	190						122.36	67.64
21	5	44	199						139.72	59.28

Figure 5.12 makes use of the regression coefficients and the intercept that were derived in Figure 5.11. The range I2:I21 uses the coefficients and the intercept to predict LDL values, by means of this formula in cell I2:

```
=A2*$F$3+B2*$E$3+$G$3
```

The formula is copied down through I21. It uses the regression coefficients and the intercept in row 3 to calculate each individual LDL prediction.

Then, in the range J2:J21, the residual values are calculated by subtracting the predicted values from the observed LDL values.

The regression sum of squares is calculated in cell E7 with this formula:

```
=DEVSQ(I2:I21)
```

Recall that Excel's DEVSQ() function subtracts the mean of a set of values from each individual value, squares each difference, and totals the squared differences. In this case, the result is the sum of squares for the regression.

Lastly, the residual sum of squares is calculated in cell F7 by means of this formula:

```
=DEVSQ(J2:J21)
```

Notice that the values calculated in this way in E7:F7 are identical to those returned by LINEST() in E15:F15.

You'll see statistical shorthand terms such as *regression sum of squares* and *sum of squares residual* for these quantities. Those terms can sound mysterious at first. These two are the important points to remember:

5

- We obtain a set of predicted values by multiplying the predictors by the coefficients and adding the intercept. We get the coefficients' values by regressing LDL onto Age residuals (with Diet removed) and also onto Diet residuals (with Age removed). So the sum of the squared deviations of the predicted values from their own mean is the *sum of squares regression*.
- We obtain another set of values by subtracting the predicted LDL values from the actual LDL values. These residual LDL values—values that remain after subtracting the predictions due to the regression—also define a sum of squared deviations, the *sum of squares residual*.

R^2 and Standard Error of Estimate

Figure 5.13 shows how to get the statistics that bear on the reliability of the regression equation. These are the values that LINEST() returns in its third and fourth rows.

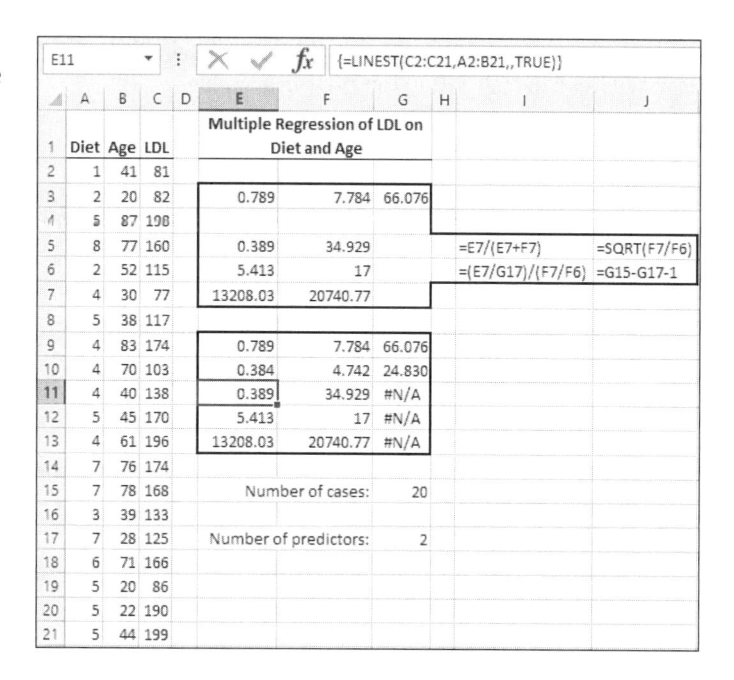

E11 | {=LINEST(C2:C21,A2:B21,,TRUE)}

	A	B	C	D	E	F	G	H	I	J
1	Diet	Age	LDL		Multiple Regression of LDL on Diet and Age					
2	1	41	81							
3	2	20	82		0.789	7.784	66.076			
4	5	87	198							
5	8	77	160		0.389	34.929			=E7/(E7+F7)	=SQRT(F7/F6)
6	2	52	115		5.413	17			=(E7/G17)/(F7/F6)	=G15-G17-1
7	4	30	77		13208.03	20740.77				
8	5	38	117							
9	4	83	174		0.789	7.784	66.076			
10	4	70	103		0.384	4.742	24.830			
11	4	40	138		0.389	34.929	#N/A			
12	5	45	170		5.413	17	#N/A			
13	4	61	196		13208.03	20740.77	#N/A			
14	7	76	174							
15	7	78	168		Number of cases:		20			
16	3	39	133							
17	7	28	125		Number of predictors:		2			
18	6	71	166							
19	5	20	86							
20	5	22	190							
21	5	44	199							

Figure 5.13
The F-ratio and the R^2 are both based on the sum of squares regression and the sum of squares residual.

Let's start with R^2, which is easier to calculate than is the F-ratio. R^2 expresses the proportion of the total variability in the predicted variable that is shared with the variability in the best combination of the predictor variables—that is, the composite variable created by applying the regression coefficients and intercept to the original observations. In the present example, LDL shares 38.9% (see cell E11 in Figure 5.13) of its variance with the composite variable that's created by applying the regression equation to the predictor variables.

The regression analysis, whether single- or multiple-predictor, calculates R^2 as follows:

1. Find the sum of squares for the predicted values (the sum of squares regression, in cell E7 of Figure 5.13).
2. Find the sum of squares for the residual values (the sum of squares residual, in cell F7 of Figure 5.13).
3. Find the total sum of squares of the predicted variable. One way to do so is to add the sum of squares regression to the sum of squares residual. In Figure 5.13 you could use this formula:

   =E7+F7

   Another way, perhaps more satisfying, to get the total sum of squares is to calculate it directly using Excel's DEVSQ() function:

   =DEVSQ(C2:C21)

4. Form the ratio of the sum of squares regression, obtained in step 1, to the total sum of squares, obtained in step 3. You can use this formula:

=E7/(E7+F7)

That's the approach taken in cell E5 in Figure 5.13. The formula is repeated as text in cell I5.

If you prefer, you can find the R^2 value with this formula:

=E7/DEVSQ(C2:C21)

The formulas are equivalent because the total sum of squares for the outcome measure, LDL, equals the regression sum of squares plus the residual sum of squares.

NOTE

That equivalence holds unless you for some reason tell LINEST( ) to force the intercept to 0.0, by setting its third, *const*, argument to FALSE. If you do so, you are requiring LINEST( ) to center the total sum of squares on 0.0 rather than on the mean of the LDL observations. Then you must use the SUMSQ( ) function rather than the DEVSQ( ) function.

So it's easy to calculate R^2, but it's just as easy to calculate the standard error of estimate, after you have your hands on the sum of squares residual. Recall that one definition of the sum of squares residual is the total sum of squares less the sum of squares regression—so the sum of squares residual represents the variability in the LDL values that *cannot* be predicted by use of the observations in A2:B21.

Therefore, we can employ the usual formula for the variance, the sum of squared deviations divided by their degrees of freedom, to get the variance of the unpredicted, residual portion of the observed LDL values. We already have the sum of squares residual. The degrees of freedom equals the number of cases (here, 20) less the constraints for the two predictors (2) less the constraint for the grand mean (1): that is, 20 – 2 – 1, or 17. Note that LINEST() agrees and returns 17 as the residual degrees of freedom in cell F12 of Figure 5.13.

Merely divide the sum of squares residual by the residual degrees of freedom. That calculation gets us the error variance: the variance of the unpredicted portion of the LDL values. It remains to take the square root of the error variance, to get its standard deviation, also termed the *standard error of estimate*. That statistic represents the standard deviation of the errors of prediction—that is, the differences between the predicted values for LDL and the actual LDL observations, or the errors of estimate. The result is in cell F5 of Figure 5.13, and its formula appears as text in cell J5. The value in cell F5 is identical to the one in cell F11, which is calculated and returned by LINEST().

F-Ratio and Residual Degrees of Freedom

The F-ratio, which LINEST() returns in the first column of its fourth row, is closely related to R^2. However, the F-ratio is a bit more complicated to assemble than R^2. The F-ratio also addresses the question of the reliability of LINEST()'s results, and the strength of the relationship between predictor variables and the predicted variable, somewhat differently than R^2 does.

The R^2 tells you directly the proportion of the total variability that's associated with the regression: It's the sum of squares regression divided by the total sum of squares.

In contrast, the F-ratio divides a measure of the variability due to regression by another measure of variability—not by the total variability, as in the case of R^2, but by the residual variability. Furthermore, in assembling the F-ratio, we adjust the sums of squares by their degrees of freedom. When that's done, the results are two estimates of the population variance: one due to regression and one due to residuals. In effect, when we look at an F-ratio we ask how many times greater than the residual variance is the variance based on the regression. The stronger the regression, the larger the F-ratio, and the more likely it is that the F-ratio measures a reliable, replicable relationship.

We can make probability statements about that relationship, such as "Assume that the F-ratio would be 1.0, if it were calculated on the full population from which we got this sample. In that case, the probability of obtaining an F-ratio as large as this one is less than 1%. That's so unlikely that we prefer to reject the assumption that the F-ratio in the population is 1.0."

Here's how to assemble the F-ratio from the sums of squares that we already have:

1. Count the number of predictor variables involved in the regression. In the present example, there are two: Age and Diet. That's the number of degrees of freedom associated with the regression sum of squares.
2. Divide the sum of squares for the regression by its degrees of freedom. The result is the numerator of the F-ratio, often termed the *mean square regression*.
3. Count the number of cases in your analysis. In the present example, that's 20. Subtract the number of predictor variables, and then subtract 1. As you saw in the discussion of the standard error of estimate, the result is 20 – 2 – 1 or 17. LINEST() supplies this value in the fourth row, second column of its results.
4. Divide the sum of squares residual by the result of step 3, the degrees of freedom for the residual. The result is often termed the *mean square residual* or, sometimes, the *mean square error*.
5. Divide the result of step 2 by the result of step 4. The result is the F-ratio.

The section "Understanding LINEST()'s F-ratio" in Chapter 4, "Using the LINEST() Function" discusses the interpretation of the F-ratio in some detail.

Calculating the Standard Errors of the Regression Coefficients

The standard errors of the regression coefficients and the intercept are the last pieces to bring into place in order to replicate the LINEST() function in the context of multiple regression. Let's start with the standard errors of the regression coefficients. See Figure 5.14.

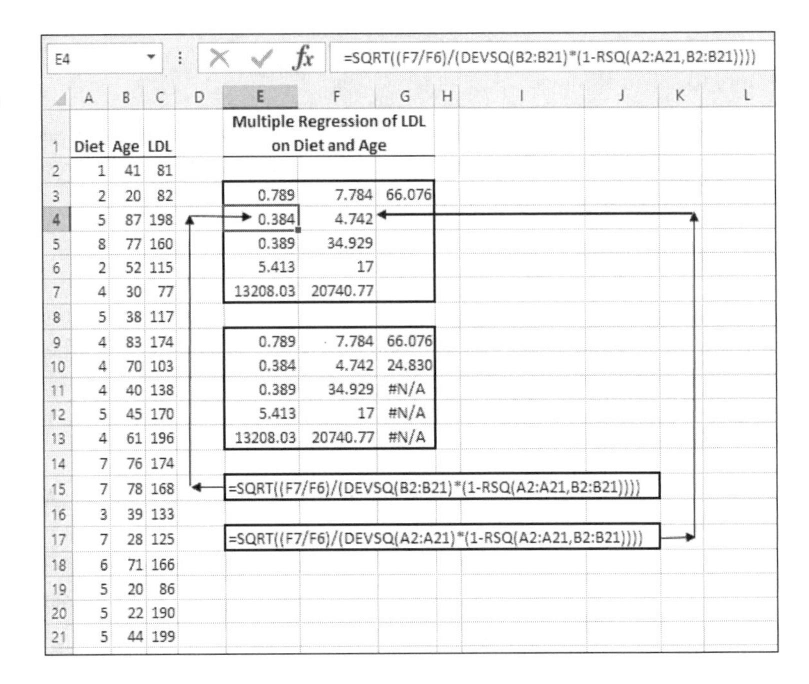

E4: =SQRT((F7/F6)/(DEVSQ(B2:B21)*(1-RSQ(A2:A21,B2:B21))))

	A	B	C	D	E	F	G
					Multiple Regression of LDL on Diet and Age		
1	Diet	Age	LDL				
2	1	41	81				
3	2	20	82		0.789	7.784	66.076
4	5	87	198		0.384	4.742	
5	8	77	160		0.389	34.929	
6	2	52	115		5.413	17	
7	4	30	77		13208.03	20740.77	
8	5	38	117				
9	4	83	174		0.789	7.784	66.076
10	4	70	103		0.384	4.742	24.830
11	4	40	138		0.389	34.929	#N/A
12	5	45	170		5.413	17	#N/A
13	4	61	196		13208.03	20740.77	#N/A
14	7	76	174				
15	7	78	168		=SQRT((F7/F6)/(DEVSQ(B2:B21)*(1-RSQ(A2:A21,B2:B21))))		
16	3	39	133				
17	7	28	125		=SQRT((F7/F6)/(DEVSQ(A2:A21)*(1-RSQ(A2:A21,B2:B21))))		
18	6	71	166				
19	5	20	86				
20	5	22	190				
21	5	44	199				

Figure 5.14
The formulas for the standard errors are more complex than for the other LINEST( ) results.

To get the standard error for the Age predictor, we need access to several statistics that we've already calculated, and one that we haven't. These are:

- The sum of squares residual and its degrees of freedom
- The sum of squares for the Age variable, calculated using the DEVSQ() function
- The R^2 for Age with Diet

For example, the formula in cell E4 of Figure 5.14 is as follows (see also cell E15 of Figure 5.14):

=SQRT((F7/F6)/(DEVSQ(B2:B21)*(1–RSQ(A2:A21,B2:B21))))

In that formula, the fragment F7/F6 returns the variance of the residuals—that is, the variance of the errors made in predicting LDL from Age and Diet. Recall that a variance is the sum of a set of squared deviations from the mean of the set, divided by the set's degrees of freedom. That's exactly what the fragment F7/F6 is: the sum of the squared deviations of the residuals in F7, divided by the residual degrees of freedom in F6.

(Don't let this take your eye off the ball, but it's an important aside. The variance of the residual errors is the square of the standard deviation of the residuals, which is also termed the *standard error of estimate*. You can find that in cell F5. Notice that the standard error of estimate, 34.929, is the square root of F7/F6, which equals 1220.045.)

The next fragment in the prior formula, DEVSQ(B2:B21), returns the sum of the squared deviations of the Age variable, the basis of the regression coefficient whose standard error we're calculating. So DEVSQ(B2:B21) finds the difference of each value of Age from the mean of Age, squares each difference and totals the squared differences. The result is one measure of the degree of variability in the predictor variable: in this case, Age.

The third and final fragment in the formula is 1–RSQ(A2:A21,B2:B21). Start with the RSQ() portion, which calculates R^2, the square of the correlation between these Diet and Age values. With its arguments set to the Diet values in A2:A21, and the Age values in B2:B21, it returns the proportion of the variance in Age that's shared with the variance in Diet. Then, when you subtract that proportion from 1, you get the proportion of the variance in Age that is *not* shared with Diet.

NOTE

With just two predictor variables, you can use RSQ( ) to determine the proportion of variance in Age that's shared with Diet (and, equivalently, the proportion of variance in Diet that's shared with Age). However, RSQ( ) can deal with only two variables at a time. With three or more predictors, you would need to rely on LINEST( ) to get the R^2 between the predictor in question and *all* the remaining predictors. Suppose that you have three predictors in A2:A21, B2:B21, and C2:C21. If you're working on the standard error of the regression coefficient for the predictor in A2:A21, you could use this to get the pertinent R^2:

=INDEX(LINEST(A2:A21,B2:C21,,TRUE),3,1)

If you grasp the reason that this formula—and its use of LINEST( )—works where RSQ( ) wouldn't, it's clear that you're following this discussion.

Let's reassemble the three fragments. Multiplying the second and third fragments, we get:

(DEVSQ(B2:B21)*(1–RSQ(A2:A21,B2:B21)))

That's the variability, measured by the sum of squared deviations, of Age times the proportion of the variability in Age not shared with Diet. The result of multiplying the two fragments together is the sum of squares in Age that's *not* shared with Diet.

You would get the same result if you used the DEVSQ() function on the residual values of regressing Age onto Diet.

Now put the first fragment back in the mix:

(F7/F6)/(DEVSQ(B2:B21)*(1–RSQ(A2:A21,B2:B21)))

The expression F7/F6 is the variance of the residuals. Recall that we get the residual values in part by multiplying a predictor variable by the regression coefficient. We're looking for an estimate of the variability in the regression coefficient that we would calculate from many samples like the one we use in Figure 5.14. Those hypothetical samples would include different cases, with different values for the predictor variables and the predicted variable. Therefore, they would result in differing values for the regression coefficients.

We can derive an estimate of the multi-sample variability in the regression weights by taking the variance of the residuals and removing the variability due to the predictor variable itself. In the regression equation, we get the residuals by multiplying the predictor variable by the regression coefficient, and subtracting the results from the predicted variable. Because we multiply the predictor variable by its coefficient, we can estimate the variability of the coefficient by reversing the multiplication: We divide the variance of the residuals by a measure of the variability in the predictor variable.

That's just what the previous formula does:

- It divides the variability in the residuals (F7/F6)
- By the variability in the predictor (DEVSQ(B2:B21))
- Times the proportion of variability in the predictor *not* shared with the other predictor (1–RSQ(A2:A21,B2:B21))

which results in an estimate of the sample-to-sample variability in the regression coefficient.

That estimate of variability constitutes the portion of the residual variance that's attributable to sample-to-sample variation in the regression coefficient. To convert the variance estimate to an estimate of the standard deviation—that is, the standard error—of the regression coefficient, we just take its square root. Repeating the formula from the beginning of this section, the standard error of the regression coefficient is:

=SQRT((F7/F6)/(DEVSQ(B2:B21)*(1–RSQ(A2:A21,B2:B21))))

It's worth noting that the only fragment that would vary if we want to calculate the standard error of the Diet predictor instead of the Age predictor is the second of the three:

- The (F7/F6) fragment is unchanged. Regardless of which predictor we're looking at, the variance of the residuals of the predicted variable remains the same and continues to equal the square of the standard error of estimate.
- There are just two predictors in this equation, Diet and Age. The Diet variable shares the same proportion of its variance with Age as the Age variable does with Diet. Therefore, the third fragment remains the same whether we're calculating the standard error for Diet or for Age. Matters would be a touch more complicated if the regression involved three or more predictors instead of just two.
- The second fragment, the sum of squares for the predictor, is the only one that changes as you move from one predictor to the next in a two-predictor analysis. Substituting DEVSQ(A2:A21) for DEVSQ(B2:B21) changes which source of variability we isolate when we multiply by the proportion of variance that's *not* shared by the two predictors.

Some Further Examples

Before moving on to other methods of calculating regression equations and their ancillary statistics, I want to cover a few further examples to illustrate how the standard errors of the regression coefficients act in different situations. To begin, let's back up to a single-predictor analysis. See Figure 5.15.

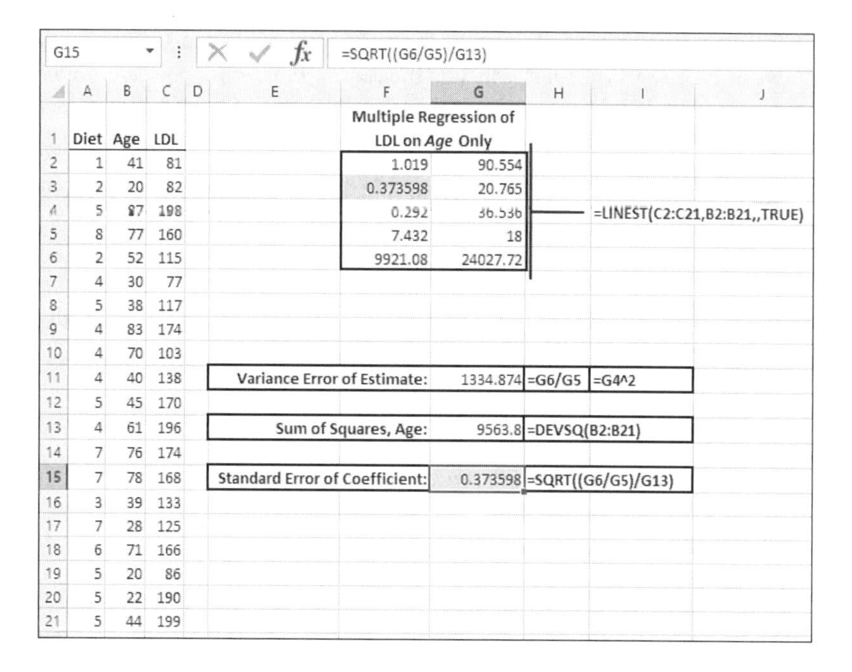

G15 =SQRT((G6/G5)/G13)

	A	B	C	D	E	F	G	H	I	J
1	Diet	Age	LDL			Multiple Regression of LDL on *Age* Only				
2	1	41	81			1.019	90.554			
3	2	20	82			0.373598	20.765			
4	5	87	198			0.292	36.536	=LINEST(C2:C21,B2:B21,,TRUE)		
5	8	77	160			7.432	18			
6	2	52	115			9921.08	24027.72			
7	4	30	77							
8	5	38	117							
9	4	83	174							
10	4	70	103							
11	4	40	138		Variance Error of Estimate:		1334.874	=G6/G5	=G4^2	
12	5	45	170							
13	4	61	196		Sum of Squares, Age:		9563.8	=DEVSQ(B2:B21)		
14	7	76	174							
15	7	78	168		Standard Error of Coefficient:		0.373598	=SQRT((G6/G5)/G13)		
16	3	39	133							
17	7	28	125							
18	6	71	166							
19	5	20	86							
20	5	22	190							
21	5	44	199							

Figure 5.15
When there's just one predictor, it cannot share variance with another predictor.

Figure 5.15 continues with the same data set used in this chapter's other figures. But the LINEST() analysis in the range F2:G6 ignores the Diet variable and predicts LDL on the basis of Age alone.

In the prior section I pointed out that the formula for the standard error of a regression coefficient adjusts its denominator to remove any variability that the predictor shares with other predictors. Here's the full denominator of the formula for the standard error of the regression coefficient once again, for the two-predictor case:

DEVSQ(B2:B21)*(1–RSQ(A2:A21,B2:B21))

It finds the proportion of variance shared by the two predictors, returned by the RSQ() function. It subtracts that value from 1 to get the portion of the variance *not* shared by the two predictors. When you multiply that quantity by the sum of squares of the values in B2:B21, you restrict the sum of squares expressed by the denominator to the amount associated with the predictor variable in question—here, the Age variable in B2:B21.

However, the analysis in Figure 5.15 uses one predictor only, Age. Therefore there's no other predictor in the equation to share variance with Age, and the denominator might as well be this:

DEVSQ(B2:B21)

omitting the 1–RSQ(A2:A21,B2:B21) fragment entirely. Then the formula for the standard error of the regression coefficient for Age, omitting Diet from the analysis, is:

=SQRT((G6/G5)/G13)

That's the formula used in cell G15 in Figure 5.15. I've shaded that cell for easy comparison with cell F3, also shaded, which is the standard error for the Age predictor as calculated and returned by LINEST(). Notice that the two values are identical.

So that's one way to avoid the necessity of adjusting the sum of squares for one predictor so that it's unique to the predictor, and not shared with another predictor: Don't use another predictor. Another, more practical way is to use predictors that are not correlated. When you use predictor variables to quantify the design of an experiment with balanced group sizes, using one of the coding systems discussed in Chapter 7, "Using Regression to Test Differences Between Group Means," that's what you're doing. When the analysis involves variables such as Age and Diet that your subjects bring along with them, it's harder to find useful predictors that are correlated with the predicted variable but are not correlated with one another. Then, you need to isolate unique variance, as is done by the denominator of the formula for the standard error of the regression coefficient discussed earlier in this section.

Still, it's worthwhile to look at what happens with two uncorrelated predictors. See Figure 5.16.

Figure 5.16
Uncorrelated predictors make the attribution of variance unambiguous.

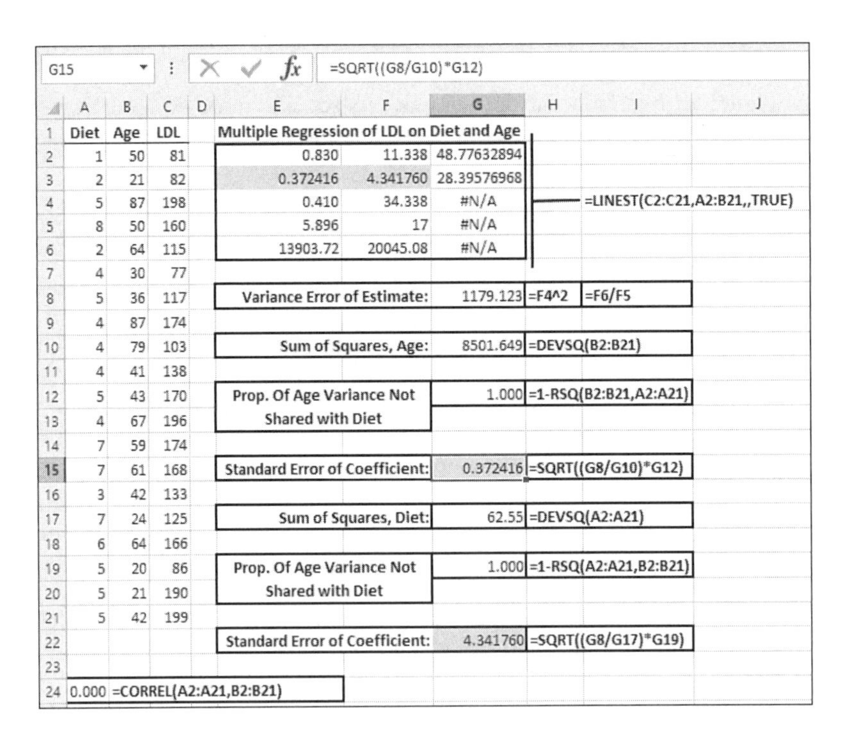

G15 =SQRT((G8/G10)*G12)

	A	B	C	D	E	F	G	H	I
1	Diet	Age	LDL		Multiple Regression of LDL on Diet and Age				
2	1	50	81		0.830	11.338	48.77632894		
3	2	21	82		0.372416	4.341760	28.39576968		
4	5	87	198		0.410	34.338	#N/A	=LINEST(C2:C21,A2:B21,,TRUE)	
5	8	50	160		5.896	17	#N/A		
6	2	64	115		13903.72	20045.08	#N/A		
7	4	30	77						
8	5	36	117		Variance Error of Estimate:		1179.123	=F4^2	=F6/F5
9	4	87	174						
10	4	79	103		Sum of Squares, Age:		8501.649	=DEVSQ(B2:B21)	
11	4	41	138						
12	5	43	170		Prop. Of Age Variance Not Shared with Diet		1.000	=1-RSQ(B2:B21,A2:A21)	
13	4	67	196						
14	7	59	174						
15	7	61	168		Standard Error of Coefficient:		0.372416	=SQRT((G8/G10)*G12)	
16	3	42	133						
17	7	24	125		Sum of Squares, Diet:		62.55	=DEVSQ(A2:A21)	
18	6	64	166						
19	5	20	86		Prop. Of Age Variance Not Shared with Diet		1.000	=1-RSQ(A2:A21,B2:B21)	
20	5	21	190						
21	5	42	199						
22					Standard Error of Coefficient:		4.341760	=SQRT((G8/G17)*G19)	
23									
24	0.000	=CORREL(A2:A21,B2:B21)							

In Figure 5.16, I have reinstated the Diet variable as a predictor in the regression equation. The LINEST() analysis in the range E2:G6 uses both Diet and Age to predict LDL. However, I changed the values of Age so that it is no longer correlated with the other predictor, Diet. The CORREL() function, used on the Diet and Age variables, returns a correlation of 0.0 in cell A24.

With two predictor variables back in the regression, there are two standard errors of the coefficients to evaluate. Let's start with Age. Its sum of squares appears in cell G10 as 8501.649. (That's a different value than shown in cell G13 of Figure 5.15: Remember that I altered the Age values to bring about a correlation of 0.0 between the predictors.)

Because the correlation between Age and Diet is now 0.0, due to the alterations I made to the values in Age, the two predictor variables share no variance. In that case, this fragment:

RSQ(A2:A21,B2:B21)

Returns 0.0, and the complete fragment:

1 – RSQ(A2:A21,B2:B21)

Returns 1.0. Now the denominator for the standard error of the Age predictor is as follows:

=DEVSQ(B2:B21) * (1 – RSQ(A2:A21,B2:B21))

Because the portion after the asterisk evaluates to 1.0, the result is simply the sum of squares for Age, just as it is when Age is used as the sole predictor variable.

Again, the formula for the standard error of the coefficient divides the residual variance by the unique sum of squares for the predictor. When you divide the residual variance by the sum of squares for Age and take the square root, as is done in cell G15 of Figure 5.16, you get the standard error. Notice that the value in cell G15 is identical to that returned by LINEST() in cell E3.

A similar analysis appears in the range G17:G22 of Figure 5.16, for the Diet predictor. Again, the standard error is returned by the ratio of the residual variance to the sum of squares for Diet. Again, there is no need in this case to adjust Diet's sum of squares for variance shared with Age, because the two predictors are uncorrelated. However, unless the predictor values express a design with balanced group sizes, the absence of correlation between predictors is an unusual situation. I have discussed it in some detail because it helps to explain the relationship between the value of a coefficient's standard error and the correlation, or lack thereof, between predictor values. It also lays some groundwork for the issues to be discussed in Chapter 7.

Another comparison is notable more for what doesn't change than for what does. In Figure 5.17 you see two data sets. One, in the range A2:C21, is identical to the data set in Figures 5.11 through 5.15. The other, in the range J2:L21, is identical to that in A2:C21 except that the values for the Diet predictor have each been multiplied by 16.

Figure 5.17 also displays two instances of LINEST() results. The range F2:H6 analyzes the data set in A2:C21, with LDL as the predicted variable. The range N2:P6 analyzes the data set in J2:L21, again with LDL as the predicted variable.

Figure 5.17
Compare the Diet values in column A with those in column J.

O2 {=LINEST(L2:L21,J2:K21,,TRUE)}

	A	B	C	D	E	F	G	H	I	J	K	L	M	N	O	P	Q
1	Diet	Age	LDL			Multiple Regression of LDL on Diet and Age				Diet	Age	LDL		Multiple Regression of LDL on (16 * Diet) and Age			
2	1	41	81			0.789	7.784	66.076		16	41	81		0.789	0.486	66.076	
3	2	20	82			0.384	4.742	24.830		32	20	82		0.384	0.296	24.830	
4	5	87	198			0.389	34.929	#N/A		80	87	198		0.389	34.929	#N/A	
5	8	77	160			5.413	17	#N/A		128	77	160		5.413	17	#N/A	
6	2	52	115			13208.03	20740.77	#N/A		32	52	115		13208.03	20740.77	#N/A	
7	4	30	77							64	30	77					
8	5	38	117							80	38	117					
9	4	83	174			*Diet*	*Age*	*LDL*		64	83	174			*Diet*	*Age*	*LDL*
10	4	70	103		Diet	1				64	70	103		Diet	1		
11	4	40	138		Age	0.364	1			64	40	138		Age	0.364	1	
12	5	45	170		LDL	0.487	0.541	1		80	45	170		LDL	0.487	0.541	1
13	4	61	196							64	61	196					
14	7	76	174							112	76	174					
15	7	78	168							112	78	168					
16	3	39	133							48	39	133					
17	7	28	125							112	28	125					
18	6	71	166							96	71	166					
19	5	20	86							80	20	86					
20	5	22	190							80	22	190					
21	5	44	199							80	44	199					

Although I made a fundamental change to the Diet values in column J, notice that:

- Both LINEST() analyses return the same sum of squares regression and the same sum of squares residual: Compare cells F6 with N6 and G6 with O6.
- Therefore, the relationship of the sum of squares regression to the sum of squares residual and to the total sum of squares is the same in both analyses. In turn, the F-ratio, degrees of freedom for the residual, R^2, and the standard error of estimate are identical in both analyses. These statistics must remain the same in the two analyses because the sums of squares remain the same despite the change to the Diet values.
- Furthermore, the regression coefficient for Age, and its standard error, are the same in both analyses. The same is true of the intercept, and its standard error, in H2:H3 and P2:P3.

So, none of the statistics I just cited change. Simply multiplying a predictor variable by a constant (or adding a constant to a predictor variable) has no effect on the correlations between that variable and the other variables in the analysis. You can confirm that by comparing the correlation matrix in the range F10:H12 with the matrix in O10:Q12.

However, both the regression coefficient for Diet and its standard error change in response to multiplying the Diet values by a constant (16 in this case). When the predictor's values increase by a factor of 16, its regression coefficient is reduced by a factor of 16. So does its standard error. Compare the shaded cells in Figure 5.17, G2:G3 with O2:O3.

You wind up with the same effect on the full regression's predicted values in either case. If, as here, the magnitude of the predictor values increases 16 times, the size of the regression coefficient becomes 16 times smaller. The result, the predictor's contribution to the set of predicted values that minimize the sum of the squared errors, is the same in both cases.

Using the Standard Error of the Regression Coefficient

The prior section examines the standard error of the regression coefficient in some depth: how it's calculated, what the calculation has to say about its meaning, and how it accounts for its predictor variable's relationship to other predictors. All that background deserves some discussion about how to use the standard error of the regression coefficient once you've got your hands on it.

Suppose that understanding the quantitative relationships between humans' ages, diets, HDL levels and LDL levels were of such compelling interest that hundreds, even thousands of researchers repeated the observational experiment that's been used as an example in this chapter. Suppose that all those researchers used the same variables and measured them in the same ways, that they selected the same number of subjects, that they analyzed the data in the same way—in short, that they conducted the research in exactly the same way.

In that case, all those researchers would get different results. Many of the results would be very similar to one another, with the regression coefficients and the R^2 values differing from one another by just an eyelash. An appreciable number of the studies would get results that were substantially different from most others. If you collected and compared the results of all those studies, you would notice that the regression coefficient for, say, Diet differs from study to study. (The same would be true, of course, for Age and HDL.) Those differences in the regression coefficients, as I've sketched this fantasy, would be due to simple sampling error: of particular interest here, differences from sample to sample in how Diet and LDL are quantitatively related.

We care about this because we want to know, principally, whether Diet's regression coefficient in the population from which we got those thousands of samples is 0.0. If it were, there would be some health-maintenance implications: Care providers and humans in general would be less concerned about low density lipoproteins as a function of diet. Also, if Diet's true regression coefficient were 0.0, those whose interests center on understanding the relationships between behaviors and cholesterol levels would drop Diet from their equations—at least those equations that seek to predict or explain cholesterol levels on the basis of personal attributes.

After all, consider what happens to the regression equation if the regression coefficient for Diet equals 0.00. It might look like this:

LDL = 0.437*Age + 0.00*Diet – 1.823*HDL + Intercept

That equation tells you that Diet has no effect on LDL levels, whether to increase them or to decrease them. You might as well leave Diet out of the equation.

But suppose that your own results are those shown in Figure 5.18.

Figure 5.18
The regression coefficient for Diet is far from 0.0.

H10 =T.DIST.2T(ABS(H9),$I$5)

	A	B	C	D	E	F	G	H	I	J	K
1	Age	HDL	Diet		LDL			Coefficient for Diet	Coefficient for HDL	Coefficient for Age	Intercept
2	41	65	1		81			12.274	-1.823	0.437	167.287
3	20	55	2		82			5.113	1.016	0.410	61.012
4	87	40	5		198			0.492	32.846	#N/A	#N/A
5	77	70	8		160			5.155	16	#N/A	#N/A
6	52	45	2		115			16686.539	17262.261	#N/A	#N/A
7	30	64	4		77						
8	38	61	5		117			Diet	HDL	Age	
9	83	52	4		174		t-ratio	2.40	-1.80	1.07	
10	70	45	4		103		Probability of t-ratio	0.02888	0.09147	0.30227	
11	40	60	4		138						
12	45	51	5		170						
13	61	54	4		196						
14	76	67	7		174						
15	78	56	7		168						
16	39	60	3		133						
17	28	69	7		125						
18	71	61	6		166						
19	20	62	5		86						
20	22	63	5		190						
21	44	42	5		199						

The regression coefficient for Diet that you actually got from your sample and from Excel's calculations based on that sample is 12.274. Should you leave it in the equation or throw it out? You might think that it can't hurt to leave it in, and it might hurt to discard it. There's something to that, but what if your intention is only partly to *predict* LDL levels? What if your main purpose is to *explain* the potential causes of LDL levels? In that case, retaining a variable that really doesn't belong in the equation is confusing at best and could even be misleading. The regression coefficient you calculated for the Diet predictor isn't 0.0, so there's no prima facie case to delete Diet from the equation. But is the actual value, 12.274, weak or strong?

That's where the standard error of the regression coefficient comes in. If you divide the regression coefficient by its standard error, the result tells you how many standard errors the observed value is from 0.0. In this case, here's the result you would get:

=12.274 / 5.113

or 2.40. So, the obtained value of 12.274 is 2.40 standard errors greater than 0.0. A standard error is a standard deviation and if you've been around standard deviations much you know that a distance of 2.40 standard deviations is quite large. It's probably a meaningful result.

Of course, that's a subjective assessment. There's another way, one that's more objective, if not entirely so. It relies heavily on the standard error of the regression coefficient. When you divide a regression coefficient by its standard error, the result is termed a *t-ratio*. (You might have encountered t-ratios in other contexts such as the difference between two means divided by the standard error of the difference between two means: That's also a t-ratio.)

Excel has several worksheet functions that assess the probability of obtaining a given t-ratio from a population in which the actual t-ratio is 0.0. In Figure 5.18, cells H10:J10 use the T.DIST.2T() function, along with the residual degrees of freedom from the regression analysis, to return that probability for each regression coefficient. For example, cell H10 contains this formula:

=T.DIST.2T(H9,$I$5)

where H9 contains the t-ratio for Diet's regression coefficient and I5 contains the residual degrees of freedom from the full regression. The reference to cell I5 is fixed by the dollar signs so that the formula could be copied and pasted into I10:J10.

NOTE

You'll find a good bit more information about the T.DIST( ) and T.INV( ) families of functions in subsequent sections, "Arranging a Two-Tailed Test" and "Arranging a One-Tailed Test." In Figure 5.18, cell I10 uses the formula =T.DIST.2T(ABS(I9),$I$5). The ABS( ) function, which returns an absolute value, is necessary because—unaccountably—the T.DIST.2T( ) function requires that its first argument be a positive number. The whole point of a two-tailed t-test is to work with either a positive or a negative t-ratio. I find it helpful to use the ABS function routinely in the T.DIST.2T( ) function, whether the t-ratio is positive or negative, and have done so in Figure 5.18.

It's a good bit easier to grasp the meaning of those probabilities by graphing them. Let's start by looking at a chart of all those thousands of hypothetical analyses that replicate your snapshot study of 20 subjects.

Suppose first that the true regression coefficient for Diet in the population is 0.0, and therefore that knowing a subject's eating habits adds no useful information to your prediction of the subject's LDL. In that case the thousands of samples and analyses would, if charted, look like the curve on the left in Figure 5.19.

The curve's height represents the relative frequency with which different values of t-ratios come about. The horizontal axis of the chart in Figure 5.19 displays the values of the regression coefficient that are associated with the t-ratios. So, the vertical axis in Figure 5.19 shows the relative frequency of the t-ratios—equivalently, the relative frequency of the regression coefficients—and the horizontal axis shows the values of the regression coefficients used to calculate the t-ratios.

The horizontal axis for the chart in Figure 5.19 (and subsequent figures in this chapter) requires a little further explanation. I have chosen to show the possible values of the regression coefficients along the horizontal axis, because our main purpose at this point is to evaluate the probability associated with a given regression coefficient. It's also true that the probability is not formally an attribute of the regression coefficient itself, but of the t-ratio that you calculate using the regression coefficient. Therefore, in discussing the charts and the probabilities they display, I refer at times to the regression coefficient itself, and at times to its associated t-ratio, depending on the nature of the point I'm trying to make.

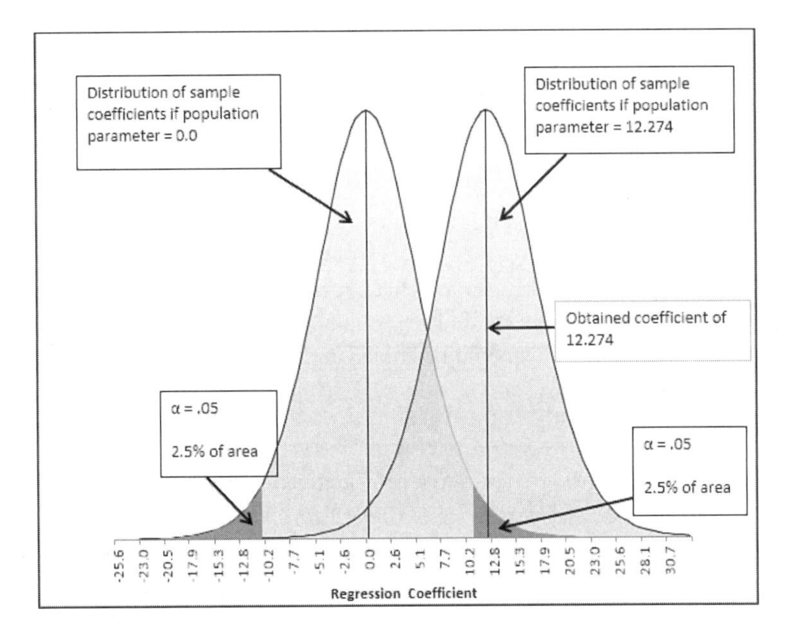

Figure 5.19
The two curves represent two different realities.

Notice that whether we're thinking in terms of regression coefficients or t-ratios, in Figure 5.19 the left-hand curve's mean is 0.0. Its mode is also 0.0. That value appears more frequently than any other value if the left-hand curve represents reality—that is, if the regression coefficient for Diet in the full population is 0.0.

But it's quite possible that the population value of Diet's regression coefficient is not 0.0 but 12.274. After all, that's the value that your own sample and analysis return. Suppose that all those thousands of other samples and analyses had resulted in the findings represented by the figure's right-hand curve. (For present purposes that assumption isn't necessary but it helps in visualizing the outcomes.)

Notice where the regression coefficient of 12.274 occurs in the left-hand curve, the one that's centered on a coefficient of 0.0. A value as high as 12.274 is well within the darker, shaded wedge in the right tail of the left-hand curve. That wedge represents 2.5% of the area under that curve. Another 2.5% is represented by the shaded wedge in the *left* tail of the same curve. Together the two areas represent what's termed *alpha*.

In the evaluation of statistical probabilities, the term *alpha* refers to the probability that you'll decide a finding does not belong to a given distribution when in fact it does. In this example, alpha refers to the probability that you'll decide the regression coefficient of 12.274 did not come from the left-hand distribution, if in fact it did. I have conventionally and arbitrarily set alpha to 5% so that we'll have a specific value for alpha that we can discuss.

In this case I have also divided alpha equally between the two tails of the full distribution, with 2.5% in each tail. (As you'll see, the value of alpha is always directly under your control. You can set it to whatever value you regard as a satisfactory criterion for what's "sufficiently unusual.")

The regression coefficient that you obtained, 12.274, and its standard error of 5.113, together result in a t-ratio of 12.274/5.113 or 2.40. The actually obtained ratio of 2.40 lies in the range of possible t-ratios that define the location of those two wedges, which, taken together, equal alpha.

Any t-ratio smaller than −2.12 or larger than +2.12 is within those wedges, and therefore has a probability of occurrence of at most 5%, *assuming that the population parameter is 0.0*. The placement of the left-hand curve on the horizontal axis assumes that its mean value is 0.0, thus building into the curve the assumption of a parameter of 0.0 for the t-ratio. Under that assumption (and another that I'll discuss before this chapter is out), we would obtain a t-ratio of 2.40, or a coefficient of 12.274 paired with a standard error of 5.113, one time at most in 20. That could certainly happen if you get an unusual sample. Just by chance you'll occasionally get an unusual result from a perfectly typical population. In that case, if your decision rule tells you that the result comes from a different population, then your results will have misled you. The probability of that occurring is equal to the value you choose for alpha.

There are many other possible realities, of course. One of them is represented by the right-hand curve in Figure 5.19. Notice how much more likely a regression coefficient of 12.274 is in the context of a curve whose mean is itself 12.274, than it is in a curve whose mean is 0.0.

So here's your choice: You can assume that the population parameter for the regression coefficient of LDL regressed onto Diet is 0.0. In that case, a person's diet makes no difference to his or her LDL level and you can omit Diet from the regression. It also means that you do *not* regard an outcome that has less than a 5% probability of occurring as very unusual—certainly not so unusual as to consider an alternative.

Your other option is to decide that a 5% or smaller likelihood, that of getting a coefficient of 12.274 from a population where the regression coefficient is actually 0.0, *is* unusual. You might decide that it's unusual enough to conclude that the assumption of a coefficient of 0.0 in the population is illogical. Many people would regard a 5% likelihood as sufficiently unusual, but many would regard it as merely a little bit abnormal.

That's what keeps subjectivity in the realm of inferential statistical analysis. The border between what's mildly unusual and what's really off the wall is a personal assessment. The usual advice, and it's good advice, is to be guided by the relative costs of different mistakes in adopting a decision rule such as this one. While that's useful input, at some point someone has to decide what's too costly to risk, and we're back to a subjective judgment.

Suppose that you wanted to make your criterion more stringent. You might want to use 2%, rather than 5%, as the borderline likelihood of a chance finding from a population with a 0.0 regression coefficient. The charts in Figure 5.20 show what that would look like.

Compare Figure 5.20 with Figure 5.19. The first difference you notice is that the wedges—which collectively constitute the alpha rate of 2%—are smaller. That's because I've changed the definition of what's unusual enough from 5% in Figure 5.19 to 2% in Figure 5.20.

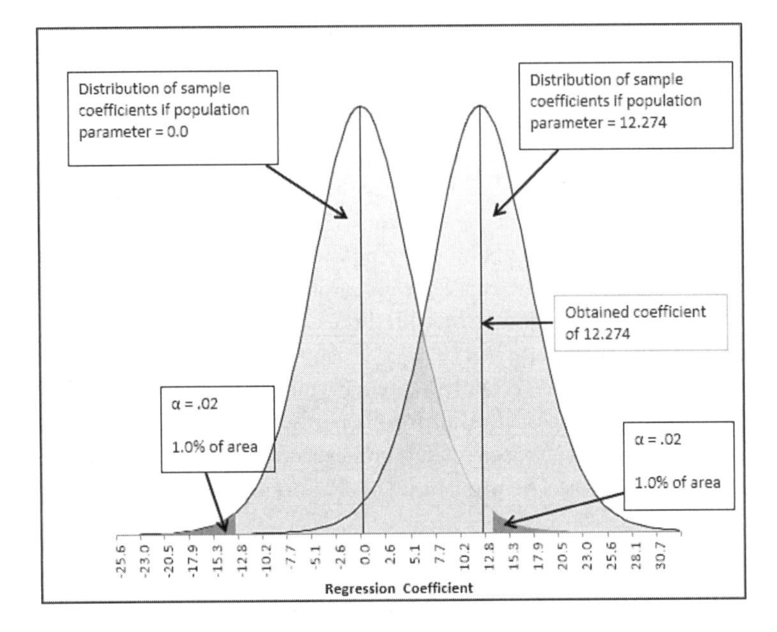

Figure 5.20
Here, alpha is 2% instead of 5%.

The second difference to notice about Figure 5.19 and Figure 5.20 is that in Figure 5.20 the regression coefficient you obtained is no longer within either of the two wedges. The total area defined by those wedges is 2% and if that's what you regard as sufficiently unusual, then you need to maintain the position that a t-ratio of 12.274 is just a chance outcome from a population whose coefficient for LDL on Diet is actually 0.0. You cannot conclude, as you did with the situation shown in Figure 5.19, that the alternate reality in fact exists, one in which the population has a regression coefficient different from 0.0.

> **NOTE** I've used the term *alpha* to mean the area under the curve defined by the two wedges in the curve's tails. That's a standard usage: For example, when statisticians change the value of alpha from, say, 2% to 5%, they speak of *relaxing alpha*. A concept that's closely related to that of alpha is termed *Type I error*. You make a Type I error when your decision rule causes you to believe that a sufficiently unusual result is evidence that a distribution such as the left-hand curve in Figure 5.19 does not in fact exist—but it actually does. In this case, you would by chance (a chance with probability equal to alpha) obtain a regression coefficient of 12.274 that really does come from a population where the regression coefficient is 0.0. That's a Type I error. There are also Type II (and even Type III) errors, and I'll discuss them in Chapter 7.

Arranging a Two-Tailed Test

Way back at the start of this chapter I defined the Diet variable, as used in the example we've focused on, as the number of times each week that a subject in the study eats red meat. Although I personally believe that the relationship between consumption of red meat and LDL levels is established, and the research settled, you might not. (If with me you do believe that the more red meat you eat the greater your LDL cholesterol level, suspend your belief for a few paragraphs. It's a little easier to grasp the concept with an open mind.)

This fictional research study finds that the relationship between Diet (as measured) and LDL is strong and positive: the more red meat consumed by the subjects, the higher their LDL levels. However, before you collected the data, you didn't know that. You took no position with respect to the relationship between Diet and LDL: For all you knew, the more red meat a subject eats, the *lower* that person's LDL. Therefore you wanted to allow for the possibility that the regression coefficient would be not 12.247 but (for example) -12.247, with a standard error of that regression coefficient of 5.113. In that case you need to arrange for alpha to occupy both ends of the distribution. The regression coefficient could turn out to be, say, either –12.247 or +12.247. You need to know the *critical values* for the regression coefficient that define the limits of the wedges at either end of the distribution.

With alpha set to 5% and half of alpha in each tail of the distribution—as in Figure 5.19—the critical values are –10.737 and +10.737. So you compare the obtained regression coefficient to those values and find that 12.247 is greater than 10.737—therefore, you reject the hypothesis that the regression coefficient in the population is 0.0, and retain the Diet variable in your regression equation.

Had you instead found that the regression coefficient obtained from your sample was –12.247 instead of +12.247, you would still have rejected the hypothesis of a 0.0 regression coefficient in the population, because –12.247 is well into the left-hand tail's portion of the alpha area.

Where do the critical values of +/–10.737 come from? Excel's T.INV() functions help provide them for you. Recall that you form a t-ratio by dividing a statistic such as a regression coefficient and dividing by the standard error of that statistic.

So in a regression analysis, a t-ratio is calculated as follows:

$$t = b / s_b$$

where t is the t-ratio, b is the regression coefficient, and s_b is the standard error of the regression coefficient. Simple rearrangement gets us this equation for finding a specific regression coefficient, given a value for the t-ratio:

$$b = t * s_b$$

LINEST() has already calculated s_b, the standard error, as 5.113. We'll get the critical t-ratio from one of Excel's T.INV() functions

NOTE

Excel offers, among many others, worksheet functions that provide information about t distributions and F distributions. The functions with the letters *DIST* in them, such as T.DIST( ) and F.DIST( ), take t-ratios or F-ratios as arguments—you supply them—and return the probability of observing the ratio you specify. The functions with the letters *INV* in them, such as T.INV( ) and F.INV( ), take a probability as an argument and return the value of the ratio associated with that probability.

Because we're looking for the critical values of the t-distribution that cut off the wedges in the tails of the distribution, we'll pass 5% off to one of the T.INV() functions and in return get back the associated value of the t-ratio. In this case, we can use this formula:

=T.INV.2T(.05,16)

The use of the T.INV.2T function requires some explanation. First, the characters ".2T" are appended to "T.INV" to inform Excel that we're after the *two tailed* version of the t-ratio. We want to divide the alpha level equally between the two tails of the distribution.

Second, the first argument in this case is .05, or 5%. That's the probability we're willing to live with as a sufficiently unusual outcome. In the present example, if we get a t-ratio that would come from a population with a 0.0 regression coefficient in only 5% of the possible samples, we're willing to reject the hypothesis that the regression coefficient is 0.0 in the parent population. (Sometimes, we'll get that unusual t-ratio from a population that in fact has a 0.0 regression coefficient. In that case we make a Type I error, and we just have to pay off to that possibility.)

Third, we supply the value 16 as the function's third argument. That's the degrees of freedom. (Unlike the normal distribution, both the t-distribution and the F-distribution are really *families* of distributions whose shapes differ according to the degrees of freedom on which the distribution is based.) When you're testing a regression coefficient, as we are here, to decide whether it is actually different from 0.0, you use the residual degrees of freedom for the full regression. In this case that's 16: See cell I5 in Figure 5.18. The residual degrees of freedom is always found in LINEST()'s fourth row, second column.

Fourth, we need two cutoff points—critical values—for a two-tailed test: We need one for the wedge in the upper tail and one for the wedge in the lower tail. The T.INV.2T() function returns only the positive value, the critical value for the upper wedge. You can get the cutoff for the lower wedge by multiplying the upper wedge's critical value by −1.0. Because the t-distribution is symmetric, the lower critical value is just the negative of the upper critical value.

The formula as just given returns the value 2.12, the cutoff value for a t-ratio when you set alpha to 5%, when the degrees of freedom is 16, and when you're dividing alpha between the two tails of the distribution. It remains to convert the critical value for the t-ratio to the metric used by the regression coefficient:

$$b = t * s_b$$

and therefore:

$$10.737 = 2.12 * 5.113$$

The result is the critical value to separate the alpha area from the remainder of the distribution in its right tail. The critical value for the portion of alpha in the left tail is simply −10.737.

As I mentioned earlier in this section, there's a minor problem with T.INV.2T(): Although it's specifically tailored to return the critical values for a two-tailed t-test, it returns only the positive critical value. You have to remember to get the negative critical value by sticking a minus sign in front of the value Excel returns.

Here's another way. You can calculate the critical values separately. If you want half of alpha, or 2.5%, in the left tail, you can use the T.INV() function instead of T.INV.2T():

=T.INV(.025,16)

to get –2.12, and the same function after adjusting the probability, as follows:

=T.INV(.975,16)

To get +2.12.

In both cases, the first argument specifies the area under the t distribution that lies to the *left* of the value that Excel returns. With a two-tailed test and with alpha set to 5%, you want 2.5% of the area to the left of the lower critical value, and 97.5% to the left of the upper critical value.

Of course it's easier to enter the T.INV.2T() function just once, use its result as the positive critical value, and multiply it by –1 to get the negative critical value. But if you try out both T.INV.2T() with a 5% probability and compare the result to T.INV() with both a 97.5% and a 2.5% probability, you might find that doing so clarifies the relationship between the T.INV() and the T.INV.2T() functions. It's also likely to clarify the relationship between two-tailed and one-tailed tests. Speaking of which ...

Arranging a One-Tailed Test

5

What if you want all the alpha area in just one tail? That can come about when you believe you know before you run the study that a regression coefficient can be only positive (or only negative). If, in this study, you believe that frequency of eating red meat can bear only a positive relationship to LDL, you can put all the alpha into one tail of the distribution and, as it turns out, thereby increase statistical power. See Figure 5.21.

Suppose that your primary purpose in doing this study is to *explain* the antecedents of unhealthy levels of LDL, in addition to simply predicting LDL levels. If diet exerts an effect on LDL, you want to account for diet in your equation. If your belief prior to running the experiment is that the more red meat in a person's diet, the higher the LDL, you want to focus on the location of Diet's regression coefficient in the *right* tail of the *left-hand* distribution in Figure 5.21.

Compare Figure 5.21 with Figure 5.19. Both figures set alpha at 5%. The difference between the two figures is that in Figure 5.19, alpha is split equally between the two tails of the left-hand distribution. The allocation of alpha assumes that the regression coefficient could be either negative or positive. The critical values, +/–2.12 if expressed as t-ratios rather than as regression coefficients, cut off 2.5% in each tail. They form the boundaries separating the two alpha areas from the central portion of the distribution.

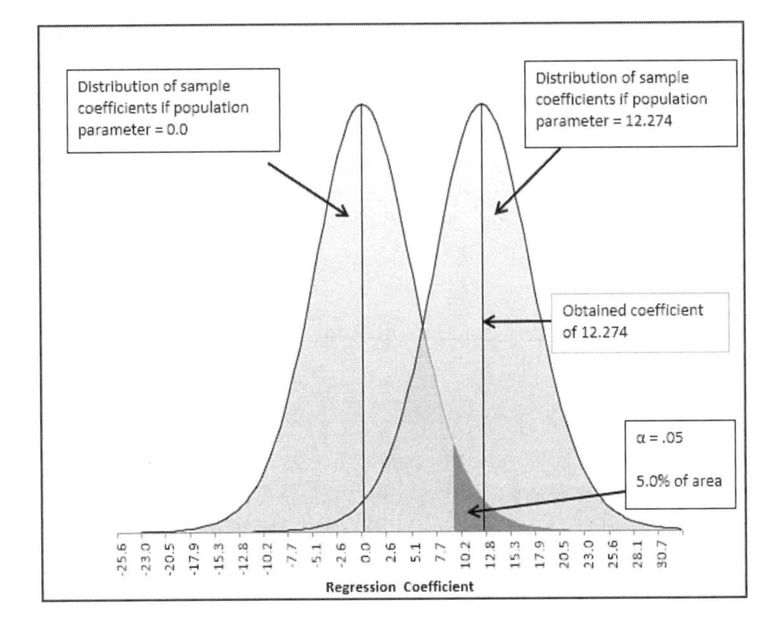

Figure 5.21
Here, alpha is 5% but it occupies only one tail of the distribution.

If the regression coefficient in your actual experiment results in a negative t-ratio farther from zero than –2.12, you have decided to regard it as evidence that the population regression coefficient is not only different from zero but is also negative. If instead your obtained t-ratio is positive and greater than +2.12, you'll decide that the population's regression coefficient is not 0.0 and that it's positive.

By splitting the 5% alpha between the two tails of the distribution as in Figure 5.19, you have arranged for decision rules that will apply whether the obtained regression coefficient is negative or positive.

Now consider the location of alpha in Figure 5.21. It's allocated entirely to the right tail of the distribution. We determine the location of the critical value to separate alpha from the rest of the distribution just as for the upper portion of the two-tailed test. Determine the critical value for the t-ratio:

=T.INV(0.95,16)

to get 1.746.

> **NOTE** Note that because in this case we're planning a one-tailed test, we use the T.INV( ) function instead of the T.INV.2T( ) function.

Multiply that result, 1.746, by the standard error, 5.113, to express the critical value in the regression coefficient's metric: The result is 8.926. A regression coefficient for Diet that is larger than 8.926 will cause you to reject the hypothesis of a regression coefficient of 0.0 in the full population.

By arranging to place all of alpha in the distribution's upper tail, you take the position that even a regression coefficient of –200 in your experiment will not contradict the hypothesis that the regression coefficient in the population is 0.0. You admit only the possibility that if the population regression coefficient is non-zero, it is positive. If your obtained coefficient is greater than the critical value—which in Figure 5.21 is 8.926—you'll decide that the population coefficient is non-zero and you'll keep Diet in the equation.

If, unaccountably, you get a negative regression coefficient, so that the *more* red meat eaten is associated with *lower* LDL in your sample, you cannot reject the hypothesis that the population regression coefficient is 0.0. Even if LINEST() tells you that the coefficient for regression LDL on Diet is –200 with a standard error of 5.11, you have to obey your decision rule and retain the hypothesis that the population coefficient is 0.0. (However, your next step should be to double-check your raw data and the way that you set it up for analysis. If all looks good, you'll have to re-examine your basic assumptions.)

Nevertheless, there's a strong argument for putting all your alpha in one tail instead of splitting it between two tails. Notice that the upper critical value in Figure 5.19 is 10.737, whereas it's 8.926 in Figure 5.21. The reason is that placing the entire 5% alpha in the right tail pushes the critical value down, to the left, to make room for the entire 5% in one tail. Therefore, your obtained regression coefficient needs only to exceed 8.926 (equivalent to a t-ratio of 1.746), not 10.737 (equivalent to a t-ratio of 2.12), in order for you to reject the hypothesis that the regression coefficient equals 0.0 in the population. Whatever the actual population value, a t-ratio of 1.746 is a more likely outcome than 2.12. In establishing a one-tailed decision rule you are increasing the statistical power of the test. If, going into the experiment, you are sure of your ground in expecting that you'll get a positive (or a negative) regression coefficient, you should establish a one-tailed test. You're more likely to reject the 0.0 population value when it's correct to do so.

By the way, statisticians tend to prefer the term *directional test* to *one-tailed test* and *nondirectional test* to *two-tailed test*. If you put all the alpha into one tail of the t-distribution, you expect that the actual result will be either positive or negative, and that's a directional test. If you split alpha between the distributional tails, you avoid taking a position on the outcome, and that's a non-directional test.

This isn't just terminological fussiness. F distributions, which we'll start to explore before this chapter is finished, have two tails just as t distributions do. However, in an F-test the focus is always on the right tail of the distribution, so the distinction implied by the terms *one-tailed* and *two-tailed* is misleading. *Directional* and *non-directional* are the preferred terms because they place the emphasis on the type of hypothesis rather than on the nature of the distributional tails.

Using the Models Comparison Approach to Evaluating Predictors

The two previous sections discuss the use of the t distribution to evaluate the effect of a given predictor variable in the regression equation. The idea is to determine the number of standard errors between the obtained regression coefficient and 0.0, where the coefficient has no effect on the equation's predicted values.

Together, the t-ratio and the degrees of freedom tell you the likelihood of obtaining that regression coefficient through sampling error, when the coefficient's value in the parent population is 0.0. If that likelihood is small, you keep the variable in the equation. If the likelihood is large, you discard it from both your equation and from your theory of how the variables interact.

There's another route to the same end, termed the *models comparison* approach, that uses the F rather than the t distribution. In practice there's little reason to use both the t-ratio and the models comparison approaches, unless you're unusually scrupulous about checking your work. In Chapter 7 I'll use the models comparison approach extensively. Furthermore, the models comparison approach puts the use of t-tests on regression coefficients in the broader context of the full regression analysis. So even if you'll never use models comparison in practice, there are good reasons to be familiar with it.

Regression analysis in general and Excel's LINEST() function in particular use the equation's R^2, or (perhaps with somewhat greater rigor) its F-ratio, to address the question of whether the equation is representative of the state of nature in the population. You will have applied LINEST(), or any linear regression application, to a sample of subjects. Among other issues, you would like to know whether you can anticipate similar results from another sample, similarly drawn.

The idea behind the models comparison approach is to derive the regression equation at least twice: once with (for example) Diet, Age, and HDL as predictors, and once with (again, for example) only Age and HDL as predictors. Doing so assumes one model with Diet, and one model without Diet. We can compare the results of the two models to quantify the change in R^2, or the change in the F-ratio. We also can test the statistical significance of that change. If the increment is statistically significant, we conclude that the variable in question (in this example, Diet) contributes a meaningful increment of information regarding the predicted variable (here, LDL). Hence the term *models comparison*.

Obtaining the Models' Statistics

Let's have a look at how the models comparison approach works with the example data set we've been using. The regression analyses in Figure 5.22 once again test the Diet variable's contribution to the equation.

Figure 5.22
This approach compares two models to test the difference in their results.

L16 | =F.DIST.RT(K16,I16,I17)

	A	B	C	D	E	F	G	H	I	J	K	L
1	Age	HDL	Diet		LDL			Coefficient for Diet	Coefficient for HDL	Coefficient for Age	Intercept	
2	41	65	1		81			12.274	-1.823	0.437	167.287	
3	20	55	2		82			5.113	1.016	0.410	61.012	
4	87	40	5		198			0.492	32.846	#N/A	#N/A	
5	77	70	8		160			5.155	16	#N/A	#N/A	
6	52	45	2		115			16686.539	17262.261	#N/A	#N/A	
7	30	64	4		77							
8	38	61	5		117			HDL	Age			
9	83	52	4		174			-0.631	0.943	130.462		
10	70	45	4		103			1.002	0.399	66.815		
11	40	60	4		138			0.308	37.164	#N/A		
12	45	51	5		170			3.790	17	#N/A		
13	61	54	4		196			10468.543	23480.257	#N/A		
14	76	67	7		174							
15	78	56	7		168		Source	R^2	df	MS	F	Prob of F
16	39	60	3		133		Difference	0.183	1	0.183	5.763	0.02888
17	28	69	7		125		Residual	0.508	16	0.032		
18	71	61	6		166							
19	20	62	5		86			Diet	HDL	Age		
20	22	63	5		190		t-ratio	2.40	-1.80	1.07		
21	44	42	5		199		Probability of t-ratio	0.02888	0.05	0.30		

The LINEST() results in the range H2:K6 pertain to what I'll call the full model: that is, LDL regressed onto Diet, Age, and HDL. You've seen this specific analysis before, most recently in Figure 5.18. It's returned by this array formula:

=LINEST(E2:E21,A2:C21,,TRUE)

For present purposes, our interest centers on the contribution of the Diet variable to the regression equation. Using models comparison, the first step to isolating Diet's effect on LDL is to use LINEST() once again but omitting Diet, and using only Age and HDL as predictors. I'll call that the restricted model. So, the array formula used to obtain that analysis, in the range H9:J13, is as follows:

=LINEST(E2:E21,A2:B21,,TRUE)

The only difference in the inputs to the two instances of LINEST() in Figure 5.22 is that the analysis in H2:K6 includes Diet as a predictor, and the analysis in H9:J13 does not. Therefore we can attribute any differences in the results to Diet's presence or absence as a predictor.

In these situations I prefer to use proportions of variance in preference to sums of squares. When I see that the R^2 is 0.492, I know exactly what it means: the composite of the predictor variables shares almost half its variance with the predicted variable. In contrast, traditional analysis of variance usually begins with sums of squares. However, knowing that the sum of squares due to the regression is 16686.539 tells me nothing at all.

Regression analysis partitions the total sum of squared deviations in the predicted variable into a sum of squares due to regression and a residual sum of squares. The sum of squares regression, divided by the total sum of squares, equals R^2. Therefore we can use R^2 in place of a measure of the sum of squares regression, and $1 - R^2$ in place of the sum of squares residual.

Notice the R^2 value in cell H4, 0.492. Just a little less than half the variability in LDL is shared with the best combination, as determined by the regression coefficients, of the three predictors.

Contrast that result with the R^2 value in cell H11, 0.308. Not quite a third of the variability in LDL is associated with the best combination of Age and LDL. The only change that might bring about the difference in the two values of R^2, 0.492 and 0.308, is the presence or absence of the Diet variable. We can test whether its presence or absence is significant in a statistical sense.

The R^2 in the full model, 0.492, consists to some degree of variance that is shared by Age, Diet, HDL, and the predicted variable LDL. The R^2 in the restricted model, 0.308, has *no* variance that is shared by Age, Diet, HDL, and LDL because Diet isn't in the model. Any difference between the two values for R^2 must therefore be due to variance shared by Diet with Age, HDL, and LDL.

So, begin by subtracting the R^2 for the restricted model from the R^2 for the full model. The result is the additional proportion of variance in LDL that can be attributed to the presence of Diet in the model. That's done in Figure 5.22 in cell H16:

```
=H4 – H11
```

The loss of R^2 due to omitting Diet from the model is 0.183.

Besides the amount of R^2 attributable to Diet, we need an amount that represents the residual variability—that portion *not* accounted for by variables in the equation. We need that in order to arrange a statistical test of the increment, and we get it by subtracting the R^2 for the full model from 1.0. So the formula in cell H17 is this:

```
=1 – H4
```

which returns 0.508, the proportion of variability in LDL that the predictors *fail* to account for in the full model.

We can also use subtraction to get the degrees of freedom for the difference in R^2. The degrees of freedom for regression in the full model is 3, the number of predictors. In the restricted model, it's 2—again, the number of predictors. We're left with 1 as the degrees of freedom for the Difference, in cell I16.

We're using $1 - R^2$ as the residual proportion of variance for the models comparison test. (We can also reach this value by dividing the full model's residual sum of squares by the total sum of squares.) Therefore 16, the residual degrees of freedom for the full model, is the one to use for the residual component of the models comparison test.

Cells J16:J17 divide the R^2 values in H16:H17 by the associated degrees of freedom in cells I16:I17. The results are not true mean squares, as implied by the label MS in J15, but they stand in for mean squares in an analysis that begins with proportions of variance instead of sums of squares.

Finally, an F-ratio is calculated in cell K16 by dividing the MS Difference in cell J16 by the MS Residual in cell J17. We can evaluate the result, 5.763, using Excel's F.DIST.RT() function. That's done in cell L16 using this formula:

=F.DIST.RT(K16,I16,I17)

The F.DIST.RT() function returns the proportion of the area under the F distribution that lies to the right of the value of its first argument—here, the value 5.763 in cell K16. The F.DIST() function, in contrast, returns the proportion of the area under the F distribution that lies to the left of an F-ratio. So you could also use this formula:

=1 – F.DIST(K16,I16,I17,TRUE)

which, because it subtracts the result of the F.DIST() function from 1, returns the proportion of the area under the F distribution that lies to the *right* of the value in cell K16.

Both 1 – F.DIST() and F.DIST.RT() return the value 0.02888 as a probability. The probability represents the likelihood of getting an F-ratio of 5.763 from a population where the actual F-ratio, the parameter, is 1.0—that is, where the variance due to regression in the F-ratio's numerator is equal to the residual variance in the F-ratio's denominator. Then, the regression provides no useful information over and above the residual variance.

Be sure to compare the probability, 0.02888, returned by the F.DIST.RT() function, with the probability returned by the T.DIST.2T() function, way back in Figure 5.18, in cell H10. The two probabilities are identical, out to the fifteenth decimal place.

In Figure 5.18, we calculated the likelihood that a t-ratio of 2.40, the regression coefficient for Diet divided by its standard error, comes from a population in which the actual regression coefficient is 0.0. In Figure 5.22, we tested the likelihood that 0.183 of the LDL variance, associated with the presence of Diet in the equation, comes from a population in which the actual proportion of the variance shared by Diet and LDL is 0.0.

These are simply two different ways of asking the same question. It takes some thought, but not an overwhelming amount, to see why they provide the same answer.

It's important to note that either method, using the t-ratio on the coefficient or using the F-ratio on the incremental variance, adjusts the test of a particular predictor by removing any variance that it shares with the remaining predictors. The t-ratio approach does so, in this chapter's example, by regressing Diet onto Age and HDL, and removing that shared variance from the standard error of the regression coefficient.

The F-ratio approach does so by subtracting the R^2 due to Age and HDL from the full model's R^2, leaving only that portion of R^2 due uniquely to Diet. We'll return to this theme in Chapter 7.

Using Sums of Squares Instead of R^2

Let's finish this discussion by demonstrating that the analysis in the prior section, which is based on proportions of variability or R^2, returns the same results as a more traditional analysis which is based on actual values of the sums of squares. See Figure 5.23.

Figure 5.23
It makes no difference to the outcome whether you use proportions of variance or sums of squares.

H16 =H6-H13

	A	B	C	D	E	F	G	H	I	J	K	L
1	Age	HDL	Diet		LDL			Coefficient for Diet	Coefficient for HDL	Coefficient for Age	Intercept	
2	41	65	1		81			12.274	-1.823	0.437	167.287	
3	20	55	2		82			5.113	1.016	0.410	61.012	
4	87	40	5		198			0.492	32.846	#N/A	#N/A	
5	77	70	8		160			5.155	16	#N/A	#N/A	
6	52	45	2		115			16686.539	17262.261	#N/A	#N/A	
7	30	64	4		77							
8	38	61	5		117			HDL	Age			
9	83	52	4		174			-0.631	0.943	130.462		
10	70	45	4		103			1.002	0.399	66.815		
11	40	60	4		138			0.308	37.164	#N/A		
12	45	51	5		170			3.790	17	#N/A		
13	61	54	4		196			10468.543	23480.257	#N/A		
14	76	67	7		174							
15	78	56	7		168		Source	SS	df	MS	F	Prob of F
16	39	60	3		133		Difference	6217.996	1	6217.996	5.763	0.02888
17	28	69	7		125		Residual	17262.261	16	1078.891		
18	71	61	6		166							
19	20	62	5		86			Diet	HDL	Age		
20	22	63	5		190		t-ratio	2.40	-1.80	1.07		
21	44	42	5		199		Probability of t-ratio	0.02888	0.05	0.30		

Every formula and value in Figure 5.23 is identical to the corresponding formula and value in Figure 5.22 with these exceptions:

- The proportions of variance in cells H16 and H17 of Figure 5.22 are replaced in Figure 5.23 by the sums of squares that they represent. In Figure 5.23, the sum of squares for the difference between the two models, in cell H16, is the difference between the regression sum of squares for the full model in cell H6, and the regression sum of squares for the restricted model in cell H13.
- The mean squares in cells J16 and J17 are true mean squares, the associated sum of squares divided by the associated degrees of freedom.

The resulting F-ratio and its probability of occurrence are identical to those calculated in Figure 5.22. The two approaches to models comparison are completely equivalent. Again, I find the proportions of variance approach more useful because I'm working with proportions of the total sums of squares, and I know that a value such as 0.750 means three quarters of the total variation in the predicted variable. If I'm working with sums of squares, I don't have a clue what 92848.62 means.

Estimating Shrinkage in R^2

There's a problem inherent in the whole idea of regression analysis, and it has to do with the number of observations in your data set and the number of predictor variables in the regression equation.

Suppose you have a random sample of 50 potential customers and 5 predictor variables. You use those predictor variables to predict how much money each customer will spend at a particular website. The regression analysis yields five regression coefficients, a constant, and an R^2 of 0.49.

Now you take another random sample of 50 people and measure them on the same five variables. You apply the earlier regression coefficients to the new sample data and calculate the R^2 between the predicted values and the actual observations, and find that the new R^2 is only 0.43.

That's a fairly typical outcome. It's *normal* to get a lower R^2 when you apply the coefficients calculated on one sample to another sample. In fact, it's a typical enough outcome to have its own name: R^2 shrinkage. Here's what happens:

The math that's the basis of regression analysis is designed to maximize R^2, the proportion of variance shared by Variable Y's actual values and its predicted values. Regression acts as if the individual correlations between the variables—between each predictor and the predicted variable, and between each pair of predictor variables—from the sample are the same as they are in the population. However, a statistic from a sample equals a parameter from a population only very rarely, and even then it's just by chance.

So, due to the errors that are nearly inevitable in the sample correlations and to the way that the process maximizes R^2, regression capitalizes on chance. When you repeat your experiment with a second sample, using regression coefficients from the first sample, you're using coefficients that are based on errors that exist in that first sample but not in the second sample. As a result, the R^2 calculated on the second data sample tends strongly to be lower than the R^2 from the first sample. The R^2 statistic is in fact positively biased, just as the sample standard deviation is negatively biased as an estimator of the population standard deviation (although the reasons for the bias in the two statistics are very different).

The most reliable solution is to increase the sample size. As is the case with all statistics, the larger the number of cases on which the statistic is based, the closer the statistic comes to the value of the population parameter.

The relationship between the number of variables and the number of cases helps determine the degree of bias. A formula called the *shrinkage formula*, reported by most statistics packages that include a regression procedure, can give you a sense of how much bias might exist in the R^2 based on your sample:

$$1 - (1 - R^2)\ (N - 1) / (N - k - 1)$$

where N is the size of the sample used to calculate R^2 and k is the number of predictor variables in the regression equation. The result of the formula is an estimate of the value of the R^2 if calculated on a population instead of on a sample.

It's worth trying out the shrinkage formula on a worksheet to see what happens as you vary the sample size and the number of predictor variables. The larger the sample size, the less the shrinkage, and the larger the number of predictors, the greater the shrinkage.

Other methods of testing the variability of the R^2 statistic involve resampling. Although I can't cover those methods here, it's worth looking up techniques such as the jackknife and bootstrapping. Also cross-validation, involving a training data set and a validation data set, is frequently an excellent idea, especially now that networked access to large data sets is widespread.

There's much in regression analysis that's straightforward, on which statisticians are largely in accord, but the approach entails some aspects that require experience and good judgment to handle properly. I'll start to explore some of those aspects in Chapter 6, "Assumptions and Cautions Regarding Regression Analysis."

Assumptions and Cautions Regarding Regression Analysis

6

IN THIS CHAPTER

About Assumptions 199

The Straw Man 204

Coping with Nonlinear and Other Problem Distributions 211

The Assumption of Equal Spread 213

Unequal Variances and Sample Sizes 220

If you bear with me through this entire book, or even just some of it, you'll find that I refer occasionally to assumptions that your data may need to meet if your analysis is to be credible. Some sources of information about regression analysis—really *outstanding* sources—spend less than a page discussing those assumptions. Other sources spend pages and pages warning about violating the assumptions without explaining if, how, and why they can be violated with impunity.

The truth of the matter is that assumptions in regression analysis are a complicated matter. You'll be better placed to understand the effects of violating them if you understand the reasons that the assumptions exist, and what happens to the numbers when your data doesn't follow them. Hence this chapter.

About Assumptions

When I first started learning about regression analysis, and about multiple regression analysis in particular, I nearly gave up before I started because it seemed that regression was based on a whole raft of assumptions—too many assumptions, so many assumptions that I had no hope of complying with them all. I figured that if I didn't comply with them, I would get called out for poor planning or sloppy management. ("No! Bad nerd!")

So where I could, I avoided tasks where the experimental design unavoidably involved multiple regression, and steered other tasks toward what I regarded as simpler, less assumption-wracked analyses such as analysis of covariance. The irony in that, of course, is that analysis of covariance *is* multiple regression analysis.

Eventually, through exposure to better sources of information, I came to realize that meeting all those assumptions wasn't necessary to producing a quality piece of research. Furthermore, none of them prevents you from running a regression analysis. There's no assumption that can prevent you from *calculating* correlations, standard errors, R^2 values, and the other elements of a regression analysis. You'll want to check some aspects of the underlying data, and although some writers cast them as assumptions, they're more accurately characterized as traps to avoid.

In preparing to write this book I reviewed several sources, published during the last 30 years, that focus on regression analysis, and quite a few websites, prepared during the last 10 years or so, that also discuss regression analysis. I'm sorry to report that many writers and other content experts still adopt a Mrs. Grundy, finger-wagging approach to the discussion of regression analysis.

Here's an example of what you can expect. I've changed several terms to conceal its source somewhat, but I haven't modified its sense or its tone one whit:

> If you violate any of these assumptions (e.g., if the relationship between the predictor and the predicted variable is curvilinear, or the residuals are correlated, or the variances aren't equal across the predictors values, or the predicted variable isn't normally distributed), then the predicted values, confidence intervals, and inferences might be uninformative in the best case and biased in the worst.

Sounds pretty dire, doesn't it? We might just as well unplug and see what the tea leaves have to say.

One problem is that the term *assumptions* is used loosely in the literature on regression analysis. Assumptions can be strong or weak, and it's a continuum, not simply a situation with some strong assumptions and some weak ones.

Another problem is that different sources cite different assumptions, but most sources cite these:

- The Y variable's residuals are normally distributed at each X point along the regression line.
- The Y values bear a linear relationship to the X values. This does not mean that the Y values fall directly on a straight line when plotted against the X values. The assumption is that the residuals have a mean of zero at each X point. Figure 6.1 illustrates this concept. (However, if the plot of the means describes a curve rather than a straight line, the solution often involves simply adding a predictor such as the square of the existing X variable.)
- The variance of the residuals is about the same at each X point. You'll see this assumption termed *homoscedasticity*, or *equal spread*. This chapter takes particular care examining the assumption of equal spread because of its importance in testing differences between means. (You're by no means out of options if the assumption is not met.)

- The residuals are independent of one another. The rationale for this assumption is that dependency among the residuals can change your estimate of the variability in the population. Positive residual correlation can spuriously reduce that estimate. In that case, the regression effect can look stronger than it actually is. Particularly from the viewpoint of statistical inference and probabilities, this is an important assumption. If your data don't meet it, it's a good idea to consider other methods of analysis that are specifically designed to account for and deal with the sources of dependency in the data. Examples include Box-Jenkins analysis and various types of exponential smoothing, which anticipate and manage the serial correlations among residuals that are common in time series analysis.

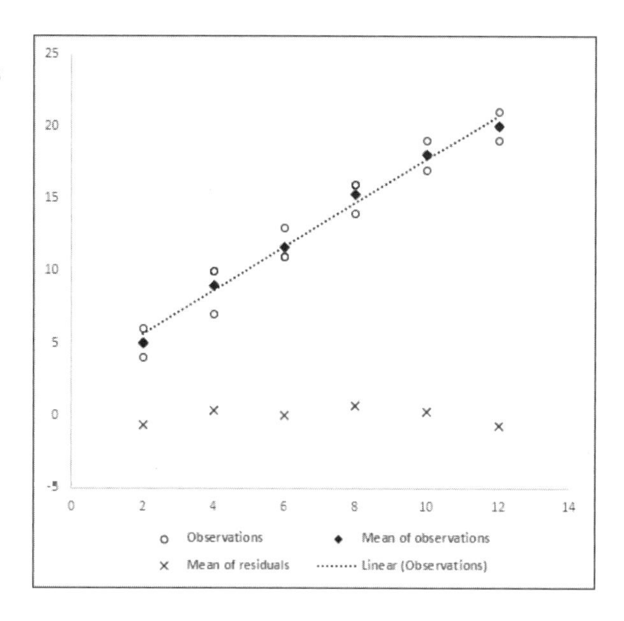

Figure 6.1
The means of the Y values lie close to the regression line, and the means of the residuals are all close to 0.0. The linearity assumption is met.

That final assumption, of the independence of residuals, is a particularly important assumption and a strong one—strong in the sense that regression analysis generally and its statistical inferences in particular are sensitive to its violation. Different methods for diagnosing dependency between residuals exist. Excel makes it so easy to calculate residuals and plot them on charts that there's really no good reason not to chart residuals against actuals, and against each predictor variable. If you're working with time series, and predicting future values by means of their relationship to measures of elapsed time, then tests of correlated errors such as autocorrelation functions and partial autocorrelation functions are useful. I'll return to the assumption of independence of residuals from time to time in the remainder of this book.

However, not all these are either/or issues. It is *not* true that, for example, having different variances in the Y values at different points in the X variable invalidates any regression analysis. If you are using regression analysis to test differences in group means, it frequently happens that violating the equal spread assumption makes no difference: Regression is said to be *robust* with respect to violation of the equal spread assumption. Although we make the assumption, we do so in the knowledge that the consequences of violating it can be tested and that we can manage them if need be.

Furthermore, the term *assumption* isn't as precise as we might want. The word itself, as well as the way that it's used, suggests that if an assumption is not met then the entire analysis is at least suspect and perhaps utterly invalid. That's simply not true, although certainly cases exist in which violating the assumption makes an important difference. The assumption of linearity provides good examples. See the section later in this chapter, "Coping with Nonlinear and Other Problem Distributions," for more on this issue.

So, we have a set of assumptions that are of different degrees of importance, and a set of precautions that it's just due diligence to observe.

The assumptions group roughly into several types, which I discuss in later sections.

Robustness: It Might Not Matter

Some of the assumptions are made in order to derive the equations underlying regression analysis. However, when it comes to actually carrying out the analysis, they turn out not to matter, or matter only in some extreme situations. Statisticians say in these cases that the analysis is *robust with respect to violations of the assumption*. The body of research that reached these conclusions is collectively termed the *robustness studies*.

For example, the assumption of equal spread states that the Y values have the same variance at each level of the X variable. If you regress weight onto height, you assume that the variance of the weights of all those who are 60 inches tall is the same as the variance of the weights of all those who are 61 inches tall, 62 inches tall, and so on.

But the assumption has a fudge factor built into it. First, we recognize that it's extremely unlikely that the variances on Y will be identical at every value of X—particularly with small sample sizes. So it's typical to plot the Y values against the X values and look for evidence of highly discrepant variances in Y at different values of X. When X is an interval variable, such as height or age or temperature, you generally don't worry too much about moderate differences in the variances, but you do keep in mind that in such a situation the standard error of estimate might not be really standard. If you want to put confidence intervals around the regression line to show where, say, 95% of the Y values will cluster for a particular X value, you're aware that the interval might be too large where the variance in Y is small, and too small where the variance in Y is large.

It's the same issue when X is a nominal rather than an interval variable: when, for example, a 1 on the predictor X means "treatment group" and a 0 means "control group." Now it's typical to pay close attention to the possibility that the variance in Y is very different in the two groups. It's a straightforward comparison. And if the group sizes are equal, it's been shown that the analysis is robust even to large differences in the variances. Even when both the variances *and* the group sizes are different, you can modify the statistical test so that the resulting probability statements are credible.

Particularly when you are using regression to test differences in group means, the assumption that the Y values—and thus, the residuals—are normally distributed is of little importance: The tests are robust with respect to violating the assumption of normality.

Even so, it's important to take a look at the data in a chart. The Data Analysis add-in contains a Regression tool. I use it infrequently because the additional analyses it provides do not compensate for the fact that the results are given as static values. However, the Regression tool does include a reasonable approximation of a normal probability plot, also termed a quantile plot. Figure 6.2 shows an example.

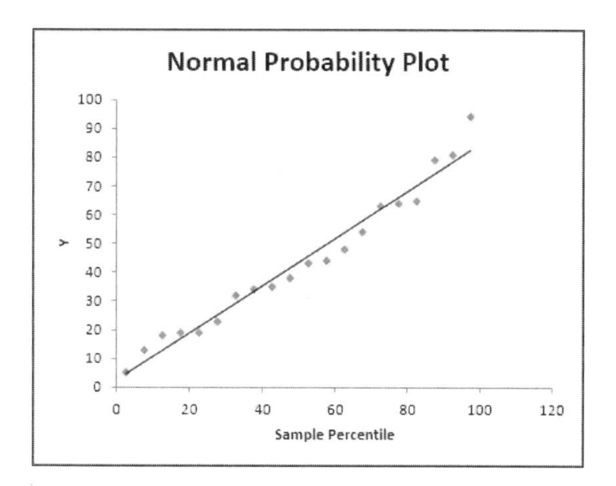

Figure 6.2
The closer the individual observations lie to the regression line, the closer their distribution is to the normal.

The normal probability plot as provided by the Data Analysis add-in's Regression tool does not include a trendline. To get one, simply right-click the data series in the chart and choose Add Trendline from the shortcut menu. The default trendline is Linear.

The observations shown in Figure 6.2 come from a normal distribution. It's not at all unusual to find a data set in which the plotted points depart somewhat from the trendline, perhaps in a sigma pattern. You don't usually worry too much about that, but Figure 6.3 shows what the normal probability plot might look like if you applied least squares regression analysis to a Y variable that has just two values, 1 and 0.

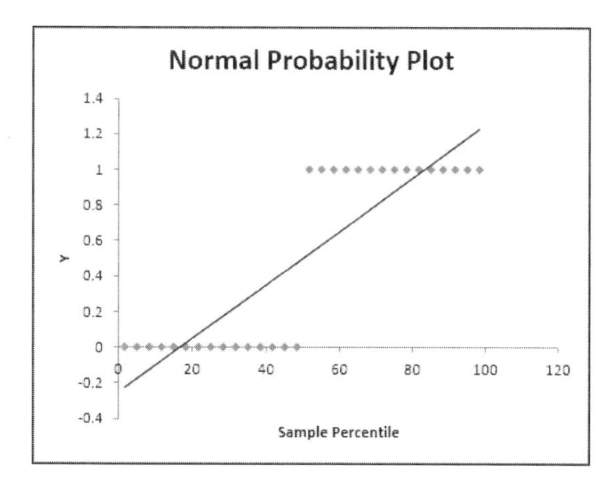

Figure 6.3
The normal probability plot of a dichotomous Y variable.

What you see in Figure 6.3 is a major departure from what you see in Figure 6.2. It's true that statisticians often ignore, with good reason, moderate departures from the normal distribution in the predicted variable. However, what you see in Figure 6.3 does not represent a moderate departure from the normal distribution in the underlying data. Least squares regression analysis, which is what this book is about, is not an appropriate method for analyzing Y values that just don't resemble a bell curve at all. As is often the case, it's not the regression method all by itself that causes the problem. It's the nature of the data we bring to the regression framework that alters the nature of the analysis.

I'll go into these matters in greater depth later in this chapter, in the section titled "Coping with Nonlinear and Other Problem Distributions."

Assumptions and Statistical Inference

Some of the assumptions regarding regression pertain only to situations in which you want to use statistical inference. That is, you might have selected subjects at random from a population of interest and assigned each subject, again at random, to a treatment and to a control group. You collect data that quantifies the treatment outcomes and run it through a regression analysis in preparation for a t-test.

The t-test is the inferential test that, in this example, tells you how probable the t-ratio is if the treatment had no effect compared to the control when applied to the entire population instead of just a sample. You need to meet the assumptions made by the t-test to make an accurate probability statement.

Otherwise, you need not meet those assumptions. If you fail to meet the t-test's assumption that the variance is the same in both groups, you'll want to do some additional checking before you bother with calculating the probability that the null hypothesis is wrong. If you're not concerned with, and you don't intend to discuss, the probability of getting your t-ratio from a population where the treatment and the control group have the same outcome, then you can ignore LINEST()'s F-ratio. It'll be accurate but depending on your purposes, you might find it irrelevant as an inferential statistic, and it's less useful than R^2 as a descriptive measure. The rest of the analysis will also be accurate, and very likely useful.

The Straw Man

The sort of thing that you often find online, in blogs and in essays sponsored by publishers of statistical software, all too often jumps directly from the authors' keyboard to your screen, with no apparent technical edit along the way. Here's an example:

A blog discusses how you would probably prefer your regression analysis to turn out: with a high R^2 value and what the blog terms a "low p value." By the term "low p value," the blog means the probability returned by a t-test of a regression coefficient. To run that t-test, you divide the coefficient by its standard error. (See Chapter 5, "Multiple Regression" for the details of setting up this sort of test.) Assume that the coefficient in the population is 0.0.

In that case, the probability of getting a large t-ratio from the data in your sample is very low—hence the author's term "low p value."

The blog states that a large R^2 (meaning a strong relationship between the predictor and the predicted variable) and a low p value (meaning a "statistically significant" regression coefficient) seem to go together. The implication is that we would be slack-jawed with astonishment should we encounter an analysis in which the R^2 is weak and the statistical significance of the regression coefficient is strong. Charts such as those in Figure 6.4 and 6.5 are provided to help us "... reconcile significant variables with a low R-squared value!"

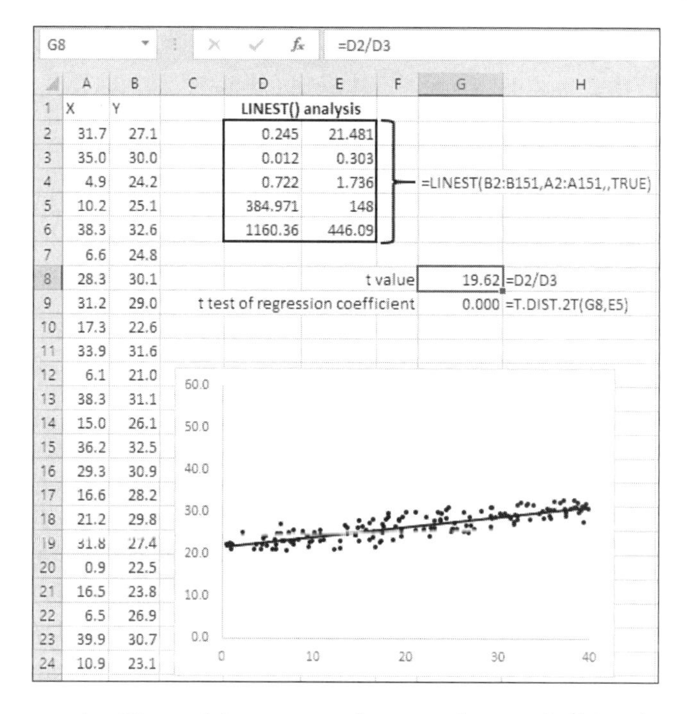

G8 | f_x =D2/D3

	A	B	C	D	E	F	G	H
1	X	Y		LINEST() analysis				
2	31.7	27.1		0.245	21.481			
3	35.0	30.0		0.012	0.303			
4	4.9	24.2		0.722	1.736	=LINEST(B2:B151,A2:A151,,TRUE)		
5	10.2	25.1		384.971	148			
6	38.3	32.6		1160.36	446.09			
7	6.6	24.8						
8	28.3	30.1			t value		19.62	=D2/D3
9	31.2	29.0	t test of regression coefficient				0.000	=T.DIST.2T(G8,E5)
10	17.3	22.6						
11	33.9	31.6						
12	6.1	21.0						
13	38.3	31.1						
14	15.0	26.1						
15	36.2	32.5						
16	29.3	30.9						
17	16.6	28.2						
18	21.2	29.8						
19	31.8	27.4						
20	0.9	22.5						
21	16.5	23.8						
22	6.5	26.9						
23	39.9	30.7						
24	10.9	23.1						

Figure 6.4
The actual observations cling closely to the trendline, and the R^2 is fairly high.

The blog suggests that Figure 6.4 represents the normal state of affairs when the relationship between the predictor and the predicted variable is strong. The observed values are all close to the trendline. Furthermore, the regression coefficient is quite some distance from 0.0, in terms of its standard errors. The coefficient of .245 divided by its standard error of .012 results in a t-ratio of 19.62: The coefficient is almost 20 standard errors greater than zero.

In contrast, the situation depicted in Figure 6.5 has a relatively low R^2 paired with a regression coefficient that is still improbably high if its value in the population is 0.0.

The blog states that the reader is probably surprised to find a low value for R^2 (0.067, shown in cell D4 of Figure 6.5) *and* a regression coefficient so many standard errors from zero in the same analysis. It's certainly true that a test of the t-ratio returns a probability that meets most reasonable people's conception of what constitutes statistical significance.

Figure 6.5
The actual observations stray well away from the trendline and the R^2 is low.

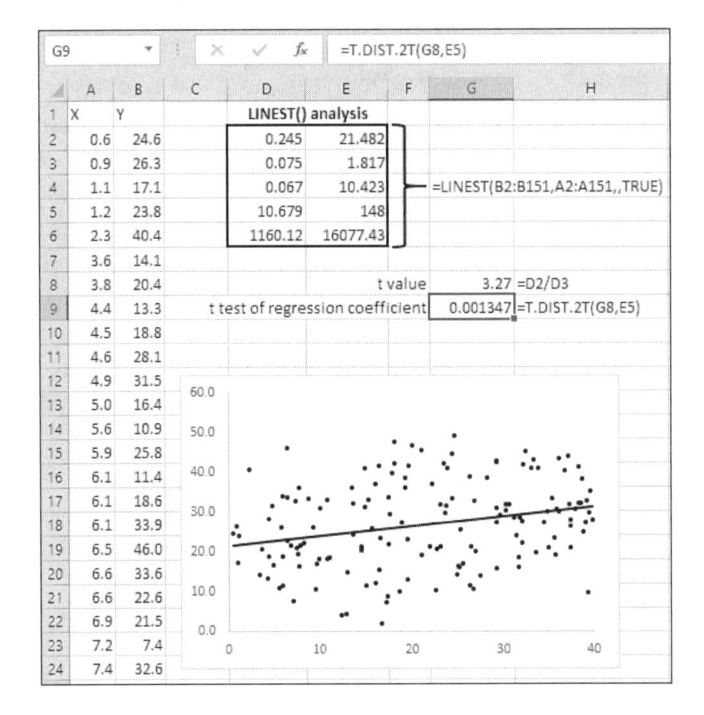

	A	B	C	D	E	F	G	H
1	X	Y		LINEST() analysis				
2	0.6	24.6		0.245	21.482			
3	0.9	26.3		0.075	1.817			
4	1.1	17.1		0.067	10.423	=LINEST(B2:B151,A2:A151,,TRUE)		
5	1.2	23.8		10.679	148			
6	2.3	40.4		1160.12	16077.43			
7	3.6	14.1						
8	3.8	20.4			t value		3.27	=D2/D3
9	4.4	13.3		t test of regression coefficient			0.001347	=T.DIST.2T(G8,E5)
10	4.5	18.8						
11	4.6	28.1						
12	4.9	31.5						
13	5.0	16.4						
14	5.6	10.9						
15	5.9	25.8						
16	6.1	11.4						
17	6.1	18.6						
18	6.1	33.9						
19	6.5	46.0						
20	6.6	33.6						
21	6.6	22.6						
22	6.9	21.5						
23	7.2	7.4						
24	7.4	32.6						

Bear in mind that more than 99.8% of a t distribution with 148 degrees of freedom lies to the left of a t-ratio of 3.0, and even more lies to the left of a t-ratio of 19.0. So a t-ratio of either 19.62 (Figure 6.4, cell G8) or 3.27 (Figure 6.5, cell G8) would occur by chance in at most 1 of 1,000 samples when the population parameter for the coefficient is actually 0.0. That's shown by means of the T.DIST.2T() function in cell G9 of both Figures 6.4 and 6.5. (I discuss this function, as well as the closely related T.DIST() and T.DIST.2T() functions, later in this chapter.)

So, still referring to Figure 6.5, how can you get such powerful evidence that the regression coefficient is non-zero from the same data set and analysis that show the R^2 to be only 0.067? Well, a statistically significant regression coefficient has no special need for a large R^2. The two statistics measure entirely different things. The blog I'm discussing sets up a straw man—the apparent contradictory pairing of a weak R^2 with a statistically significant regression coefficient—and then tries to knock it down with advice such as "Low R^2 values can be a problem when you need precise predictions." (Exactly when is it that we need imprecise predictions?)

Start by considering the definition of R^2. Regression divides the total sum of squares into two additive components: the regression sum of squares and the residual sum of squares. R^2 is the ratio of the regression sum of squares to the total sum of squares, and so can be thought of as a proportion or percentage. The closer each predicted value is to each actual observation, the larger the regression sum of squares, the lower the residual sum of squares, and the higher the R^2.

Compare the chart in Figure 6.4 with that in Figure 6.5. The observations in Figure 6.4 are clearly much closer to the regression line than they are in Figure 6.5. That difference is confirmed numerically by the sums of squares residual in cell E6 of each figure. The sum of squares regression is very nearly the same in both figures at a touch over 1160 (see cell D6). But the sum of squares residual is 446.09 in Figure 6.4 versus 16077.43 in Figure 6.5 (see cell E6). That increase in the sum of squares residual causes the decrease in the R^2 from Figure 6.4 to Figure 6.5.

At the same time, the value of the regression coefficient is identical in both figures (see cell D2). This example shows that it's entirely possible for the R^2 to weaken while the value of the coefficient remains unchanged. It's an unusual situation, yes. But I've made it unusual to better illustrate this point: that the accuracy of the regression equation's predicted values can weaken substantially without changing the value of the regression coefficient.

Now let's look more closely at the question of the statistical significance of the regression coefficient. As I just mentioned, the regression coefficient doesn't change as the sum of squares residual increases between Figure 6.4 and 6.5. However, the increase in the sum of squares residual does have an effect on the coefficient's standard error. Notice that the standard error in cell D3 grows from 0.012 in Figure 6.4 to 0.075 in Figure 6.5, more than a six-fold increase.

NOTE

The variance error of estimate is the square of the standard error of estimate in cell E4 of both figures. The variance error of estimate equals the average squared deviation between the actual observations and the predicted values. It is also the numerator of the standard error of the regression coefficient. Therefore, other things being equal, when the variance error of estimate increases, two of the results are a smaller R^2 and a larger standard error of the regression coefficient.

So in the process of increasing the sum of squares residual from Figure 6.4 to Figure 6.5, we have also increased the standard error of the regression coefficient—and the t-ratio *decreases*. In cell G8 of Figure 6.4, the t-ratio is an enormous 19.62—the regression coefficient is almost 20 standard errors above 0.0. In cell G8 of Figure 6.5, the t-ratio is 3.27: still quite large when the df is 148, but not truly enormous.

A t-ratio of 19.62 with 148 degrees of freedom comes about by chance, from a population where the true value of the regression coefficient is 0.0, an almost immeasurably small percent of the time. A t-ratio of 3.27 from the same population, although still very unlikely, is measurable and occurs about once in 1,000 replications.

Somewhat disingenuously, the blog I'm discussing doesn't mention that increasing the residual sum of squares both reduces R^2 and reduces the inferential t-ratio from 19.62 to 3.27. That's a dramatic difference, but it's hidden by the glib statement that both "p values" are equal to or less than .001.

This section began by noting that there's nothing contradictory in the pairing of a weak R^2 and a statistically significant regression coefficient. The two statistics measure different aspects of the regression and coexist happily, as shown in Figure 6.5. Still, it's worth looking more closely at what is being measured by the regression coefficient and R^2.

R^2 is clearly a descriptive—as distinct from an inferential—statistic. It tells you the proportion of variability in the predicted variable that's shared with the predictor variable. (To keep this discussion straightforward, I'll continue to assume that we're working with just one predictor, but the same arguments apply to a multiple regression analysis.)

The regression coefficient is also a descriptive statistic, but (especially in a multiple predictor situation, where a single coefficient cannot be regarded as a slope) it's more profitably thought of as a contributor to the accuracy of the predictions. In that case, the question is whether the variable contributes *reliably* to the prediction. If, in hundreds of replications of a study, a variable's regression coefficient stays close to 0.0, it's not helping. One way to test that is to calculate a t-ratio by dividing the coefficient by its standard error. You can then compare that result to a known distribution of t-ratios: So doing will tell you whether you can expect the outcome of a single study to repeat itself in subsequent, hypothetical studies.

Another way to make the same assessment is by way of an F test, which you can set up by rearranging some of the components that make up the R^2. LINEST() provides the value of the F ratio directly, and that's useful, but it's not much help in understanding what the F ratio tells you. (In contrast, the value of the R^2 has virtually immediate intuitive meaning.) Later chapters take a much closer look at the F ratio in regression analysis, but for the moment let's see how it can act as a measure of the reliability of the regression's results.

As you would expect, the F ratio has a numerator and a denominator. The numerator is itself a ratio:

MS Regression = SS Regression / df Regression

The abbreviation *MS* is shorthand for *m*ean *s*quare. It's an estimate of the population variance based on the sum of the squared deviations of the predicted values from their mean. The MS regression has as many degrees of freedom as there are predictor variables in the equation, so in the case of simple single predictor regression, the df for MS Regression is 1. In the case of the analysis in Figure 6.5, the MS regression is the SS Regression of 1160.12 (cell D6) divided by 1, or 1160.12.

The denominator of the F ratio is yet another ratio:

MS Residual = SS Residual / df Residual

In Figure 6.5, the SS Residual is 16077.43 (cell E6). The df Residual is given by this formula:

df Residual = $n - k - 1$

where n is the number of observations and k is the number of predictor variables. So, in both Figures 6.4 and 6.5, the df Residual is 150 – 1 – 1, or 148. LINEST() provides you that figure in its fourth row second column—cell E5 in both figures.

Following the formulas just given for the MS Regression and the MS Residual, the LINEST() results in Figure 6.5 return:

MS Regression = SS Regression / df Regression = 1160.12 / 1 = 1160.12

MS Residual = SS Residual / df Residual = 16077.43 / 148 = 108.63

You won't find either of those two mean squares in the LINEST() results: It doesn't return them. However, the next step in this analysis is to calculate the F ratio, which LINEST() does supply. The F ratio is the ratio of the MS Regression to the MS Residual, or, in this case:

1160.12 / 108.63 = 10.679

Notice that's the same value as LINEST() returns in cell D5 of Figure 6.5. LINEST() always returns the F ratio for the full regression in the fourth row, first column of the worksheet range its results occupy.

It remains to evaluate the F ratio, to calculate the probability of getting an F ratio as large as 10.679, with 1 and 148 degrees of freedom, when the relationship between the predictor and the predicted variable is simply random in the full population.

Figure 6.6 shows the frequency distribution of the F ratio with 1 and 148 degrees of freedom, when the relationship between the predictor and the predicted variable is random. Due to sampling error, it occasionally happens that you get an F ratio large enough to make you wonder if the underlying hypothesis of a random relationship between the predictor values and the predicted values is wrong.

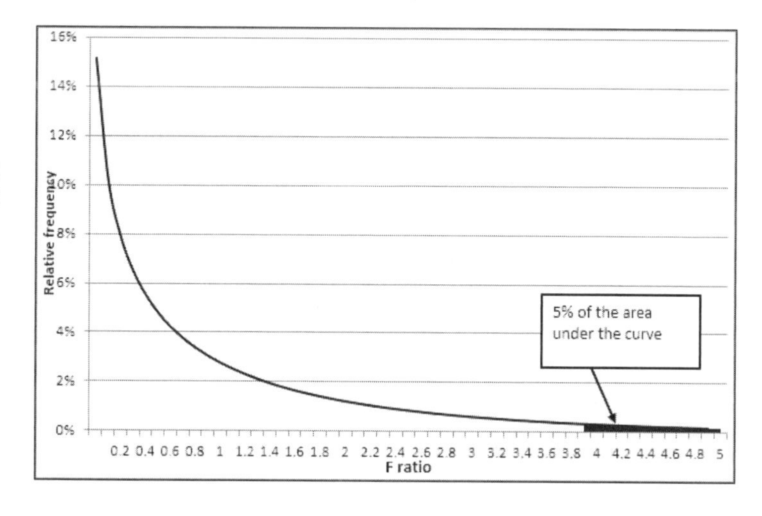

Figure 6.6
The distribution shown is termed a *central F distribution*. It represents samples from a population where the MS Regression equals the MS Residual.

When that relationship is a random one, both the MS Regression and the MS Residual estimate the same quantity: the simple variance of the predicted variable.

NOTE

If you're relatively new to the terms *mean square regression* and *mean square residual*, it can help to remember what a variance is: It's the average of squared deviations. The term *mean square* is just verbal shorthand for "average squared deviation." The term *mean square regression*, then, just means "an estimate of the total variance in the predicted variable, based on the differences between the predicted observations and their mean."

What you deviate a value from depends on the context. For example, if you're calculating an MS Residual, you subtract the predicted value from the actual value and square the result. If you're calculating the variance of a sample, you subtract the sample mean from each observation, and—as usual—square the result.

When only one predictor variable appears in a regression equation, it turns out that testing the significance of its regression coefficient is equivalent to the significance of the regression equation itself. In one case you're testing whether the regression coefficient in the population is 0.0. In the other case you're testing whether the equation's R^2 in the population is 0.0.

When there's just one predictor variable, the probability attached to its regression coefficient is identical to the probability attached to the equation's R^2. With more than one predictor, the probabilities are different because the value of R^2 depends on more than just one predictor.

Figure 6.7 shows how the probability estimate that results from a t-test of a regression coefficient is equivalent to the probability estimate that results from the F-test of the regression equation.

Figure 6.7
With just one predictor, the t-test and the F-test necessarily reach the same probability estimate.

G8 =1-F8

	A	B	C	D	E	F	G	H
1		Regression Coefficient	Standard Error of Regression Coefficient	df	t value	T.DIST()	1 - T.DIST()	2 * (1- T.DIST())
2								
3		0.245	0.075	148	3.27	0.99933	0.00067	0.00135
4								
5								
6		SS Regression	SS Residual	df	F value	F.DIST()	1 - F.DIST()	
7								
8		1160.12	16077.43	148	10.679	0.998653	0.00135	

The values used in Figure 6.7 are taken from the LINEST() analysis in Figure 6.5. Notice these two points in particular:

- The value of the t-ratio is the square root of the F-ratio. This is always true when the test is of a regression equation with one predictor only, equivalent to an ANOVA or a t-test of the difference between two group means.

- The probability returned by the nondirectional t-test in cell H3 is identical to the probability returned by the F-test in cell G8. (For a discussion of the meaning of "nondirectional" in this context, see the section titled "Understanding the Differences Between the T.DIST( ) Functions," later in this chapter.)

Coping with Nonlinear and Other Problem Distributions

Earlier in this chapter, in "Robustness: It Might Not Matter," I noted that the way that regression analysis works often depends on the nature of the data that we bring to it. One such aspect is the distribution of residuals. The normal distribution assumption holds that the residuals—that is, the difference between the actual values on the Y variable and the predicted Y values—themselves are normally distributed at each value of the X variable.

The linearity assumption holds that the mean of the residuals is zero at every point along a straight regression line. In linear regression, the predicted values lie precisely on the regression line. So the residuals are assumed to have a mean of zero when X equals 5, when X equals 17.4, and so on.

If the data do not conform to those assumptions, approaches other than linear regression analysis are available to you. One involves curvilinear regression analysis, in which you raise the values of an X variable to the second, third, and conceivably higher powers. That approach often fits beautifully in the overall structure of multiple regression analysis, and this book discusses it in greater detail in Chapter 7, "Using Regression to Test Differences Between Group Means."

However, problems can arise with curvilinear regression, in particular when you approach the observed limits of the predictor variables. Figure 6.8 illustrates the problem.

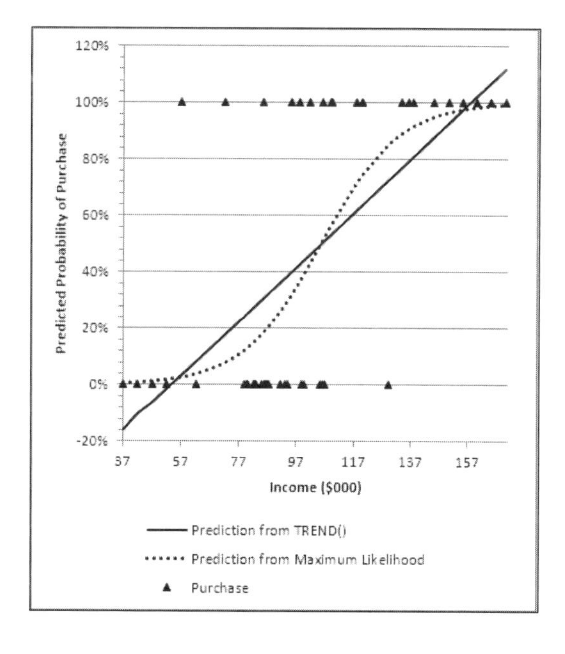

Figure 6.8
At the upper and lower limits the predicted values from linear regression are unrealistic.

6

Figure 6.8 depicts the results of two ways of analyzing (fictional) data on the relationship between annual income and the probability of purchasing a house. Clearly, the likelihood that a purchase will occur increases as income increases—you don't need a chart to tell you that. The chart does clarify the differences between the results of predicting the probability of purchasing a house by means of linear regression, and the results you might get from a different approach. In this case, Figure 6.8 contrasts linear regression with logistic regression.

Correlations form the basis for linear (and curvilinear, for that matter) regression analysis. Odds and odds ratios form the basis for logistic regression. Despite the use of the word *regression* in both methods' names, the two approaches are very different. One result of those differences is that logistic regression does not necessarily produce a straight regression line, as does linear regression. That aspect of logistic regression means that its maximum likelihood line, shown as a sigmoid-shaped series of dots in Figure 6.8, adjusts its slope as it moves across the continuum of X values (here, that's income).

Therefore, logistic regression doesn't wind up with negative purchase probabilities at the low end of the income scale, nor with purchase probabilities greater than 100% at its upper end. Logistic regression has other advantages as compared to linear regression, but the issues of scale are among the clearest.

Figure 6.8 also shows the actual probabilities, 1.0 and 0.0, of purchase for each of the observations that make up the linear and the logistic regression analyses. It's clear that they are not normally distributed across the income values, and that their residuals cannot be normally distributed at each income value. So in this case, because of the nature of the available observations, two of the assumptions made by linear regression are not met.

Figure 6.9 shows that you can turn a strictly linear regression into a curvilinear regression by raising an X variable to a power: 2 for a quadratic relationship, 3 for a cubic relationship, and so on.

You can arrange for a cubic component (or other nonlinear component such as the quadratic) simply by right-clicking the original charted data series, choosing Add Trendline from the shortcut menu, selecting the Polynomial option, and choosing an order of 3 (for cubic), 2 (for quadratic), and so on.

Although plenty of cases exist in which adding a quadratic or cubic component can improve the accuracy of regression predictions, as well as their reliability, the situation shown in Figures 6.8 and 6.9 isn't one of them. Notice that the cubic component turns up at the low end of the income scale, and turns down at the upper end. That outcome does not align with what we know of the relationship between income and purchasing behavior.

These situations typically come up when you have a nominal or qualitative variable as the Y variable and interval variables as the X variable or variables. Besides logistic regression, discriminant function analysis is another credible way to analyze such a data set, although many statisticians have moved away from the discriminant function and toward logistic regression in recent years.

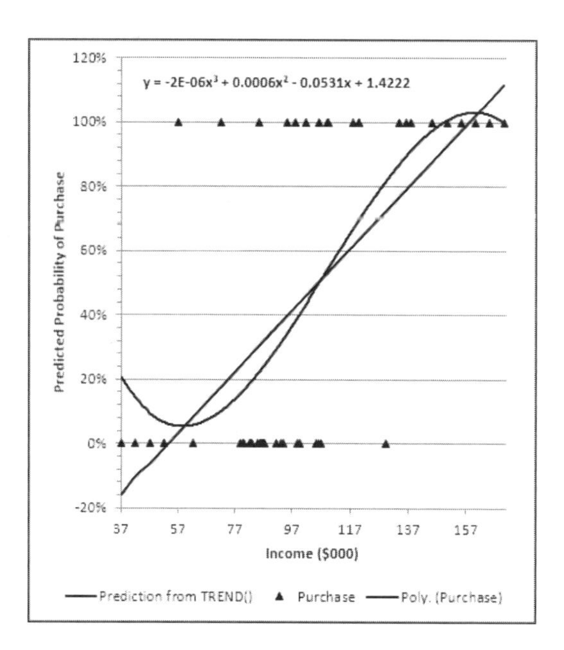

Figure 6.9
A cubic component with two changes in direction can provide a rough approximation of the logistic regression's maximum likelihood trendline.

I have included the present section of this chapter primarily to emphasize that the issues of normal distributions and linearity do not present hard-and-fast criteria. It's not easy to come up with a naturally occurring distribution that's further from the normal curve than a dichotomy such as Buys versus Doesn't Buy, and this section has suggested that you consider options other than linear regression in that sort of case.

Still, the litany of assumptions has shades of gray. Although the curvilinear solution shown in Figure 6.9 is not ideal, it does dispense with the assumption of linearity, quite neatly in fact, by the expedient of supplying a curved predictor.

The Assumption of Equal Spread

One of the assumptions cited for regression analysis is that of *homoscedasticity*. You can see why it's more typically called *equal spread*. For clarity here, I'll refer to the predictor variable (or the least squares combination of the predictor variables in a multiple regression) as the *X* variable, and the predicted variable as the *Y* variable.

The equal spread assumption states that, at every point along the continuum defined by the X variable, the variance of the Y variable is constant. Put another way: Suppose that those cases that have the score 5.5 on the X variable have a variance of 31 on the Y variable. The assumption states that the variance on Y of the cases with the score 92 on the X variable also have a variance of 31 on the Y variable. The same assumption is made for cases with the score –42 on the X variable, and for cases with the score 3.1416 on the X variable: We assume that all those subsets have the same variance, 31.

Figure 6.10 illustrates this assumption.

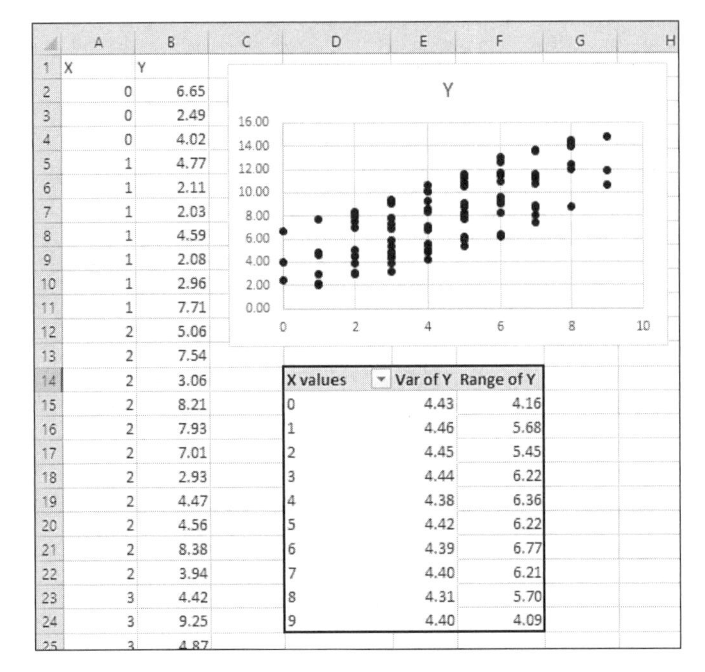

X	Y
0	6.65
0	2.49
0	4.02
1	4.77
1	2.11
1	2.03
1	4.59
1	2.08
1	2.96
1	7.71
2	5.06
2	7.54
2	3.06
2	8.21
2	7.93
2	7.01
2	2.93
2	4.47
2	4.56
2	8.38
2	3.94
3	4.42
3	9.25

X values	Var of Y	Range of Y
0	4.43	4.16
1	4.46	5.68
2	4.45	5.45
3	4.44	6.22
4	4.38	6.36
5	4.42	6.22
6	4.39	6.77
7	4.40	6.21
8	4.31	5.70
9	4.40	4.09

Figure 6.10 The variances of Y at each X point are all nearly the same.

The data shown in Figure 6.10 complies with the assumption of equal spread. You can see from the chart that the spread of the Y values at each X value is quite similar, but a much more objective measure appears in the pivot table that starts in cell D14. There, the variances at each point are called out specifically, and no variance is more than 0.15 from another.

It's useful to keep these points in mind:

- The assumption of equal spread does not impose a requirement that the variances all precisely equal one another. *That* sort of assumption is so restrictive that the technique of regression analysis would become completely infeasible. The variances should be fairly close to one another if possible, but methods exist to deal with situations in which there are appreciable differences (this chapter discusses those methods). The data in Figure 6.10 results in variances that surely qualify as "fairly close."
- When you look at a chart similar to the one shown in Figure 6.10, you'll often see a cigar-shaped scatter of charted values. That's almost the case in Figure 6.10, where the center of the scatter is thicker than at its tails. However, your eye is reacting to the *ranges* of values, and the assumption has to do with their *variances*. The range is sensitive to the number of observations, whereas the variance is not: It's the average squared deviation. So although you'll notice that the tails of the scatter diagram tend to be thinner than its center, keep in mind that the important statistic in this case is the variance, not the range.

Now, to clarify the rationale for the assumption of equal spread, I have to anticipate some of the material in Chapter 7. The next section provides a brief overview of that usage.

Using Dummy Coding

The final two chapters in this book detail the use of regression as an analysis technique that's superior to and more informative than traditional approaches such as t-tests, the Analysis of Variance (ANOVA), and the Analysis of Covariance (ANCOVA). You reach the same end point using regression as you do using t-tests, ANOVA, or ANCOVA, but using regression you gain additional insight regarding the relationships among the variables.

The usual (if not the only) purpose of a t-test is to assess the probability of obtaining a difference between two mean values from samples, under the assumption that there is no difference between the means in the full populations from which the samples were taken. Traditionally this requires that you separate the data that applies to Group 1, calculate its mean and variance, and then do the same for Group 2. You maintain the separation of the groups by running the calculations separately.

The same is true of ANOVA, but normally there are more than just two groups to worry about and therefore the number of calculations increases. The t-test you run with two groups tells you whether you can regard the group means as "significantly different." The F-test you run in an ANOVA is sometimes termed an *omnibus* F-test, because it tells you whether *any* of three or more comparisons between two means is a significant difference.

Regression analysis of the same data that you might subject to a t-test (with two groups) or to an ANOVA (with three or more groups) takes a different tack. Suppose that you have just two group means to consider. Figure 6.11 shows how you might lay the data out for analysis by the t-Test: Two-Sample Assuming Equal Variances tool in Excel's Data Analysis add-in.

Figure 6.11
The data laid out in two separate lists.

	A	B	C	D	E	F
1	Treatment	Control		t-Test: Two-Sample Assuming Equal Variances		
2	125	95				
3	94	116			*Treatment*	*Control*
4	113	111		Mean	107.944	98.667
5	92	105		Variance	190.291	191.647
6	95	103		Observations	18	18
7	121	89		Pooled Variance	190.969	
8	100	117		Hypothesized Mean Difference	0	
9	109	103		df	34	
10	124	80		t Stat	2.014	
11	86	86		P(T<=t) one-tail	0.026	
12	112	76		t Critical one-tail	1.691	
13	118	105		P(T<=t) two-tail	0.052	
14	101	104		t Critical two-tail	2.032	
15	106	86				
16	117	102		Mean Difference	9.278	=AVERAGE(A2:A19)-AVERAGE(B2:B19)
17	84	97		Pooled Variance	190.969	=(DEVSQ(A2:A19)+DEVSQ(B2:B19))/(E6+F6-2)
18	128	123		Variance Error of the Mean Diff.	21.219	=E17*(1/E6+1/F6)
19	118	78		Standard Error of the Mean Diff.	4.606	=SQRT(E18)
20				t-ratio	2.014	=E16/E19

The data to be analyzed is in the range A2:B19, in two columns that keep the groups segregated. The result of running the t-test tool on the data appears in the range D1:F14. Please note in particular the value of the t-ratio, shown both in cell E10 (calculated by the t-test tool) and in cell E20, calculated by a simple sequence of math operations in cells E16:E20.

> **NOTE** Assuming that Excel's Data Analysis add-in has been installed on your computer, you can expect to find it in the Analysis group on the Ribbon's Data tab. You might need to make it available to Excel by clicking the Ribbon's File tab, clicking the Options item on the navigation bar, and choosing the Add-Ins item. Choose Excel Add-ins in the Manage dropdown and click Go.

You can take a very different approach by running the data through a regression analysis. Start by creating a new variable that contains information regarding which group a given subject belongs to. For example, you might code a subject with a 1 on a Group variable if she belongs to a Treatment group, and with a 0 on that variable if she belongs to a Control group. (This coding scheme is termed *dummy coding*.) See Figure 6.12.

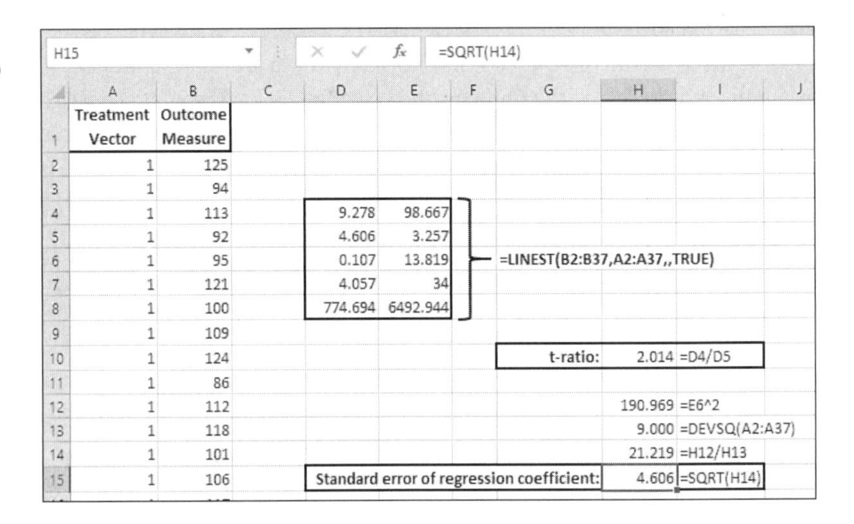

Figure 6.12
Information about group membership appears in column A.

In Figure 6.12, a vector of information regarding group membership is found in the range A2:A37. Measures that quantify the result of being assigned to either the Treatment or the Control group appear in the range B2:B37. Note that except for the arrangement of the outcome measures, the data is exactly the same as in Figure 6.11.

6

> **NOTE** For reasons of space, we have omitted the final 20 rows of data in columns A and B of Figure 6.12. The full data set in the Excel workbook for this chapter is available from the publisher's website.

LINEST() can now treat the group membership data in A2:A37 as a predictor variable, and the measures in B2:B37 as a predicted variable. The results returned by LINEST() appear in the range D4:E8.

LINEST() does not calculate a t-ratio directly but it's easy to do so. Just divide the regression coefficient by its standard error. That's done in cell H10 of Figure 6.12. Notice that the result is exactly the same as the t-ratios shown in Figure 6.11, in cells E10 and E20.

Comparing the Regression Approach to the t-test Approach

How does that come about? Well, notice first the regression coefficient returned by LINEST() in cell D4 of Figure 6.12. It is 9.278, identical to the difference between the two group means shown in cell E16 of Figure 6.11. One way of interpreting the regression coefficient is that it expresses the change in the Y value that accompanies a change of one unit in the X value: that is, the rise over the run. So, as the Group code moves from 0 to 1 in Figure 6.12, the outcome measure moves from 98.667 to 107.944, a distance of 9.278 units.

Second, note that the standard error of the regression coefficient, 4.606 in cell D5 of Figure 6.12, is identical to the standard error of the mean difference, shown in cell E19 of Figure 6.11. Whether you take the traditional t-test approach as in Figure 6.11 or use the regression approach shown in Figure 6.12, you wind up dividing the difference between the two group means by the standard error of the difference between two means.

Let's take a look at the two different routes that lead to the same standard error.

The Standard Error of the Mean Difference

The pertinent calculations are shown in Figure 6.11, cells E16:E19. (The formulas in each of those three cells are shown as text in F16:F19.) Cell E17 pools the variances of the two groups in A2:A19 and B2:B19 by adding their sums of squares (returned by the DEVSQ() function) and dividing by their combined degrees of freedom. Then, cell E18 converts the pooled variance to the variance error of the mean difference: It does so by multiplying the pooled variance by the sum of the reciprocals of the two group sizes. This is analogous to calculating the variance error of the mean, when you divide a group's variance by the number of observations in the group. (See "The F-ratio in the Analysis of Variance" in Chapter 4, "Using the LINEST() Function.") Finally, the variance error of the mean is converted to the standard error of the mean by taking the square root in cell E19. The standard error of the mean difference is the result you would calculate if you repeatedly drew two samples from parent populations, calculated each sample mean, found the difference between the two means, and then calculated the standard deviation of the mean differences.

The Standard Error of the Regression Coefficient

Notice that we just calculated the standard error of the difference between two means, and now we're about to calculate the standard error of the regression coefficient. We just finished demonstrating that the regression coefficient of 9.278 is identical to the observed mean difference of 9.278. Once again, we're calculating the same quantity via two different routes.

In Chapter 5, the section titled "Calculating the Standard Errors of the Regression Coefficients," went into some detail regarding how the standard error of a regression coefficient is calculated. I'll recap that here, just briefly. The idea is to take the standard error of estimate (the standard deviation of the differences between the actual and the predicted Y values) and square it. The result is the variance error of estimate. Cell H12 of Figure 6.12 does that. It's interesting to note that the variance error of estimate in cell H12 is equal to the pooled variance shown in Figure 6.11, cells E7 and E17.

The size of the variance error of estimate is in part a function of the variability in the predictor variable. The next step is to remove its effect on the variance error of estimate by dividing by the sum of the squared deviations of the predictor. That's done in cell H13 (to calculate the sum of the squared deviations) and cell H14 (to divide it out of the variance error of estimate).

If this were a multiple regression analysis, we would also divide by $1 - R^2$, where R^2 refers to the proportion of variance shared by the predictor of interest and the other predictors. However, in this example there is only one predictor, so we skip that step.

Finally, cell H15 takes the square root of the value in cell H14, and the result is the standard error of the regression coefficient—which is identical to the standard error of the mean differences, calculated in cell E19 of Figure 6.11.

Two Routes to the Same Destination

The minutiae of these calculations aren't the important point to take from this section (although it can be illuminating to note that several intermediate results are identical in the two approaches). Before moving on to the relationship between these matters and the assumption of equal spread, it's helpful to step back and take a somewhat longer view.

The t-test approach works directly with means and variances: calculating one group's mean separately from another group's mean, calculating one group's variance separately from another group's variance. The variances are pooled by averaging and the means are contrasted by subtraction. The mean difference is evaluated in terms of the standard error of the difference between means and the resulting t-ratio is compared to a reference distribution: the t distribution that results from a population mean difference of 0.0 for a given number of degrees of freedom.

Regression analysis does not segregate groups in its calculations. It takes note of the fact that it's working with more than one group of scores by means of a dummy-coded vector of, in this example, 0's and 1's. Regression works with correlations between the X and Y variables to determine the regression coefficient (recall that the numeric value of the coefficient equals the mean difference calculated by the t-test). Regression also works directly with the variance of the Y variable and the variance of the errors of estimate: It doesn't need to pool the group variances because it never disaggregated them.

Yet the two approaches yield results that are identical. At root the reason is that both approaches depend on the general linear model, which posits three components for each observation of the outcome or predicted variable (or more than three, when you include additional predictors):

- The grand mean of all the observations of the outcome variable
- The mean of each group, if two or more groups are present, expressed as a deviation from the grand mean of the outcome variable

- The individual observed values, expressed as deviations from the mean of the observation's group or, with just one group, from the grand mean of the outcome variable. (You might have one group only if your purpose is to test whether a group mean differs from a specified constant.)

Both regression and t-tests (and ANOVA and ANCOVA, for that matter) take the general linear model as the basis of their calculations, and it's no great surprise that the two approaches bring about identical results.

However, there *are* major differences between the two approaches. Any application that includes multiple regression, and that performs the calculations correctly, can provide a complete analysis of designs such as the following, among various others:

- Relationships between one *or more* predictor variables and a predicted variable, when all the variables are measured on interval scales. For example, the combined effect of age and cardiovascular activity on cholesterol levels.
- Differences between the means of two groups (independent and dependent groups t-tests).
- Differences between the means of more than two groups (ANOVA).
- Differences between the means of groups when there are two or more factors—for example, sex *and* ethnicity—and the interactions between the factors.
- Differences between the means of two or more groups while removing bias and increasing statistical power by including one or more covariates (ANCOVA).
- Assessment of *curvilinear* relationships between two or more variables.

In contrast, t-tests are limited to the assessment of differences between the means of two groups (which may be either independent of one another or dependent, such as two groups comprising siblings).

In Excel's case, LINEST() returns a regression coefficient and a standard error for each predictor, and one each of the following: R^2, the standard error of estimate, the F-ratio, the residual degrees of freedom, the sum of squares regression, and the sum of squares residual. Given that brief set of statistics, LINEST() can return any variation on the list of six analyses I just presented.

It's all in how you present the predictor variables to LINEST() on the worksheet. In prior chapters, you've already seen how to put two or more interval variables into the mix as predictors. In this book's final chapters you'll see more regarding the analysis of nominal predictors such as treatment and comparison groups.

This section of the current chapter has discussed enough of the use of dummy coding and t-tests to address this question: How can unequal variances cause you to revise your probability statements about the population of interest?

Unequal Variances and Sample Sizes

When you use a regression utility such as LINEST() to analyze the differences between group means, the effect of unequal variances on the analysis takes on a different dimension than when you analyze the relationship between X variables measured on an interval scale and a Y variable measured similarly.

I've used the example of predicting cholesterol levels from age and diet several times in this book, but it's pertinent here too. Unless you have a sample of thousands of subjects, it's highly unlikely that you'll have two or more subjects of precisely the same age *and* precisely the same amount of fats consumed daily.

Therefore, it's highly unlikely that you'll have any groups of the sort discussed in the prior section on t-tests. In any manageable sample you're unlikely to have a "group" of two or more people who are exactly 29 years and 3 weeks old, who also consume 5 grams of fat weekly. You may have one such person, but not two or more in each of the range of possible values of age and fat consumption. In that case the presence of unequal variances still matters, but it matters in a different way than it does when you have two or more groups that represent different sexes or ethnicities or job categories or type of medication. In the latter case, when you have (say) 20 men to contrast with 20 women, half of whom are carnivores and half of whom are vegans, you can calculate the variance of each group and decide whether their variances on the Y variable are different enough to worry about. That's not feasible when an X variable is measured in an interval scale such as ounces.

But the presence of different variances at different points along the X scale still matters. The next two sections discuss first, the problem of heterogeneous variances when you're working with groups—and therefore you're using a nominal scale for your X values—and second, the same problem when the X scale represents interval measurement.

Unequal Spread: Conservative Tests

Suppose that you have two groups of subjects—say, broccoli plants—growing on two adjacent plots of land. As a test, you apply a relatively expensive organic fertilizer to the plants on one plot and a less expensive inorganic fertilizer to the plants on the other plot. Your research hypothesis is that you'll get a higher yield from the costlier organic fertilizer than from the cheaper inorganic alternative. Your null hypothesis is that no difference in average yield exists between the population that gets the organic fertilizer and the one that gets the inorganic fertilizer—with respect to type of fertilizer and plant yield, the samples come from the same population.

You started out with 20 plants on each plot. Unfortunately, not long before harvesting all those heads of broccoli, a truck backed over one of the plots and wiped out 10 of its plants. You press on with your project anyway, because you know that the statistical test you have in mind does not require an equal number of observations in each group.

You harvest the broccoli and weigh the number of ounces of vegetable matter in each plant. The results appear in columns A and B of Figure 6.13.

Figure 6.13
The organic group now has twice as many observations as the inorganic group.

H23 =G17/G18

	A	B	C	D	E	F	G	H	I
1	Organic	Inorganic		Group	Score		t-Test: Two-Sample Assuming Equal Variances		
2	129	95		1	129				
3	87	117		1	87			*Organic*	*Inorganic*
4	111	111		1	111		Mean	104.25	100.2
5	87	104		1	87		Variance	247.671	154.622
6	90	101		1	90		Observations	20	10
7	121	89		1	121		Pooled Variance	217.763	
8	96	115		1	96		Hypothesized Mean Difference	0	
9	106	103		1	106		df	28	
10	120	79		1	120		t Stat	0.709	
11	81	88		1	81		P(T<=t) one-tail	0.242	
12	109			1	109		t Critical one-tail	1.701	
13	116			1	116		P(T<=t) two-tail	0.484	
14	98			1	98		t Critical two-tail	2.048	
15	102			1	102				
16	115			1	115				
17	76			1	76		4.05	100.2	
18	129			1	129		5.715	4.667	
19	116			1	116		0.018	14.757	
20	107			1	107		0.502	28	
21	89			1	89		109.35	6097.35	
22				0	95				
23				0	117		t ratio:	0.709	

In Excel's Data Analysis add-in, there's a tool named *t-test: Two-Sample Assuming Equal Variances*. You open that tool, select the options it offers (principally, where the data is located and where you want the results written), and get the results shown in the range G1:I14 of Figure 6.13. Looking the results over, you see that you have a problem.

First, the two groups have different numbers of broccoli plants. You knew before you ran the test that you were going to be dealing with unequal n's, but the cause was unrelated to the nature of the treatment that you applied, so you're not concerned with that aspect of the unequal n's.

> **NOTE** It can be a different story if the treatment might *cause* the difference in group sizes by introducing a difference in dropout rates (also termed *mortality* in the jargon of experimental design). I'll discuss that issue in Chapter 7, "Using Regression to Test Differences Between Group Means."

The problem you identify has to do with unequal group variances *in combination with* unequal group sizes. Notice from cell H11 that the probability of getting a sample t-ratio of 0.709 from a population where the group means are equal is 0.242. So the t-ratio you obtained isn't at all unlikely—it can come about by chance almost 25% of the time. For most people, that's not a rare enough occurrence to cause you to reject the null hypothesis of no difference between the population means.

And yet there's the problem of the unequal variances and the unequal n's. It turns out that when the larger group also has the larger variance, the statistical test becomes what statisticians term *conservative*. By that, they mean you'll reject a true null hypothesis less frequently that you think you will.

Statistical jargon refers to the rejection of a true null hypothesis as a *Type I error*. In the terms used by the present example, suppose that the null hypothesis is true and that there is no difference between the two fertilizers' average yields. In that case, if the plots in your experiment returned sufficiently different results that you conclude that a true difference exists, you'd be making a Type I error. You expect to control the probability of a Type I error by setting alpha to some level—say, 0.05.

But in the situation described in this example, your statistical test is more rigorous, more *conservative*, that you think. With your decision rules, you might think that your risk of a Type I error is 0.05 when it might actually be 0.03.

The opposite side of that coin is that when the larger group has the smaller variance, the statistical test is liberal. You will reject a true null hypothesis more frequently than you expect. When the sample sizes are equal, the test is robust with respect to violation of the assumption of equal variances. With equal sample sizes, violation of that assumption has at most a negligible effect on the probability statements that guide your decision about whether to retain or reject a null hypothesis.

The outcome shown in Figure 6.13, a conservative test with the larger sample having the larger variance, indicates that the actual probability of your obtained t-ratio is less than the mathematics of the t distribution indicate. The calculated probability is 0.242 (or 24.2%) as shown in cell H11 of Figure 6.13. Because we aren't in compliance with the assumption of equal spread *and* because the sample sizes are unequal, the actual probability of the t-ratio of 0.709 with 28 degrees of freedom is somewhat less than 0.242—perhaps even as low as 0.050 or 5%.

With the figures returned by the t-test, you're operating conservatively. You continue to regard as tenable the null hypothesis of no difference in the population means, because the sample means aren't far enough apart to justify rejecting the null hypothesis. That might be because the t-test's assessment of the probability of that difference has been knocked askew by the combination of the difference in sample variances and the difference in sample sizes.

Let's take a more concrete look at what's going on in the numbers. Figure 6.14 has a more detailed analysis of the conservative situation, when the larger group has the larger variance.

Figure 6.14 requires a fair amount of explanation. It shows only a pivot table. That's because there's just too much material that underlies the pivot table to show in one or two printed figures. It's all there in the Excel workbook that accompanies this chapter, but it will help to know where the relevant information came from.

I began by creating two populations of 1,000 largely random numbers. I say "largely random" because they were constrained in the following three ways:

- Both populations follow a normal distribution.
- Both populations have a mean of 0.022.
- One population has a variance of 0.944; the other has a variance of 9.98.

Figure 6.14
The pivot table shows that you are getting fewer instances of extreme t-ratios than you expect.

T2 | f_x =T.DIST(S2,28,TRUE)

	S	T	U	V	W	X	Y
1	t value	Prob of t		Prob Range for t	Count of Prob of t		
2	-0.741646	0.232		0-0.025	9		
3	1.0188859	0.842		0.025-0.05	0		
4	0.4513095	0.672		0.05-0.075	11		
5	-0.95158	0.175		0.075-0.1	13		
6	1.0828871	0.856		0.1-0.125	15		
7	-0.52714	0.301		0.125-0.15	39		
8	-0.470325	0.321		0.15-0.175	27		
9	0.3539203	0.637		0.175-0.2	23		
10	0.1390924	0.555		0.2-0.225	27		
11	1.2991104	0.898		0.225-0.25	31		Rows 12 through 31 are
32	0.394155	0.652		0.75-0.775	27		hidden to save space
33	0.363766	0.641		0.775-0.8	27		
34	-0.474589	0.319		0.8-0.825	26		
35	-0.546371	0.295		0.825-0.85	12		
36	0.1969179	0.577		0.85-0.875	21		
37	-1.025316	0.157		0.875-0.9	16		
38	-0.663251	0.256		0.9-0.925	15		
39	-0.447402	0.329		0.925-0.95	19		
40	0.9173693	0.817		0.95-0.975	8		
41	-0.843068	0.203		0.975-1	2		
42	-0.180175	0.429		Grand Total	1000		

NOTE

Excel's Data Analysis add-in has a handy tool for random number generation that enables the user to specify a distributional shape (such as Normal, Binomial, or Poisson), a mean, and a standard deviation.

I then wrote a VBA macro (which you can view in the workbook for this chapter, using the Visual Basic Editor) to create 1,000 samples from each of the two populations. The macro completes these tasks in a loop that executes 1,000 times:

1. Writes 1,000 random numbers next to each 1,000-member population.
2. Sorts each population using the random numbers generated in step 1 as the sort key.
3. Takes the first 20 members of the population with the larger variance, and the first 10 members of the population with the smaller variance, as two samples.
4. Calculates the mean and the sum of squares (using Excel's DEVSQ() worksheet function) of each sample, and records them as values on the worksheet.
5. Goes on to the next 1,000 samples.

Using the first instance of two sample means, two sample sums of squares, and the two sample sizes, I then manually entered formulas to calculate a pooled variance, a variance error of the difference between two means, the standard error of the difference, and finally the t-ratio for the two current samples. I copied those formulas down to make the same calculation on each of the remaining 999 pairs of samples, and stored the t-ratio's value on the worksheet.

Using the t-ratio and the degrees of freedom for the first sample, I used Excel's T.DIST() function to calculate the probability of the t-ratio in a central t distribution with that number of degrees of freedom. I copied that T.DIST() formula down through the remaining 999 samples.

The point of doing all this is to test whether the distribution of the calculated t-ratios really follows the theoretical distribution. If they do, then over 1,000 samples you would expect to see 25 samples with t-ratios whose probability is .025 (2.5%) or less—but that's not what we're looking at in Figure 6.14. We're finding only 9 of 1,000 samples whose probability of occurring is 2.5%, when we should be getting 25 of 1,000.

It's much the same story at the other end of the scale, where again we expect 25 samples to occupy the probability range of 0.975 through 1.00. Instead we get 2 (see cell W41 in Figure 6.14).

In sum, because the group variances are unequal (by a factor of 10), because the sample sizes are unequal (by a factor of 2), and because the larger sample has the larger variance, we are seeing fewer samples than we expect in the tails of the observed t distribution.

Suppose you decided beforehand that you would reject the null hypothesis of no difference between the population means, if the t-ratio turned out to have less than a 5% probability of occurring. With a non-directional research hypothesis, you would normally divide that 5% alpha between the two tails of the t distribution.

After the truck runs over your broccoli, you have 30 plants left and these two Excel formulas:

=T.INV(.025,28)

and

=T.INV(.975,28)

return the values –2.048 and 2.048, respectively. Those are the t-ratios that separate the lower and the upper 2.5% of the t distribution, with 28 df, from the middle 95%. So your decision rule is to reject the null hypothesis of no mean difference in the population if you get a t-ratio lower than –2.048 or greater than 2.048.

However, fewer samples occupy those tails than you expect. That's because the process of pooling the variances gives greater weight to the sample with the greater number of observations and in this case that's the sample with the larger variance. That inflates the denominator of the t-ratio and tends to pull the t-ratios closer to 0.0. The theoretical t distribution, which assumes equal variances, says to expect 5% of t-ratios outside +/–2.048. The simulated t distribution, which is derived using unequal group variances, results in only 1.1% of samples outside those limits. Therefore, you will reject the null hypothesis less frequently (best estimate: 1.1% of the time) than you think (in theory, 5% of the time). That's what's meant by the term *conservative test*.

Unequal Spread: Liberal Tests

What about the other side of the coin, when the larger group has the smaller variance instead of the larger variance? Then you're in the territory of liberal tests, where you'll reject a null hypothesis more often than the nominal probability says you will.

Figure 6.15 has the outcome of the same sort of simulation shown in Figure 6.14.

Figure 6.15
The pivot table shows that you are getting *more* instances of extreme t-ratios than you expect.

T2 =T.DIST(S2,28,TRUE)

	S	T	U	V	W
1	t value	Prob of t		Row Labels	Count of Prob of t
2	-4.829928	0.000		0-0.025	79
3	-4.197762	0.000		0.025-0.05	34
4	-3.893567	0.000		0.05-0.075	37
5	-3.875705	0.000		0.075-0.1	29
6	-3.676022	0.000		0.1-0.125	30
7	-3.650906	0.001		0.125-0.15	23
8	-3.622838	0.001		0.15-0.175	23
9	-3.608259	0.001		0.175-0.2	20
10	-3.553559	0.001		0.2-0.225	18
11	-3.518568	0.001		0.225-0.25	23
32	-2.776462	0.005		0.75-0.775	18
33	-2.771838	0.005		0.775-0.8	17
34	-2.756682	0.005		0.8-0.825	17
35	-2.754943	0.005		0.825-0.85	19
36	-2.701514	0.006		0.85-0.875	28
37	-2.683152	0.006		0.875-0.9	29
38	-2.66983	0.006		0.9-0.925	23
39	-2.646577	0.007		0.925-0.95	35
40	-2.646087	0.007		0.95-0.975	34
41	-2.645469	0.007		0.975-1	71
42	-2.633004	0.007		Grand Total	1000

Rows 12 through 31 are hidden to save space

Figure 6.15's pivot table is based on t-ratios, and their nominal probabilities, that are calculated from samples in which the *larger* sample size comes from the population with the *smaller* variance. That pairing leads to a liberal test—in contrast to the situation depicted in Figure 6.14, where the larger sample size paired with the larger variance leads to a conservative test.

NOTE I used the same two populations for Figure 6.15 as I did for Figure 6.14. To keep the focus on the relationship between group size and size of variance, the only difference in the sampling was that in Figure 6.15 I sampled 10 records from the population with the larger variance and 20 records from the population with the smaller variance.

The pivot table in Figure 6.15 shows that the lowest 2.5% of the 1,000 probabilities consists of 79 sample comparisons (cell W2) and the highest 2.5% consists of 71 sample comparisons (cell W41). Based on the simulation, 79 + 71 or 150 sample comparisons will have calculated t-ratios that have nominal probability of occurrence equal to 2.5% + 2.5% or 5%.

Those 150 sample comparisons will have t-ratios lower than −2.048 or greater than 2.048. If you have Excel, or any other source of statistical analysis, evaluate one of those t-ratios (for example,

```
=T.DIST(–2.1,28,TRUE)
```

which returns 0.022) you'll get a nominal probability in the lower 2.5% or the upper 2.5% of the t distribution with 28 degrees of freedom. Over 1,000 samples, then, the expectation is that you'll get 25 in the left tail and 25 in the right tail, for a total of 50. Instead we have 150, three times more than the expectation. You will reject the null hypothesis three times more often than you should.

The reason is, of course, the reverse of the reason for conservative tests. When the larger sample has the smaller variance, it contributes more than it should to the pooled variance. Its contribution therefore consists of a smaller sum of squares than it would when the larger sample has the larger variance.

Yes, you divide the total sums of squares by the degrees of freedom to arrive at the pooled variance—but that does not fully correct for the discrepancy between the population variances, and the net effect is as follows:

- The sample with the greater number of observations has the larger variance. Other things being equal, as the larger variance increases, the more conservative the eventual t-test.
- The sample with the smaller number of observations has the smaller variance. Other things being equal, as the smaller variance decreases, the more liberal the eventual t-test.

Unequal Spreads and Equal Sample Sizes

6

To put into context the preceding material on conservative and liberal tests, consider the effect of unequal variances in populations on samples with *equal* sample sizes. If that's the situation, the nominal probabilities are the same (or very nearly so) as the actual probabilities, even with great discrepancy in the population variances.

Figure 6.16 shows the results of the simulation discussed in the prior two sections, with a 10-to-1 difference in the sizes of the population variances, but with 20 observations in each sample.

In Figure 6.16, the population variances are the same as in the examples of conservative and liberal tests, but here the sample sizes are 20 in each group. The result demonstrates that with equal sample sizes the t-test is robust with respect to violation of the equal variance assumption.

Figure 6.16
The actual distributions are almost precisely what the nominal probabilities lead you to expect.

T2 | =T.DIST(S2,28,TRUE)

	S	T	U	V	W	X
1	t value	Prob of t		Row Labels	Count of Prob of t	
2	-3.69058	0.000		0-0.025	25	
3	-3.44799	0.001		0.025-0.05	30	
4	-3.056285	0.002		0.05-0.075	23	
5	-2.990417	0.003		0.075-0.1	19	
6	-2.965237	0.003		0.1-0.125	29	
7	-2.954137	0.003		0.125-0.15	25	
8	-2.629818	0.007		0.15-0.175	25	
9	-2.595319	0.007		0.175-0.2	20	
10	-2.507527	0.009		0.2-0.225	19	
11	-2.497077	0.009		0.225-0.25	22	Rows 12 through 31 are hidden to save space
32	-1.964107	0.030		0.75-0.775	27	
33	-1.96182	0.030		0.775-0.8	24	
34	-1.959436	0.030		0.8-0.825	33	
35	-1.952598	0.030		0.825-0.85	24	
36	-1.931002	0.032		0.85-0.875	25	
37	-1.912502	0.033		0.875-0.9	26	
38	-1.912187	0.033		0.9-0.925	24	
39	-1.868779	0.036		0.925-0.95	25	
40	-1.86679	0.036		0.95-0.975	26	
41	-1.865531	0.036		0.975-1	23	
42	-1.85257	0.037		Grand Total	1000	

Notice that 25 pairs of samples result in t-ratios whose probabilities put them in the lowest 2.5% of the distribution (cell W2): an exact match of the expected with the actual outcome. Another 23 pairs of samples occupy the highest 2.5% of the distribution (cell W41). So, the overall lesson is that you don't need to worry about equal variances when the sample sizes are equal. Make your null and research hypotheses, select an alpha level, run either a traditional t-test or push the data through LINEST(), check the probability of the t-ratio if the null hypothesis is true, and decide whether to reject or retain the null hypothesis accordingly.

Figure 6.17 illustrates the process. Let's walk briefly through the steps:

Your null hypothesis is that there's no difference in the population means of broccoli crop yield as a function of the use of one or the other of the two fertilizers. Your research hypothesis is that the costlier organic fertilizer will result in a greater yield than will the inorganic fertilizer. Note that this is a directional hypothesis. If you had no opinion about which fertilizer would work better, you might instead make the nondirectional research hypothesis that the yields are different (rather than that a particular fertilizer will cause a greater yield).

You arbitrarily decide to use the conventional alpha level of 0.05 as your criterion for rejecting the null hypothesis. You would prefer to choose alpha on more substantive grounds, such as tradeoffs between statistical power and the relative costs of the fertilizers, but you haven't yet read Chapters 7 and 8 of this book, so you settle for an alpha of 0.05.

Figure 6.17
Equal sample sizes let you move ahead without worrying about unequal variances.

E21 | =T.DIST.RT(E20,E9)

	A	B	C	D	E	F
1	Organic	Inorganic		t-Test: Two-Sample Assuming Equal Variances		
2	132	95				
3	90	116			Organic	Inorganic
4	114	111		Mean	107.944	98.667
5	90	105		Variance	262.173	191.647
6	93	103		Observations	18	18
7	124	89		Pooled Variance	226.910	
8	99	117		Hypothesized Mean Difference	0	
9	109	103		df	34	
10	123	80		t Stat	1.848	
11	84	86		P(T<=t) one-tail	0.037	
12	112	76		t Critical one-tail	1.691	
13	119	105		P(T<=t) two-tail	0.073	
14	101	104		t Critical two-tail	2.032	
15	105	86				
16	118	102		Mean Difference	9.278	=E4-F4
17	79	97		Pooled Variance	226.910	=(DEVSQ(A2:A19)+DEVSQ(B2:B19))/((18-1)+(18-1))
18	132	123		Variance Error of the Mean Diff.	25.212	=E17*(1/18+1/18)
19	119	78		Standard Error of the Mean Diff.	5.021	=SQRT(E18)
20				t-ratio	1.848	=E16/E19
21				Probability of t-ratio if null is true	0.037	=T.DIST.RT(E20,E9)

You successfully ward off the runaway truck and wind up with the same number of broccoli plants in each sample at the end of the growing period. You note that the two samples have different variances but, because the samples have equal n's, you don't worry about that. Instead you enter the size of each plant into the range A2:B19 as shown in Figure 6.17.

You click the Data tab on Excel's Ribbon and choose Data Analysis from the Analysis group. You scroll through the list box of available tools, click the t-Test: Two Sample Assuming Equal Variances tool, and click OK. The dialog box that enables you to define the analysis appears as shown in Figure 6.18.

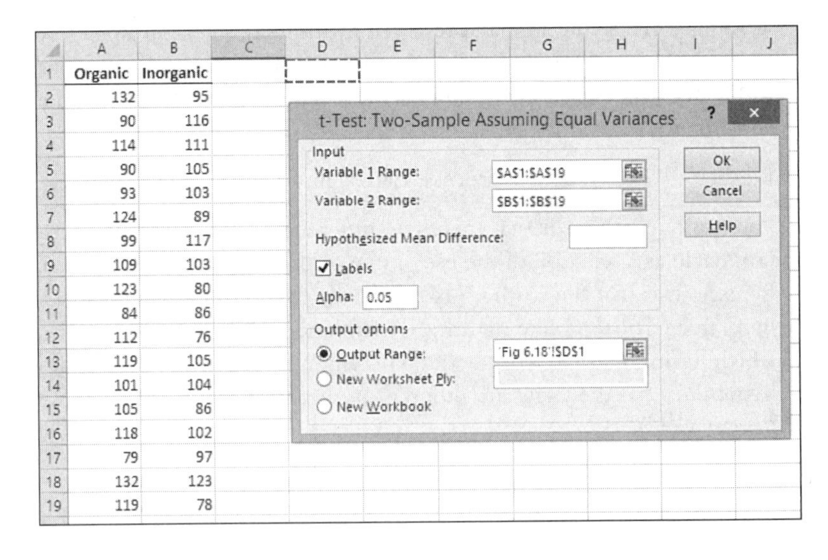

Figure 6.18
This Data Analysis tool has some problems, but it's convenient.

Enter A1:A19 for the Variable 1 range and B1:B19 for the Variable 2 range. Fill the Labels checkbox so that Excel will know it should treat the headers in A1 and B1 as labels rather than data. If the Alpha edit box doesn't show 0.05, enter that value. Click the Output Range option button, and then click in the associated edit box. Click in cell D1, or enter its address, to establish the upper-left cell of the results. Click OK to start the analysis.

You'll get the results shown in the range D1:F14 in Figure 6.17. I like to add some additional calculations, just to dot the i's and cross the t's. One of the drawbacks to most of the results provided by tools in the Data Analysis add-in is that they are provided as static values rather than as formulas. Therefore, if you want to edit the underlying data, or add or delete one or more observations, you have to run the tool again to recalculate its results. It can also be useful to document how different intermediate results are obtained. So I like to add these labels and formulas as shown in the range D16:F21 of Figure 6.17.

Mean Difference: The difference, which is what we're testing, between the means of the two groups.

Pooled Variance: Both group variances are unbiased estimators of the population variance. Each group variance is the ratio of the group sum of squares divided by its degrees of freedom (the sample size minus 1). We can pool the group variances by adding their sums of squares and dividing by the total of their degrees of freedom. The pooled variance is a more accurate estimate of the population variance than is either group variance by itself.

Variance Error of the Mean Difference: This step is directly analogous to dividing a group's variance by n, the number of observations in the group, so as to estimate the variance error of the mean. To estimate the variance of the differences between two means, taken over many similar samples, we multiply the pooled variance by the sum of the reciprocals of the sample sizes.

Standard Error of the Mean Difference: Just as the square root of the variance error of the mean returns the standard error of the mean, the square root of the variance of the mean difference returns the standard error of the mean difference.

t-ratio: Dividing the mean difference by the standard error of the mean difference results in the t-ratio.

Probability of t-ratio if null is true: We can get the theoretical probability that the t-ratio comes from a population where the difference between the two means is actually 0.0 by passing the t-ratio, along with its degrees of freedom, to a T.DIST() function. We want the area under the curve that lies to the *right* of the obtained t-ratio, because we started out by making a directional research hypothesis. Therefore, we want to know the probability that the mean for the organically fertilized broccoli is greater than the mean for the conventionally fertilized broccoli.

To get the area to the right of the obtained t-ratio, we could use either this formula:

```
=T.DIST.RT(E20,E9)
```

where the RT specifies the right tail only in the t distribution, or this formula:

```
=1−T.DIST(E20,E9,TRUE)
```

where we find the proportion of the total area under the curve to the left of the t-ratio in E20, and subtract that proportion from 1.0 to get the proportion to the *right* of the t-ratio. (The TRUE argument calls for the area under the curve left of the t-ratio rather than the height of the curve at the t-ratio. The T.DIST.RT() function does not accept a third argument.)

Using LINEST() Instead of the Data Analysis Tool

I generally prefer LINEST() to a tool from the Data Analysis add-in to analyze data. I like the fact that the formula, in contrast to the static values returned by the add-in, is volatile and will recalculate in response to changing data. If I need to add or remove observations, either of these two options is available:

- If the underlying data is in an Excel list, I just add or remove observations from the list and modify the LINEST() formula's X and Y addresses accordingly.
- If the underlying data is in an Excel table, I just add or remove observations from the table. The LINEST() formula automatically updates in response to the addition or removal of observations.

Besides the fact that the volatile function updates automatically, it's helpful that LINEST() returns statistics that the t-test doesn't: for example, the R^2 and the standard error of estimate. It's true that those two statistics are less useful in the context of a t-test, with its nominal groups, than in the context of an analysis involving variables measured on an interval scale, but as you'll see in Chapters 7 and 8, R^2 and the standard error of estimate support more complex analyses.

As this chapter discusses in the section "Comparing the Regression Approach to the t-test Approach," you can get the denominator of the t-ratio directly from the LINEST() results—it's the standard error of the regression coefficient. See Figure 6.19.

Figure 6.19
LINEST() can tell you more than the t-test tool, but you still need to supplement it.

F12 =T.DIST.RT(F11,F6)

	A	B	C	D	E	F	G
1	Group	Score					
2	1	132			=LINEST(B2:B37,A2:A37,,TRUE)		
3	1	90			9.278	98.667	
4	1	114			5.021	3.551	
5	1	90			0.091	15.064	
6	1	93			3.414	34	
7	1	124			774.694	7714.944	
8	1	99					
9	1	109			Mean Difference	9.278	=E3
10	1	123			Standard Error of the Mean Difference	5.021	=E4
11	1	84			t-ratio	1.848	=F9/F10
12	1	112			Probability of t-ratio if null is true	0.037	=T.DIST.RT(F11,F6)
13	1	119					

When you test the difference between two sample means by way of the procedures shown in Figure 6.17, you are in fact testing whether the difference between the population means is 0.0. That's also taking place in Figure 6.19.

Compare the value returned by LINEST() in cell E4 of Figure 6.19 with the value in cell E19 of Figure 6.17, where it's termed the "standard error of the mean difference." In Figure 6.17, you have to calculate that value by pooling the group variances, converting the pooled variance to the variance error of the mean differences, and then converting the variance error to the standard error of the mean differences—all because the Data Analysis tool declines to show it explicitly.

LINEST() does require that you calculate the t-ratio, as shown in cell F11 of Figure 6.19, and the probability of the t-ratio, as shown in cell F12. Calculating the t-ratio is clearly quite easy, given the statistics returned by LINEST(). Calculating the probability by means of one of Excel's T.DIST() functions is a bit more difficult, largely because you must choose from among four different versions of the basic T.DIST() function.

Understanding the Differences Between the T.DIST( ) Functions

Since Excel 2010, you have had available three different forms of the function that tells you the proportion of the t distribution found to the left and/or to the right of a given t-ratio. The functions are T.DIST(), T.DIST.RT(), and T.DIST.2T(). Another function, T.TEST(), has been available in Excel since much earlier than 2010. I'll discuss T.TEST() later in this chapter, because its principal use has to do with unequal group variances and unequal sample sizes, not the placement of a proportion of the area under the curve.

To use the T.DIST() functions, you must provide the t-ratio that you want evaluated. You must also provide the number of degrees of freedom: At its most basic, that's the total of the observations in both groups, less 1 degree of freedom for each of the two groups. The shape of the t distribution varies according to its degrees of freedom, and you must supply that value so that Excel can evaluate the distribution properly.

The T.DIST() function returns the proportion of the area under the curve that lies to the *left* of the t-ratio you specify. For example, suppose that the t-ratio returned by LINEST() or the Data Analysis tool is –1.69.

> **NOTE** The negative t-ratio comes about when the Data Analysis tool subtracts the larger mean from the smaller, or when the data you submit to LINEST( ) assigns a code of 0 to the group with the higher mean, and a code of 1 to the group with the lower mean.

Suppose you have 36 observations and the degrees of freedom for the analysis is therefore 34. In that case, this formula:

```
=T.DIST(–1.69,34,TRUE)
```

returns 0.05, or 5%. Figure 6.20 shows the location of that 5%.

NOTE The third argument to the T.DIST( ) function can be either TRUE or FALSE. If TRUE, the function returns the cumulative area under the curve to the left of the t-ratio, expressed as a proportion. If FALSE, the function returns the relative height of the curve at the point of the t-ratio. The other versions of the T.DIST( ) function, T.DIST.RT( ) and T.DIST.2T( ), do not require a third argument, just the t-ratio and the degrees of freedom.

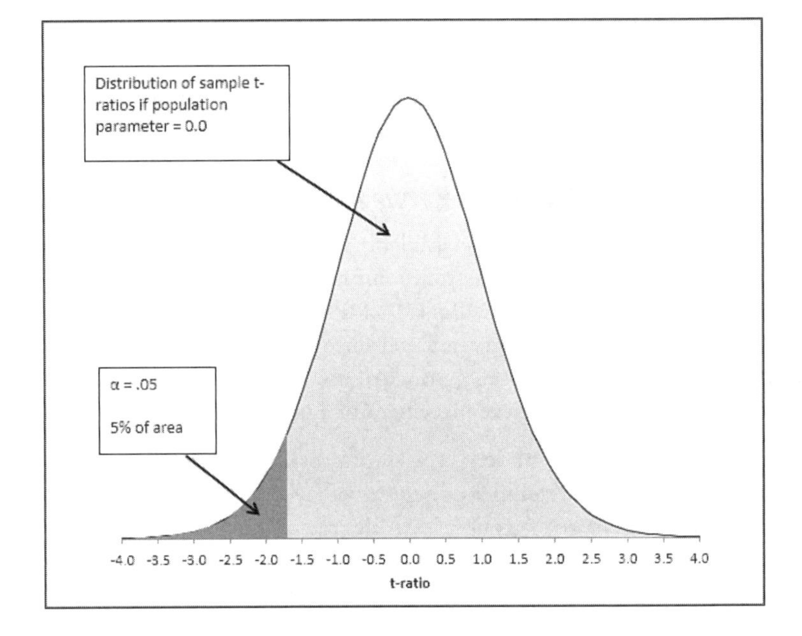

Figure 6.20
The darkly shaded area in the curve's left tail corresponds to the probability returned by the T.DIST( ) function.

The situation shown in Figure 6.20 represents a negative t-ratio, which occurs when you subtract the larger mean from the smaller. It would come about if you arranged for Variable 1 in the Data Analysis tool to contain the values with the lower mean, not (as in Figure 6.17) the values with the greater mean. Or, laying things out for LINEST(), it would come about if you assigned a code of 0 to the values with the greater mean, not (as in Figure 6.19) the values with the smaller mean. In either of those two cases, you would wind up with a negative t-ratio and the proportion of the area under the curve would look something like the chart in Figure 6.20.

Whether you get a positive or a negative t-ratio is often a matter of how you choose to lay out or code the underlying data. For various reasons it's normal to arrange the layout or coding so that you wind up with a positive t-ratio. If you have a positive t-ratio and you deploy the T.DIST() function, like this:

```
=T.DIST(1.69,34,TRUE)
```

you'll get 0.95 as a result. Just as 5% of the area under the curve falls to the left of a t-ratio of –1.69 with 34 degrees of freedom, so 95% of the area under the curve falls to the left of a t-ratio of 1.69. See Figure 6.21.

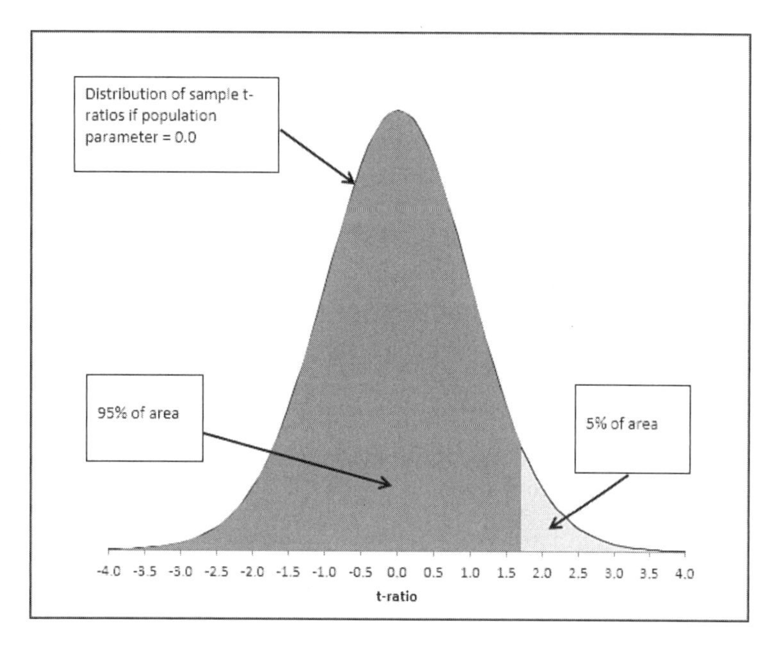

Figure 6.21
The darkly shaded area corresponds to the proportion of the curve to the left of a large, positive t-ratio.

But it's almost certain that you're not interested in that 95%. You're very likely interested in the probability of getting a large t-ratio such as 1.69 or larger when the t-ratio in the population is 0.0. That probability is 0.05 or 5%. Figure 6.22 shows what you're looking for.

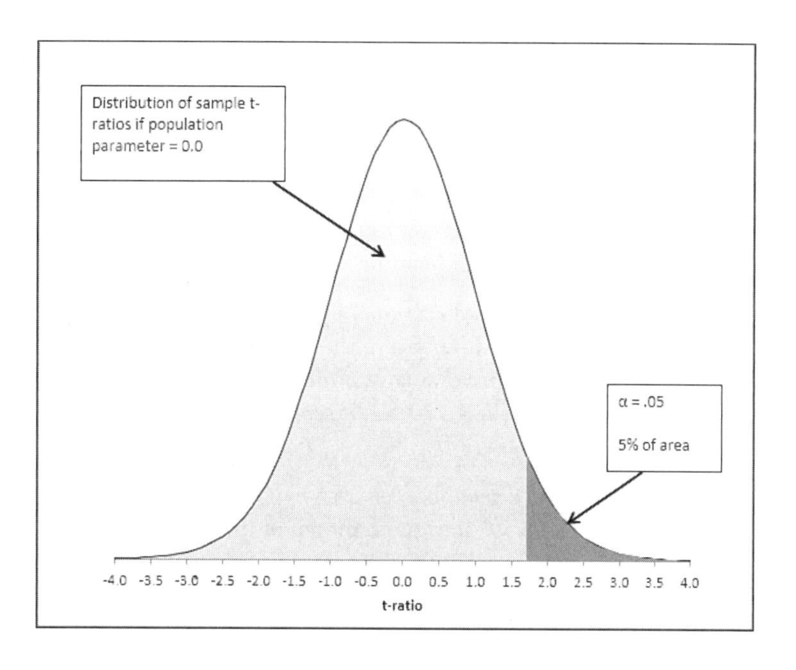

Figure 6.22
The darkly shaded area corresponds to the 5% of the curve to the *right* of the t-ratio.

6

The two most straightforward ways of getting a T.DIST() function to return the 5% (or any proportion) of the curve that's to the right of the t-ratio are to use either this formula:

=1 – T.DIST(1.69,34,TRUE)

or the T.DIST.RT() function:

=T.DIST.RT(1.69,34)

The area under the curve must equal 100%, so subtracting the proportion to the left of the t-ratio, returned by the function, must result in the proportion to the right of the t-ratio.

You might prefer to get the proportion directly, using the T.DIST.RT() function, in which the RT might stand for "right" or for "right tail." Be sure to bear in mind that the T.DIST.RT() function does not recognize the third, cumulative-area argument. (If you forget and include a third argument anyway, no harm done—Excel just complains and puts you back in editing mode.)

Using either T.DIST() or T.DIST.RT() implies that you have made a directional research hypothesis. Returning to the broccoli example, you might hypothesize that you'll get a larger yield from an organic fertilizer than with a conventional fertilizer. That's a directional hypothesis because you're positing that one particular treatment will result in a numerically larger outcome than the other.

If your research hypothesis were less specific, and stated only that you would get a different yield from the two types of fertilizer, you would no longer be positing that a particular treatment will have a numerically larger outcome—merely that the outcomes will be different.

In either case you need a decision rule: a criterion that, if it's met, tells you whether to reject the null hypothesis of no difference in the population or to retain it. Here's an example:

"I will reject the null hypothesis of no difference in the population if the organic fertilizer results in a *higher* yield than the conventional fertilizer. Additionally I want the probability of incorrectly rejecting the null hypothesis to be 5% at most."

This decision rule involves a directional hypothesis: It specifies that the organic fertilizer's yield must exceed, not just differ from, the yield for the conventional fertilizer. And it sets alpha at 5%: The probability of a false positive must not exceed 5%. If the probability of the obtained t-ratio is even 6%, you'll retain the null hypothesis for at least the time being.

The combination of the directional research hypothesis and setting alpha to 5% means that you're putting the entire 5% into the right tail of the distribution, and to get the t-ratio you will subtract the mean yield for the conventional fertilizer from the mean yield for the organic fertilizer. Only if the probability of your obtained t-ratio is 5% or less will you reject the null hypothesis.

What if you wind up with a large *and negative* t-ratio, something such as –3.5? That's a hugely unlikely outcome if the population means are identical, and you might be tempted to reject the null hypothesis. However, having already specified that you will reject the null hypothesis if the organic fertilizer outperforms the conventional, you're ethically bound to respect that position. You've made a directional research hypothesis favoring the organic fertilizer. You can't now reject it, even if you wind up with strong evidence that the conventional fertilizer works better. Even with a monstrously unlikely outcome, you have to live with both your decision rule and your null hypothesis, and start planning your next experiment (but it wouldn't hurt to verify that the data was entered correctly).

NOTE The rationale won't become clear until Chapter 7, but directional hypotheses such as the one just discussed normally result in a more sensitive statistical test (that greater sensitivity is also termed greater *statistical power*).

On another hand, perhaps your decision rule reads like this:

"I will reject the null hypothesis of no difference in the population if the organic fertilizer results in a *different* yield than the conventional fertilizer. Additionally, I want the probability of incorrectly rejecting the null hypothesis to be 5% at most."

Alpha remains the same at 5% but it's distributed differently in a nondirectional research hypothesis than with a directional research hypothesis. Because you have to allow for two different outcomes (the organic fertilizer is better, or the conventional fertilizer is better) the probability of a false positive has to be distributed between the tails of the distribution. It's normal to distribute alpha equally in the two tails, as shown in Figure 6.23.

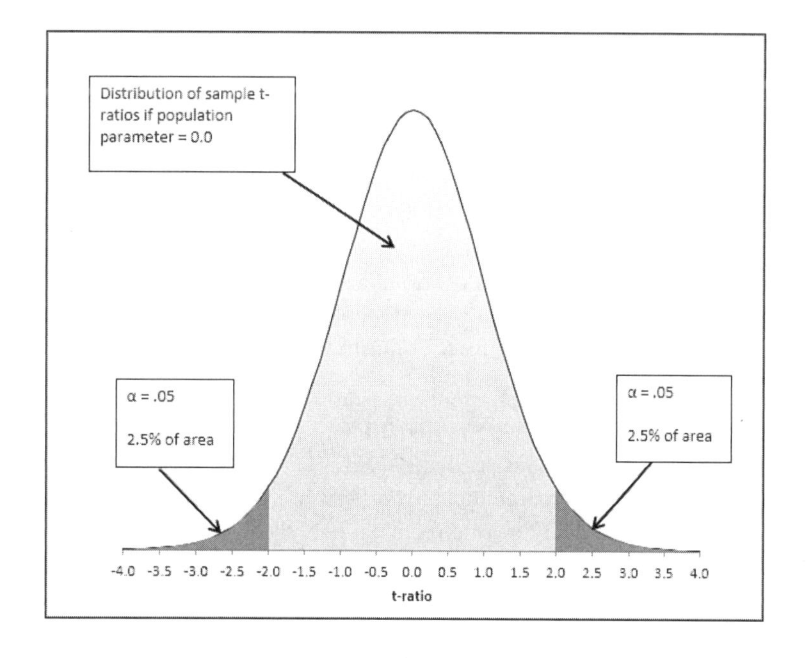

Figure 6.23
Here, the 5% alpha is divided into two 2.5% segments.

Dividing alpha as shown in Figure 6.23 allows for either possibility: Organic is better or conventional is better. This lets you off the hook: You need not specify which fertilizer you expect to outperform the other, and you need not swallow a highly unlikely outcome that doesn't jibe with a directional research hypothesis. (On the other hand, as you might expect from the prior Note, your experiment's statistical power will be lower.)

Excel's T.DIST.2T() function can help here. The "2T" portion of the function name means "two tails." Using the same calculated t-ratio and degrees of freedom as in the prior example, this formula:

=T.DIST.2T(1.69,34)

returns 0.100, or 10%. The T.DIST.2T() function finds the proportion of the curve to the right of the t-ratio *plus* the proportion of the curve to the left of the negative of the t-ratio. Put differently, T.DIST.2T() returns the result of this formula:

=T.DIST(–1.69,34,TRUE) + T.DIST.RT(1.69,34)

or this one:

=2*T.DIST(–1.69,34,TRUE)

> **NOTE** For no particularly good theoretical reason, the T.DIST.2T() function returns the #NUM! error value if you provide a negative number for the t-ratio, its first argument.

So if you have made a nondirectional research hypothesis, a calculated t-ratio of 1.69 or –1.69 does not cause you to reject the null hypothesis with an alpha of 0.05. To do so would be to say that a t-ratio that results, 10% of the time, from a population where the t-ratio is 0.0, is unlikely enough to reject the null. However, you have already stated that you require a 5%, not a 10%, false positive rate.

No, with a nondirectional research hypothesis, you have to adjust your decision rule's criterion. With a directional research hypothesis, the current example shows that a t-ratio of at least 1.69 comes about 5% of the time that the null hypothesis is true. With a nondirectional hypothesis, a t-ratio of at least 1.69 *or* less than –1.69 comes about 10% of the time. To maintain your 5% alpha level, you need to cut off 2.5% of the total distribution in each tail. Figure 6.23 illustrates those cutoffs.

With 34 degrees of freedom, it takes a t-ratio of +/–2.03 to limit the false positive rate of a nondirectional test to 5%. You find 2.5% of the curve to the left of –2.03 and the other 2.5% to the right of 2.03. Clearly, +/–2.03 is farther from the middle of the distribution than +/–1.69. So a nondirectional hypothesis has to obtain a t-ratio farther from the middle of the distribution than does a directional hypothesis in order to reject the null hypothesis. The farther a value is into the tails of a t distribution, the harder it is to encounter—and that hints at the reason that directional tests tend to have more statistical power than nondirectional tests.

TIP

You can use the T.DIST( ) function's cousin, T.INV( ), to determine a critical t-ratio. With T.DIST( ), you supply a t-ratio to determine a probability. With T.INV( ), you supply a probability to determine a t-ratio. For example, T.INV(0.95,34) returns 1.69, the t-ratio that, with 34 degrees of freedom, has 95% of the curve's area to its left and therefore 5% of the area to its right. The same relationship between DIST and INV holds in Excel for the F and the chi-square distributions—for example, F.DIST( ) and F.INV( ), and CHISQ.DIST( ) and CHISQ.INV( ).

I've spent a lot of space in this section to discuss the differences between the various T.DIST() functions, which have nothing to do—directly, at any rate—with the section's main topic, the equality of variances in different groups or at different points along the X axis. I did so in order to provide a context for the topic of the next section, using the Data Analysis add-in to manage unequal variances even when the group sizes are quite different.

Using Welch's Correction

There exists no entirely satisfactory solution to the problem of unequal group variances combined with unequal group sizes. But the approach developed by a statistician named Welch is probably the most frequently used method for approximating the correct probability of a given t-ratio under those conditions.

The method depends on two modifications to the standard t-test approach:

- Calculate what's sometimes termed a *quasi t statistic*, substituting the square root of the sum of the variance errors of the mean for the usual standard error of the difference between two means. The distribution of this quasi t approximates the usual t distribution. (The variance error of the mean is the square of the standard error of the mean.)
- Adjust the degrees of freedom for the ratio of the two variances, so that you act as though you have more degrees of freedom if the usual t-test is conservative, and fewer degrees of freedom if the usual t-test is liberal.

We'll take a look at some examples of how this works, starting with Figure 6.24.

Figure 6.24 shows two data sets that are somewhat different from those used earlier in this chapter, modified to make a somewhat clearer point regarding the quasi t approach. The data in columns A and B conform to a conservative t-test: The larger sample in Column A has the larger variance (see cells I5:J5 and I6:J6). The same two samples along with Group codes appear in columns E and F, laid out for regression analysis as in, for example, Figure 6.13.

The analysis shown in the range H1:J14, particularly cell I11, indicates that you would *retain* the null hypothesis of no difference in the population means, if you began with a directional research hypothesis and set alpha at 0.050. The probability of getting the t-ratio of 1.639 is a little larger than the 0.050 alpha that you signed up for.

Figure 6.24
Analyzing unequal variances with unequal sample sizes as though the variances were equal.

I24 | f_x =T.DIST.RT(I23,I20)

	A	B	C	D	E	F	G	H	I	J
1		Organic	Inorganic		Group	Score		t-Test: Two-Sample Assuming Equal Variances		
2		129	95		1	129				
3		87	117		1	87			*Organic*	*Inorganic*
4		111	111		1	111		Mean	104.250	94.800
5		87	84		1	87		Variance	247.671	166.400
6		90	101		1	90		Observations	20	10
7		121	89		1	121		Pooled Variance	221.548	
8		96	81		1	96		Hypothesized Mean Difference	0	
9		106	103		1	106		df	28	
10		120	79		1	120		t Stat	1.639	
11		81	88		1	81		P(T<=t) one-tail	0.056	
12		109			1	109		t Critical one-tail	1.701	
13		116			1	116		P(T<=t) two-tail	0.112	
14		98			1	98		t Critical two-tail	2.048	
15		102			1	102				
16		115			1	115		=LINEST(E2:E31,D2:D31,,TRUE)		
17		76			1	76		9.450	94.800	
18		129			1	129		5.765	4.707	
19		116			1	116		0.088	14.884	
20		107			1	107		2.687	28	
21		89			1	89		595.350	6203.350	
22					0	95				
23					0	117		t ratio:	1.639	
24					0	111		Probability of t	0.056	

The LINEST() results shown in the range H17:I21 are provided mainly so that you can compare them to the results from the Data Analysis t-test tool. Note that the t-ratio calculated in cell I23 from the LINEST() results, and the probability calculated in cell I24 from the T.DIST.RT() function's handling of the LINEST() results, are the same as those returned by the Data Analysis tool.

Now compare the information in Figure 6.24 with that in Figure 6.25.

The main difference between the analysis in Figure 6.24 and that in Figure 6.25 is that they use different Data Analysis tools. Figure 6.24 uses the tool named t-Test: Two Sample Assuming Equal Variances. Figure 6.25 uses the tool named t-Test: Two Sample Assuming Unequal Variances. The latter tool employs the Welch correction for degrees of freedom and calculates the quasi t-ratio instead of the traditional ratio.

6

You invoke the t-Test: Two Sample Assuming Unequal Variances tool just as you do the t-Test: Two Sample Assuming Equal Variances tool—they both appear in the list box when you click the Data Analysis link in the Data tab's Analyze group. You fill in the boxes and choose your options in the dialog box just as you do with the Equal Variances version of the tool.

Three cells in Figure 6.25 deserve your comparison with cells in Figure 6.24:

- Figure 6.24 shows 28 degrees of freedom in cell I9. In Figure 6.25, cell F8 shows 22 degrees of freedom. (I'll show how you adjust the degrees of freedom later in this section.)

- Figure 6.24 shows the t-ratio as 1.639 in cell I10. Figure 6.25 gives the t-ratio as 1.754 in cell F9. (Again, I'll show the formula later.)
- Figure 6.24 returns a probability for the obtained t-ratio as 0.056 in cell I11. Figure 6.25 shows the probability as 0.047 in cell F10.

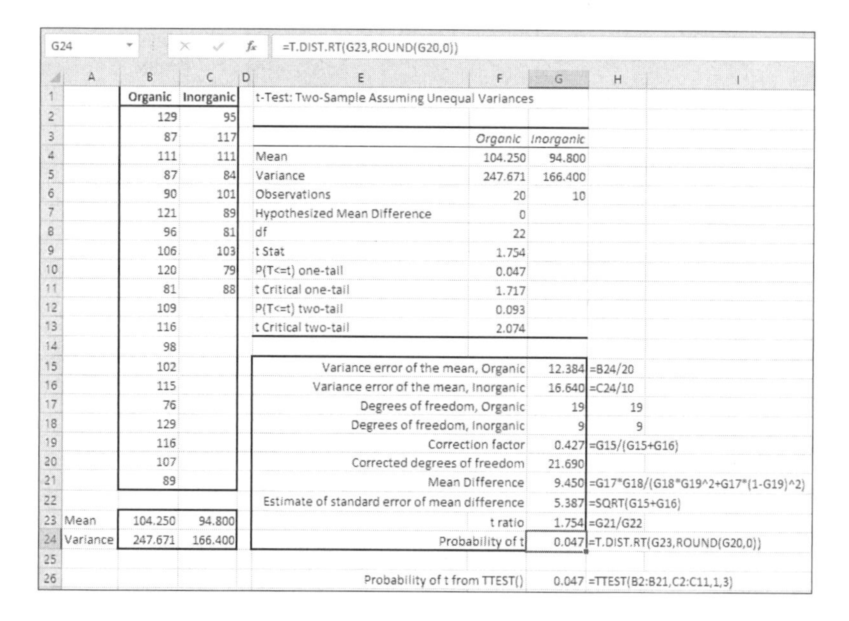

G24 =T.DIST.RT(G23,ROUND(G20,0))

	A	B	C	D	E	F	G	H
1		Organic	Inorganic		t-Test: Two-Sample Assuming Unequal Variances			
2		129	95					
3		87	117			Organic	Inorganic	
4		111	111		Mean	104.250	94.800	
5		87	84		Variance	247.671	166.400	
6		90	101		Observations	20	10	
7		121	89		Hypothesized Mean Difference	0		
8		96	81		df	22		
9		106	103		t Stat	1.754		
10		120	79		P(T<=t) one-tail	0.047		
11		81	88		t Critical one-tail	1.717		
12		109			P(T<=t) two-tail	0.093		
13		116			t Critical two-tail	2.074		
14		98						
15		102			Variance error of the mean, Organic		12.384	=B24/20
16		115			Variance error of the mean, Inorganic		16.640	=C24/10
17		76			Degrees of freedom, Organic		19	19
18		129			Degrees of freedom, Inorganic		9	9
19		116			Correction factor		0.427	=G15/(G15+G16)
20		107			Corrected degrees of freedom		21.690	
21		89			Mean Difference		9.450	=G17*G18/(G18*G19^2+G17*(1-G19)^2)
22					Estimate of standard error of mean difference		5.387	=SQRT(G15+G16)
23	Mean	104.250	94.800		t ratio		1.754	=G21/G22
24	Variance	247.671	166.400		Probability of t		0.047	=T.DIST.RT(G23,ROUND(G20,0))
25								
26					Probability of t from TTEST()		0.047	=TTEST(B2:B21,C2:C11,1,3)

Figure 6.25
Handling unequal variances with unequal sample sizes using Welch's correction.

So, the Unequal Variances tool adjusts the number of degrees of freedom from 28 to 22, a change that would normally make a test more conservative. (Reducing the degrees of freedom increases the size of the standard error of the mean. Because the standard error of the mean, or the standard error of mean differences, is the denominator of the t-ratio, a larger standard error means a smaller t-ratio—and therefore a more conservative t-test.)

However, the Unequal Variances tool also increases the size of the t-ratio, which makes the test more liberal. In this case, the net of the changes to the degrees of freedom and the t-ratio is to make the test more liberal. The data set, analyzed by the Equal Variances tool, brings about a conservative result because the larger sample has the larger variance.

The probability returned by the Equal Variances tool, 5.6%, misses the 5% criterion for alpha that you specified at the outset. In contrast, the adjustments for the conservative test result in a probability for the quasi t-ratio of 4.7%, which is within the limit you specified, and which allows you to reject the null hypothesis of no differences in the population.

Figures 6.26 and 6.27 show the reverse effect, that of making more conservative a test that is known to be more liberal than the nominal probabilities lead you to believe.

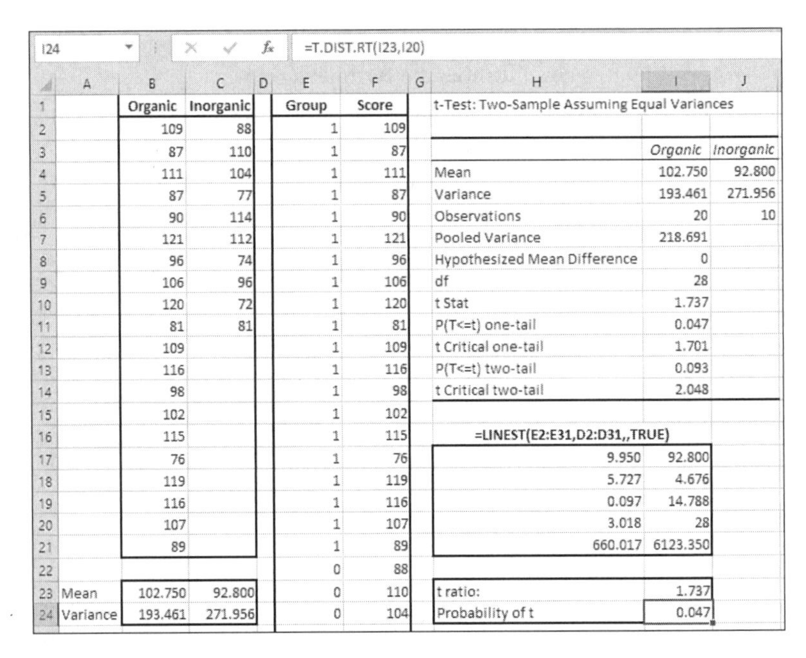

I24 | =T.DIST.RT(I23,I20)

	A	B	C	D	E	F	G	H	I	J
1		Organic	Inorganic		Group	Score		t-Test: Two-Sample Assuming Equal Variances		
2		109	88		1	109				
3		87	110		1	87			Organic	Inorganic
4		111	104		1	111		Mean	102.750	92.800
5		87	77		1	87		Variance	193.461	271.956
6		90	114		1	90		Observations	20	10
7		121	112		1	121		Pooled Variance	218.691	
8		96	74		1	96		Hypothesized Mean Difference	0	
9		106	96		1	106		df	28	
10		120	72		1	120		t Stat	1.737	
11		81	81		1	81		P(T<=t) one-tail	0.047	
12		109			1	109		t Critical one-tail	1.701	
13		116			1	116		P(T<=t) two-tail	0.093	
14		98			1	98		t Critical two-tail	2.048	
15		102			1	102				
16		115			1	115		=LINEST(E2:E31,D2:D31,,TRUE)		
17		76			1	76		9.950	92.800	
18		119			1	119		5.727	4.676	
19		116			1	116		0.097	14.788	
20		107			1	107		3.018	28	
21		89			1	89		660.017	6123.350	
22					0	88				
23	Mean	102.750	92.800		0	110		t ratio:	1.737	
24	Variance	193.461	271.956		0	104		Probability of t	0.047	

Figure 6.26
The larger sample has the smaller variance, which makes the conventional t-test more liberal than the nominal probability.

The data sets in Figures 6.26 and 6.27 differ from those used in Figures 6.24 and 6.25, in order to pair the larger sample with the smaller variance. The ranges I5:J5 and I6:J6 verify that's the data configuration. Figure 6.26 includes the results from the Data Analysis tool that assumes equal variances. Figure 6.27 includes the results from the tool that returns the quasi t-test and the correction of the degrees of freedom.

Again, to see the effects of the adjustments for the unequal variances and sample sizes, compare the degrees of freedom, the t-ratios, and the probabilities associated with a directional research hypothesis shown in Figure 6.26 with those shown in Figure 6.27.

G24 | =T.DIST.RT(G23,ROUND(G20,0))

	A	B	C	D	E	F	G	H	I
1		Organic	Inorganic		t-Test: Two-Sample Assuming Unequal Variances				
2		109	88						
3		87	110			Organic	Inorganic		
4		111	104		Mean	102.750	92.800		
5		87	77		Variance	193.461	271.956		
6		90	114		Observations	20	10		
7		121	112		Hypothesized Mean Difference	0			
8		96	74		df	16			
9		106	96		t Stat	1.639			
10		120	72		P(T<=t) one-tail	0.060			
11		81	81		t Critical one-tail	1.746			
12		109			P(T<=t) two-tail	0.121			
13		116			t Critical two-tail	2.120			
14		98							
15		102				Variance error of the mean, Organic	9.673	=B24/20	
16		115				Variance error of the mean, Inorganic	27.196	=C24/10	
17		76				Degrees of freedom, Organic	19	19	
18		119				Degrees of freedom, Inorganic	9	9	
19		116				Correction factor	0.262	=G15/(G15+G16)	
20		107				Corrected degrees of freedom	15.606	=G17*G18/(G18*G19^2+G17*(1-G19)^2)	
21		89				Mean Difference	9.950	=B23-C23	
22						Estimate of standard error of mean difference	6.072	=SQRT(G15+G16)	
23	Mean	102.750	92.800			t ratio	1.639	=G21/G22	
24	Variance	193.461	271.956			Probability of t	0.060	=T.DIST.RT(G23,ROUND(G20,0))	
25									
26						Probability of t from TTEST()	0.061	=TTEST(B2:B21,C2:C11,1,3)	

Figure 6.27
The liberal test shown in Figure 6.26 has been made more conservative by applying the Welch correction and the quasi t.

The degrees of freedom shown in cell F8 of Figure 6.27 is reduced all the way from 28 (Figure 6.26, cell I9) to 16, making what is originally a liberal test more conservative. Furthermore, the t-ratio of 1.737 in Figure 6.26 is reduced to 1.639. The net effect is to render the probability statement more conservative in Figure 6.27 (0.060) than in Figure 6.26 (0.047). So the results of the quasi t in Figure 6.27 might well convince you to retain the null hypothesis of no mean difference in the population, given that you set alpha to 0.05 at the outset.

NOTE

As I've presented the examples in Figures 6.24 through 6.27, the decisions to retain or reject the null hypothesis are (or decision is) balanced on a knife edge. I did that to show how changing the statistical test from a conventional t-test to a quasi t test can reverse the outcome. At the same time, the difference between a probability of, say, 0.047 and 0.060 is quite small. If the cost of an incorrect decision is high, I would much prefer to repeat the experiment with fresh data than to make a potentially disastrous decision on flimsy grounds.

To close the current section of this chapter, let's take a look at how the quasi t-ratio is calculated and how the degrees of freedom for the test is modified to adjust the unequal sample sizes.

The numerator of the t-ratio is unchanged in the quasi t-ratio: It's just the difference between the two means. If you use LINEST() instead of a traditional t-test to get the basic statistics, the numerator of the t-ratio is still the regression coefficient, numerically the same as the difference between the sample means.

Recall that in the conventional t-ratio, the denominator begins with the calculation of the pooled variance for the two groups: the sums of the squared deviations are totaled, and the total is divided by the sum of the two degrees of freedom.

The result, the pooled variance, is multiplied by the sum of the reciprocals of the degrees of freedom. That results in the variance error of the difference between means, and its square root is the standard error of the difference between means. So,

$$\text{Pooled Variance} = (\Sigma x_1^2 + \Sigma x_2^2)/(df_1 + df_2)$$

$$\text{Variance Error of the Mean Difference} = \text{Pooled Variance} * (1/n_1 + 1/n_2)$$

$$\text{Standard Error of the Mean Difference} = \sqrt{\text{Variance Error}}$$

And

$$\text{t-ratio} = (\overline{X}_1 - \overline{X}_2)/\mathit{Standard\ Error}$$

In contrast, to get the quasi t-ratio, replace the standard error of the difference between means with the square root of the sum of the variance errors of the mean for each group. So:

$$S_{\overline{X}_1}^2 = S_{X_1}^2/n_1$$

And

$$S^2_{\overline{X}_2} = S^2_{X_2}/n_2$$

Then:

$$\text{Quasi t-ratio} = (\overline{X}_1 - \overline{X}_2)/\sqrt{S^2_{\overline{X}_1} + S^2_{\overline{X}_2}}$$

This works out on the Excel worksheets in Figure 6.27 as follows:

- The variance for the Organic group is calculated in cell B24 with this formula:

  =VAR.S(B2:B21)

- The variance for the Inorganic group is calculated in cell C24 with this formula:

  =VAR.S(C2:C11)

  Because the VAR() functions ignore empty cells, you could also use this formula:

  =VAR.S(C2:C21)

  But although that could be convenient, it leaves you open to problems if a stray digit somehow finds its way into the range C12:C21.

- The variance error of the mean of each group is given in cells G15 and G16 with these formulas:

  =B24/20

  =C24/10

  That is, the variance of each group divided by the number of observations in each group.

- The estimate of the standard error of the difference between two means is the square root of the sum of the two standard errors of the mean. Cell G22 contains this formula:

  =SQRT(G15+G16)

The difference between the group means is in cell G21. So we can calculate the quasi t-ratio in cell G23 as 1.639, using this formula:

=G21/G22

6

Before we can calculate the probability of getting a quasi t-ratio of 1.639 from a population whose actual t-ratio is 0.0, we need to adjust the degrees of freedom for the test. That's a little more complicated than adjusting the quasi t-ratio's denominator, but not by much. Begin by calculating a correction factor *c*, using this formula:

$$c = S^2_{\overline{X}_1}/(S^2_{\overline{X}_1} + S^2_{\overline{X}_2})$$

The correction factor is calculated in cell G19 using this formula:

=G15/(G15+G16)

Then, the degrees of freedom for the quasi t-test is calculated with this formula:

$$v = v_1 v_2 / [v_2 c^2 + v_1 (1-c)^2]$$

where:

- v is the degrees of freedom for the quasi t-test.
- v_1 is the degrees of freedom for the first group.
- v_2 is the degrees of freedom for the second group.
- c^2 is the square of the correction factor.

That formula is expressed in cell G20 with this Excel syntax:

```
=G17*G18/(G18*G19^2+G17*(1–G19)^2)
```

Be sure to keep the degrees of freedom straight. If you calculate the correction factor with v_1 as its numerator, multiply it by $(1 - c)^2$ in the adjusted degrees of freedom. (It makes no difference whether Group 1 or Group 2 has the larger variance, or which group has the larger number of observations.)

It's quite likely that the calculation of the corrected degrees of freedom will result in a fractional value. In cell G20 of Figure 6.27, that value is 15.606. Although the T.DIST.RT( ) function will accept a fractional number of degrees of freedom without complaint, it's best to be explicit with the arguments. I use Excel's ROUND( ) function to get the nearest integer value for the degrees of freedom. The formula in cell G24 of Figure 6.27 is:

```
=T.DIST.RT(G23,ROUND(G20,0))
```

The Data Analysis tool, t-Test: Two Sample Assuming Unequal Variances, returns the same t-ratio and probability of that ratio as do the formulas this section has described. The Data Analysis tool does not return the adjusted denominator for the t-ratio, nor does it show how it adjusts the degrees of freedom, but you can see how the tool arrives at those values by examining the formulas this section has provided. The final results are identical.

The TTEST( ) Function

If you're confronted by the situation that this section has focused on—unequal variances and unequal sample sizes—you should know that there's a much quicker route to the eventual probability statement than trotting out the Unequal Variances tool or entering each of the from-scratch equations just discussed. Excel's TTEST( ) function can return the probability of the quasi t-ratio.

I have placed the TTEST() function in cell G26 of both Figures 6.25 and 6.27. You can compare its results with those supplied in cells F10 and G24, supplied by the Data Analysis tool and the actual formulas. If you do so, you'll note that the TTEST() result differs slightly from the other results, with the discrepancy showing up in the thousandths decimal place.

I can't think of good reason that you would want to use this function, and I'm mentioning it here primarily as an opportunity to say that its results differ unaccountably from those returned by less cryptic methods, and that it's nuts to report nothing but a probability with no supporting data such as means, group counts, and variances.

TTEST() is one of what Excel, since 2010, has termed its "compatibility" functions: Functions that remain in the application so that older workbooks can be interpreted by newer versions of Excel. The function's full syntax is as follows:

TTEST(array_1, array_2,tails,type)

Where:

- Array_1 is the worksheet address of the observations for the first group, and array_2 is the address for the second group,
- Tails (1 or 2) indicates whether to report the probability beyond the t-ratio in one or in both tails of the distribution
- Type (1, 2, or 3) indicates whether to use a t-test for groups with paired members such as siblings (1); a t-test assuming equal group variances (2); or a quasi t-test assuming unequal group variances (3).

The first of the three types of test supported by TTEST() is also termed the *dependent groups t-test*. I won't be discussing it further in this chapter, for two reasons:

- In a dependent groups t-test, each observation in one group is paired with an observation from the other group: for example, with ten pairs of twins, one group might consist of one member of each pair and the other group might consist of the other member. In that case, the group sizes must be identical and, as you've seen, the t-test is robust with respect to violation of the assumption of equal variances when sample sizes are equal.
- The dependent groups t-test is actually a simple form of the analysis of covariance, which is the topic of Chapter 8.

First, though, let's look at extending the use of coding techniques from the simple two-group design discussed in this chapter to three or more factor levels discussed in Chapter 7.

Using Regression to Test Differences Between Group Means

7

IN THIS CHAPTER

Dummy Coding 246

Effect Coding 259

Orthogonal Coding 267

Factorial Analysis 272

Statistical Power, Type I and Type II Errors ... 283

Coping with Unequal Cell Sizes 288

This is a book about regression analysis. Nevertheless, I'm going to start this chapter by discussing different scales of measurement. When you use regression analysis, your *predicted* (or *outcome*, or *dependent*) variable is nearly always measured on an interval or ratio scale, one whose values are numeric quantities. Your *predictor* (or *independent*, or *regressor*) variables are also frequently measured on such numeric scales.

However, the predictor variables can also represent nominal or category scales. Because functions such as Excel's LINEST() do not respond directly to predictors with values such as STATIN and PLACEBO, or REPUBLICAN and DEMOCRAT, you need a system to convert those nominal values to numeric values that LINEST() can deal with.

The system you choose has major implications for the information you get back from the analysis. So I'll be taking a closer look at some of the underlying issues that inform your choice.

It will also be helpful to cover some terminology issues early on. This book's first six chapters have discussed the use of regression analysis to assess the relationships between variables measured on an interval or a ratio scale. There are a couple of reasons for that:

- Discussing interval variables only allows us to wait until now to introduce the slightly greater complexity of using regression to assess differences between groups.

- Most people who have heard of regression analysis at all have heard of it in connection with prediction and explanation: for example, predicting weight from known height. That sort of usage *tends* to imply interval or ratio variables as both the predicted variable and the predictor variables.

With this chapter we move into the use of regression analysis to analyze the influence of nominal variables (such as make of car or type of medical treatment) on interval variables (such as gas mileage or levels of indicators in blood tests). That sort of assessment tends to focus on the effects of belonging to different groups upon variables that quantify the outcome of group membership (gas mileage for different auto makes or cholesterol levels after different medical treatments).

We get back to the effects of interval variables in Chapter 8, "The Analysis of Covariance," but in this chapter I'll start referring to what earlier chapters called *predicted variables* as *outcome variables*, and what I have called *predictor variables* as *factors*. Lots of theorists and writers prefer terms other than *outcome variable*, because it implies a cause-and-effect relationship, and inferring that sort of situation is a job for your experimental design, not your statistical analysis. But as long as that's understood, I think we can get along with *outcome variable*—at least, it's less pretentious than some of its alternatives.

Dummy Coding

Perhaps the simplest approach to coding a nominal variable is termed *dummy coding*. I don't mean the word "simplest" to suggest that the approach is underpowered or simple-minded. For example, I prefer dummy coding in logistic regression, where it can clarify the interpretation of the coefficients used in that method.

Dummy coding can also be useful in standard linear regression when you want to compare one or more treatment groups with a comparison or *control* group.

An Example with Dummy Coding

Figures 7.1 and 7.2 show how the data from a small experiment could be set up for analysis by an application that returns a traditional analysis of variance, or *ANOVA*.

In ANOVA jargon, a variable whose values constitute the different conditions to which the subjects are exposed is called a *factor*. In this example, the factor is Treatment. The different values that the Treatment factor can take on are called *levels*. Here, the levels are the three treatments: Medication, Diet, and Placebo as a means of lowering amounts of an undesirable component in the blood.

Figure 7.1
The Data Analysis tool requires that the factor levels occupy different columns or different rows.

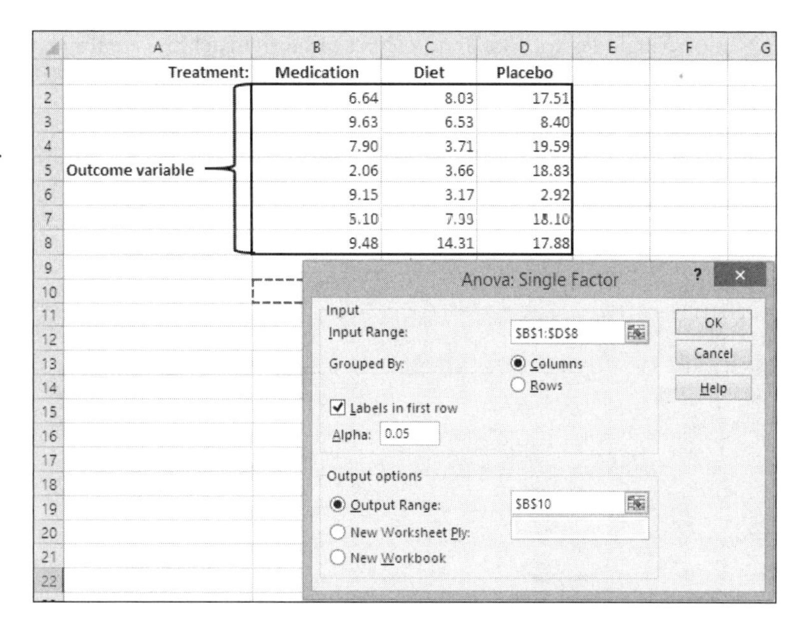

	A	B	C	D
1	Treatment:	Medication	Diet	Placebo
2		6.64	8.03	17.51
3		9.63	6.53	8.40
4		7.90	3.71	19.59
5	Outcome variable	2.06	3.66	18.83
6		9.15	3.17	2.92
7		5.10	7.33	18.10
8		9.48	14.31	17.88

Figure 7.2
If you choose Labels in First Row in the dialog box, the output associates the summary statistics with the label.

	A	B	C	D	E	F	G	H
1	Treatment:	Medication	Diet	Placebo				
2		6.64	8.03	17.51				
3		9.63	6.53	8.40				
4		7.90	3.71	19.59				
5	Outcome variable	2.06	3.66	18.83				
6		9.15	3.17	2.92				
7		5.10	7.33	18.10				
8		9.48	14.31	17.88				
9								
10		Anova: Single Factor						
11								
12		SUMMARY						
13		*Groups*	*Count*	*Sum*	*Average*	*Variance*		
14		Medication	7	49.98	7.14	7.76		
15		Diet	7	46.75	6.68	15.12		
16		Placebo	7	103.23	14.75	41.49		
17								
18		ANOVA						
19		*Source of Variation*	*SS*	*df*	*MS*	*F*	*P-value*	*F crit*
20		Between Groups	287.448	2	143.724	6.699	0.007	3.555
21		Within Groups	386.197	18	21.455			
22								
23		Total	673.645	20				

Excel's Data Analysis add-in includes a tool named *ANOVA: Single Factor*. To operate correctly, the data set must be arranged as in the range B2:C8 of Figure 7.2. (Or it may be turned 90 degrees, to have different factor levels in different rows and different subjects in different columns.) With the data laid out as shown in the figure, you can run the

ANOVA: Single Factor tool and in short order get back the results shown in the range A12:H23. The Data Analysis tool helpfully provides descriptive statistics as shown in B14:F16.

Figure 7.3 has an example of how you might use dummy coding to set up an analysis of the same data set by means of regression analysis via dummy coding.

Figure 7.3
One minor reason to prefer the regression approach is that you use standard Excel layouts for the data.

L12 =I12-I14

	A	B	C	D	E	F	G	H	I	J	K	L
1	Out-come	Treatment	Medication Vector	Diet Vector		Diet Coefficient	Medication Coefficient					
2	6.64	Medication	1	0		-8.069	-7.608	14.75				
3	9.63	Medication	1	0		2.476	2.476	1.751				
4	7.90	Medication	1	0		0.427	4.632	#N/A	=LINEST(A2:A22,C2:D22,,TRUE)			
5	2.06	Medication	1	0		6.699	18	#N/A				
6	9.15	Medication	1	0		287.448	386.197	#N/A				
7	5.10	Medication	1	0								
8	9.48	Medication	1	0		Anova: Single Factor						
9	8.03	Diet	0	1								
10	6.53	Diet	0	1		SUMMARY						
11	3.71	Diet	0	1		*Groups*	*Count*	*Sum*	*Average*	*Variance*		
12	3.66	Diet	0	1		Medication	7	49.98	7.14	7.76		-7.608
13	3.17	Diet	0	1		Diet	7	46.75	6.68	15.12		-8.069
14	7.33	Diet	0	1		Placebo	7	103.23	14.75	41.49		
15	14.31	Diet	0	1								
16	17.51	Placebo	0	0					=F.DIST.RT(F5,2,G5)		0.007	
17	8.40	Placebo	0	0		ANOVA						
18	19.59	Placebo	0	0		*Source of Variation*	*SS*	*df*	*MS*	*F*	*P-value*	*F crit*
19	18.83	Placebo	0	0		Between Groups	287.448	2	143.724	6.699	0.007	3.555
20	2.92	Placebo	0	0		Within Groups	386.197	18	21.455			
21	18.10	Placebo	0	0								
22	17.88	Placebo	0	0		Total	673.645	20				

When you use any sort of coding there are a couple of rules to follow. These are the rules that apply to dummy coding:

- You need to reserve as many columns for new data as the factor has levels, minus 1. Notice that this is the same as the number of degrees of freedom for the factor. With three levels, as in the present example, that's 3 – 1, or 2. It's useful to term these columns *vectors*.
- Each vector represents one level of the factor. In Figure 7.3, Vector 1 represents Medication, so every subject who receives the medication gets a 1 on Vector 1, and everyone else receives a 0 on that vector. Similarly, every subject receives a 0 on Vector 2 except those who are treated by Diet—they get a 1 on Vector 2.
- Subjects in one level, which is often a control group, receive a 0 on all vectors. In Figure 7.3, this is the case for those who take a placebo.

With the data laid out as shown in the range A2:D22 in Figure 7.3, array-enter this LINEST() function in a blank range five rows high and three columns wide, such as F2:H6 in the figure:

```
=LINEST(A2:A22,C2:D22,,TRUE)
```

Don't forget to array-enter the formula with the keyboard combination Ctrl+Shift+Enter. The arguments are as follows:

- The first argument, the range A2:A22, is the address that contains the outcome variable. (Because the description of this study suggests that it's a true, controlled experiment, it's not misleading to refer to the levels of a given component in the blood as an outcome variable, thus implying cause and effect.)
- The second argument, the range C2:D22, is the address that contains the vectors that indicate which level of the factor a subject belongs to. In other experimental contexts you might refer to these as *predictor variables*.
- The third argument is omitted, as indicated by the consecutive commas with nothing between them. If this argument is TRUE or omitted, Excel is instructed to calculate the regression equation's constant normally. If the argument is FALSE, Excel is instructed to set the constant to 0.0.
- The fourth argument, TRUE, instructs Excel to calculate and return the third through fifth rows of the results, which contain summary information, mostly about the reliability of the regression equation.

In Figure 7.3 I have repeated the results of the traditional ANOVA from Figure 7.2, to make it easier to compare the results of the two analyses. Note these points:

- The sum of squares regression and the sum of squares residual from the LINEST() results in cells F6 and G6 are identical to the sum of squares between groups and the sum of squares within groups returned by the Data Analysis add-in in cells G19 and G20.
- The degrees of freedom for the residual in cell G5 is the same as the degrees of freedom within groups in cell H20. Along with the sums of squares and knowledge of the number of factor levels, this enables you to calculate the mean square between and the mean square within if you want.
- The F-ratio returned in cell F5 by LINEST() is identical to the F-ratio reported by the Data Analysis add-in in cell J19.
- The constant (also termed the intercept) returned by LINEST() in cell H2 is identical to the mean of the group that's assigned codes of 0 throughout the vectors. In this case that's the Placebo group: Compare the value of the constant in cell H2 with the mean of the Placebo group in cell I14. (That the constant equals the group with codes of 0 throughout is true of dummy coding, not effect or orthogonal coding, discussed later in this chapter.)

The regression coefficients in cells F2 and G2, like the t-tests in Chapter 6, express the differences between group means. In the case of dummy coding, the difference is between the group assigned a code of 1 in a vector and the group assigned 0's throughout.

For example, the difference between the means of the group that took a medication and the group that was treated by placebo is 7.14 – 14.75 (see cells I12 and I14). That difference equals –7.608, and it's calculated in cell L12. That's the regression coefficient for Vector 1, returned by LINEST() in cell G2. Vector 1 identifies the Medication group with a 1.

Similarly, the difference between the mean of the group treated by diet and that treated by placebo is 6.68 – 14.75 (see cells I13 and I14). The difference equals –8.069, calculated in cell L13, which is also the regression coefficient for Vector 2.

> **NOTE**
>
> It's here that LINEST()'s peculiarity in the order of the coefficients shows up again. Recall that if predictor variables A, B, and C appear in that left-to-right order on the worksheet, they appear in the left-to-right order C, B, and then A in the LINEST() results.
>
> So in Figure 7.3, the vector that represents the Medication treatment is in column C, and the vector that represents Diet is to its right, in column D. However, LINEST() puts the regression coefficient for Medication in cell G2, and the regression coefficient for Diet to its *left*, in cell F2. The potential for confusion is clear and it's a good idea to label the columns in the LINEST() result to show which variable each coefficient refers to.

One bit of information that LINEST() does not provide you is statistical significance of the regression equation. In the context of ANOVA, where we're evaluating the differences between group means, that test of statistical significance asks whether *any* of the mean differences is large enough that the null hypothesis of no difference between the means in the population can be rejected. The F-ratio, in concert with the degrees of freedom for the regression and the residual, speaks to that question.

You can determine the probability of observing a given F-ratio if the null hypothesis is true by using Excel's F.DIST.RT() function. In this case, you use it in this way (it's also in cell K16):

```
=F.DIST.RT(F5,2,G5)
```

Notice that the value it returns, 0.007, is identical to that returned in cell K19 by the Data Analysis add-in's ANOVA: Single Factor tool. If there is no difference, measured by group means, in the populations of patients who receive the medication, or whose diet was controlled, or who took a placebo, then the chance of observing an F-ratio of 6.699 is 7 in 1,000. It's up to you whether that's rare enough to reject the null hypothesis. It would be for most people, but a sample of 21 is a very small sample, and that tends to inhibit the generalizability of the findings—that is, how confidently you can generalize your observed outcome from 21 patients to your entire target population.

Populating the Vectors Automatically

So: What does all this buy you? Is there enough advantage to running your ANOVA using regression in general and LINEST() in particular that it justifies any extra work involved?

I think it does, and the decision isn't close. First, what are the steps needed to prepare for the Data Analysis tool, and what steps to prepare a regression analysis?

To run the Data Analysis ANOVA: Single Factor tool, you have to arrange your data as shown in the range B1:D8 in Figure 7.2. That's not a natural sort of arrangement of data in either a true database or in Excel. A list or table structure of the sort shown in A1:D22 of Figure 7.3 is much more typical, and as long as you provide columns C and D for the dummy 0/1 codes, it's ready for you to point LINEST() at.

To prepare for a regression analysis, you do need to supply the 0s and 1s in the proper rows and the proper columns. This is *not* a matter of manually entering 0s and 1s one by one. Nor is it a matter of copying and pasting values or using Ctrl+Enter on a multiple selection. I believe that the fastest, and most accurate, way of populating the coded vectors is by way of Excel's VLOOKUP() function. See Figure 7.4.

To prepare the ground, enter a key such as the one in the range A2:C4 in Figure 7.4. That key should have as many columns and as many rows as the factor has levels. In this case, the factor has three levels (Medication, Diet, and Placebo), so the key has three columns, and there's one row for each level. It's helpful but not strictly necessary to provide column headers, as is done in the range A1:C1 of Figure 7.4.

Figure 7.4
Practice in the use of the VLOOKUP() function can save you considerable time in the long run.

G2 =VLOOKUP($F2,$A$2:$C$4,2,0)

	A	B	C	D	E	F	G	H
1	Treatment	Medication Vector	Diet Vector		Out-come	Treatment	Medication Vector	Diet Vector
2	Medication	1	0		6.64	Medication	1	0
3	Diet	0	1		9.63	Medication	1	0
4	Placebo	0	0		7.90	Medication	1	0
5					2.06	Medication	1	0
6					9.15	Medication	1	0
7					5.10	Medication	1	0
8					9.48	Medication	1	0
9					8.03	Diet	0	1
10					6.53	Diet	0	1
11					3.71	Diet	0	1
12					3.66	Diet	0	1
13					3.17	Diet	0	1
14					7.33	Diet	0	1
15					14.31	Diet	0	1
16					17.51	Placebo	0	0
17					8.40	Placebo	0	0
18					19.59	Placebo	0	0
19					18.83	Placebo	0	0
20					2.92	Placebo	0	0
21					18.10	Placebo	0	0
22					17.88	Placebo	0	0

The first column—in this case, A2:A4—should contain the labels you use to identify the different levels of the factor. In this case those levels are shown for each subject in the range F2:F22.

You can save a little time by selecting the range cells in the key starting with its first row and its *second* column—so, B2:C4. Type 0, hold down the Ctrl key and press Enter. All the selected cells will now contain the value 0.

In the same row as a level's label, enter a 1 in the column that will represent that level. So, in Figure 7.4, cell B2 gets a 1 because column B represents the Medication level, and cell C3 gets a 1 because column C represents the Diet level. There will be no 1 to represent Placebo because we'll treat that as a control or comparison group, and so it gets a 0 in each column.

With the key established as in A2:C4 of Figure 7.4, select the first row of the first column where you want to establish your matrix of 0's and 1's. In Figure 7.4, that's cell G2. Enter this formula:

```
=VLOOKUP($F2,$A$2:$C$4,2,0)
```

Where:

- $F2 is the label that you want to represent with a 1 or a 0.
- $A$2:$C$4 contains the key (Excel terms this a *table lookup*).
- 2 identifies the column in the key that you want returned.
- 0 specifies that an exact match for the label is required, and that the labels in the first column of the key are not necessarily sorted.

I've supplied dollar signs where needed in the formula so that it can be copied to other columns and rows without disrupting the reference to the key's address, and to the column in which the level labels are found.

Now copy and paste cell G2 into H2 (or use the cell's selection handle to drag it one column right). In cell H2, edit the formula so that VLOOKUP()'s third argument has a 3 instead of a 2—this directs Excel to look in the key's third column for its value.

Finally, make a multiple selection of cells G2 and H2, and drag them down into G3:H22. This will populate columns G and H with the 0's and 1's that specify which factor level each record belongs to.

You can now obtain the full LINEST() analysis by selecting a range such as A6:C10, and array-entering this formula:

```
=LINEST(E2:E22,G2:H22,,TRUE)
```

By the way, you might find it more convenient to switch the contents of columns E and F in Figure 7.4. I placed the treatment labels in column F to ease the comparison of the labels with the dummy codes. If you swap columns E and F, you might find the LINEST() formula easier to handle. You'll also want to change the first VLOOKUP() formula from this:

```
=VLOOKUP($F2,$A$2:$C$4,2,0)
```

to this:

```
=VLOOKUP($E2,$A$2:$C$4,2,0)
```

The Dunnett Multiple Comparison Procedure

When you have completed a test of the differences between the means of three or more groups—whether by way of traditional ANOVA methods or a regression approach—you have learned the probability that *any* of the means in the population is different from any of the remaining means in the population. You have not learned *which* mean or means is different from others.

Statisticians studied this issue and wrote an intimidatingly comprehensive literature on the topic during the middle years of the twentieth century. The procedures they developed came to be known as *multiple comparisons*. Depending on how you count them, the list of different procedures runs to roughly ten. The procedures differ from one another in several ways, including the nature of the error involved (for example, per comparison or per experiment), the reference distribution (for example, F, t, or q), planned beforehand (a priori) or after the fact (post hoc), and on other dimensions.

If you choose to use dummy coding in a regression analysis, in preference to another coding method, it might well be because you want to compare all the groups but one to the remaining group. That approach is typical of an experiment in which you want to compare the results of two or more treatments to a control group. In the context of dummy coding, the control group is the one that receives 0's throughout the vectors that represent group membership. One result of dummy coding, as you've seen, is that the regression coefficient for a particular group has a value that is identical to the difference between the group's mean and the mean of the control group.

These procedures tend to be named for the statisticians who developed them, and one of them is called the Dunnett multiple comparison procedure. It makes a minor modification to the formula for the t-ratio. It also relies on modifications to the reference t-distribution. In exchange for those modifications, the Dunnett provides you with comparisons that have somewhat more statistical power than alternative procedures, given that you start your experiment intending to compare two or more treatments to a control.

As you'll see, the calculation of the t-ratios is particularly easy when you have access to the LINEST() worksheet function. Access to the reference distribution for Dunnett's t-ratio is a little more complicated. Excel offers you direct access to, for example, the t-distribution and the F-distribution by way of its T.DIST(), T.INV(), F.DIST(), and F.INV() functions, and their derivatives due to the RT and 2T tags. But Excel does not have a DUNNETT() function that tells you the t-ratio that demarks the 95%, 99% or any other percent of the area beneath the distribution as it does for t and F.

> **NOTE** You cannot legitimately calculate a t-ratio using Dunnett's methods and then compare it to a standard t-distribution of the sort returned by T.DIST() and T.INV(). Dunnett's t-distribution has a different shape than the "standard" t-distribution.

Although the values for Dunnett's t are not directly available in Excel, they are available on various online sites. It's easy enough to download and print the tables (they occupy two printed pages). A search using the keywords "Dunnett," "multiple comparison," and "tables" will locate more sites than you want, but many of them show the necessary tables. The tables also appear as an appendix in most intermediate-level, general statistics textbooks in print.

Let's look at how you could conduct a Dunnett multiple comparison after running the Data Analysis ANOVA: Single Factor tool. See Figure 7.5.

Figure 7.5
When the group sizes are equal, as here, all the comparisons' t-ratios have the same denominator.

I2 =SQRT(D22*(1/B13+1/B14))

	A	B	C	D	E	F	G	H	I
1		Treatment							
2	Control	Med 1	Med 2	Med 3		Denominator of t-ratio			11.05
3	164	153	165	150					
4	141	191	168	132		t-ratio, Med 1 vs. Control			1.52
5	144	192	175	123		t-ratio, Med 2 vs. Control			2.52
6	138	126	189	155		t-ratio, Med 3 vs. Control			-0.38
7	153	162	182	159					
8							Critical value		2.23
9	Anova: Single Factor								
10									
11	SUMMARY								
12	*Groups*	*Count*	*Sum*	*Average*	*Variance*				
13	Control	5	740	148.0	111.5				
14	Med 1	5	824	164.8	769.7				
15	Med 2	5	879	175.8	97.7				
16	Med 3	5	719	143.8	241.7				
17									
18									
19	ANOVA								
20	*Source of Variation*	*SS*	*df*	*MS*	*F*	*P-value*	*F crit*		
21	Between Groups	3323.4	3	1107.8	3.63	0.04	3.2389		
22	Within Groups	4882.4	16	305.15					
23									
24	Total	8205.8	19						

The data is laid out for the Data Analysis tool in A2:D7. The ANOVA: Single Factor tool in the Data Analysis add-in returns the results shown in the range A11:G24. You'll need the group means, the Mean Square Error from the ANOVA table, and the group counts.

The first step is to calculate the denominator of the t-ratios. With equal group sizes, the same denominator is used for each t-ratio. The formula for the denominator is:

$$\sqrt{MSE\left(\frac{1}{n_1}+\frac{1}{n_2}\right)}$$

where MSE is the mean square error from the ANOVA table, shown in Figure 7.5 in cell D22. (*Mean square error* is simply another term for *mean square within* or *mean square residual*.)

When, as here, the group sizes are equal, you can also use this arithmetically equivalent formula:

$$\sqrt{2MSE/n}$$

The denominator for the t-ratios in this example is given in cell I2. It uses this formula:

=SQRT(D22*(1/B13+1/B14))

where cell D22 contains the mean square error and cells B13 and B14 contain group counts. With all groups of the same size, it doesn't matter which group counts you use in the formula. As suggested earlier, with equal group sizes the Excel formula could also be:

=SQRT((2*D22)/B13)

The next step is to find the difference between the mean of each treatment group and the mean of the control group, and divide those differences by the denominator. The result is one or more t-ratios. For example, here's the formula in cell I4 of Figure 7.5.

=(D14–D13)/I2

The formula divides the difference between the mean of the Med 1 group (D14) and the mean of the Control group (D13) by the denominator of the t-ratio (I2). The formulas in cells I5 and I6 follow that pattern:

I5: =(D15–D13)/I2

I6: =(D16–D13)/I2

At this point you look up the value in the Dunnett tables that corresponds to three criteria:

- The Degrees of Freedom Within from the ANOVA table (in Figure 7.5, the value 16 in cell C22)
- The total number of groups, including the control group
- The value of alpha that you selected before seeing the data from your experiment

Most printed tables give you a choice of 0.05 and 0.01. That's restrictive, of course, and it delights me that Excel offers exact probabilities for any probability level you might present it for various distributions including the chi-square, the binomial, the t and the F.

For the present data set, the printed Dunnett tables give a critical value of 2.23 for four groups and 16 Degrees of Freedom Within, at the 0.05 alpha level. They give 3.05 at the 0.01 alpha level.

Because the t-ratio in cell I5, which contrasts Med 2 with Control, is the only one to exceed the critical value of 2.23, you could reject the null hypothesis of no difference in the population means for those two groups at the .05 confidence level. You could not reject it at the 0.01 confidence level because the t-ratio does not exceed the 0.01 level of 3.05.

Compare all that with the results shown in Figure 7.6.

Start by noticing that cells G2, H2, and I2, which contain the regression coefficients for Med 3 Vector, Med 2 Vector, and Med 1 Vector (in that order), express the differences between the means of the treatment groups and the control group.

Figure 7.6
Much less calculation is needed when you start by analyzing the data with regression.

J8 =I2/I3

	A	B	C	D	E	F	G	H	I	J
1	Treatment	Outcome	Med 1 Vector	Med 2 Vector	Med 3 Vector		=LINEST(B2:B21,C2:E21,,TRUE)			
2	Control	164	0	0	0		-4.2	27.8	16.8	148
3	Control	141	0	0	0		11.05	11.05	11.05	7.81
4	Control	144	0	0	0		0.41	17.47	#N/A	#N/A
5	Control	138	0	0	0		3.63	16	#N/A	#N/A
6	Control	153	0	0	0		3323.4	4882.4	#N/A	#N/A
7	Med 1	153	1	0	0					
8	Med 1	191	1	0	0		t-ratio, Med 1 vs. Control			1.52
9	Med 1	192	1	0	0		t-ratio, Med 2 vs. Control			2.52
10	Med 1	126	1	0	0		t-ratio, Med 3 vs. Control			-0.38
11	Med 1	162	1	0	0					
12	Med 2	165	0	1	0				Critical value	2.23
13	Med 2	168	0	1	0					
14	Med 2	175	0	1	0					
15	Med 2	189	0	1	0					
16	Med 2	182	0	1	0					
17	Med 3	150	0	0	1					
18	Med 3	132	0	0	1					
19	Med 3	123	0	0	1					
20	Med 3	155	0	0	1					
21	Med 3	159	0	0	1					

For example, the regression coefficient in cell H2 (27.8) is the difference between the Med 2 mean (175.8, in cell D15 of Figure 7.5) and the Control mean (148.0, in cell D13 of Figure 7.5). So right off the bat you're relieved of the need to calculate those differences. (You can, however, find the mean of the control group in the regression equation's constant, in cell J2.)

Now notice the standard errors of the regression coefficients, in cells G3, H3, and I3. They are all equal to 11.05, and in any equal-cell-size situation with dummy coding, the standard errors will all have the same value. That value is also the one calculated from the mean square error and the group sizes in Figure 7.5 (see that figure, cell I2).

So, all you need to do if you start the data analysis with LINEST() is to divide the regression coefficients by their standard errors to get the t-ratios that correspond to the Dunnett procedure. Notice that the t-ratios in cells J8, J9, and J10 are identical to those calculated in Figure 7.5, cells I4, I5, and I6.

Now let's have a look at a slightly more complicated situation, one in which you have different numbers of subjects in your groups. See Figure 7.7.

In Figure 7.7, the basic calculations are the same, but instead of using just one denominator as was done in Figure 7.5 (because the groups all had the same number of subjects), we need three denominators because the group sizes are different. The three denominators appear in the range I4:I6, and use the version of the formula given earlier:

$$\sqrt{MSE\left(\frac{1}{n_1}+\frac{1}{n_2}\right)}$$

Figure 7.7
Using traditional ANOVA on a data set with unequal group sizes, you need to calculate a different denominator for each t-ratio.

I4 =SQRT($D$22*(1/B13+1/B14))

	A	B	C	D	E	F	G	H	I	J
1		Treatment								
2	Control	Med 1	Med 2	Med 3					Denominator	
3	164	153	165	160					of t-ratio	t-ratio
4	141	191	168	132		t-ratio, Med 1 vs. Control			11.13	1.62
5	144	192	175	123		t-ratio, Med 2 vs. Control			10.71	2.79
6	138	126	189	155		t-ratio, Med 3 vs. Control			10.40	-0.13
7		162	182	149						
8			181	147					Med 1 vs. Control	18.05
9				152					Med 2 vs. Control	29.92
10	Anova: Single Factor								Med 3 vs. Control	-1.32
11	SUMMARY									
12	*Groups*	*Count*	*Sum*	*Average*	*Variance*				Critical value, 0.05	2.21
13	Control	4	587	146.75	138.25				Critical value, 0.01	3.01
14	Med 1	5	824	164.8	769.7					
15	Med 2	6	1060	176.6667	82.66667					
16	Med 3	7	1018	145.4286	174.2857					
17										
18										
19	ANOVA									
20	*Source of Variation*	*SS*	*df*	*MS*	*F*	*P-value*	*F crit*			
21	Between Groups	3926.721	3	1308.907	4.757	0.013	3.1599			
22	Within Groups	4952.598	18	275.1443						
23										
24	Total	8879.318	21							

So, the formulas to return the t-ratio's denominator are:

I4: =SQRT($D$22*(1/B13+1/B14))

I5: =SQRT($D$22*(1/B13+1/B15))

I6: =SQRT($D$22*(1/B13+1/B16))

Notice that the only difference between the formulas is that they alter a reference from B14 to B15 to B16, as the number of observations in the Med 1, Med 2, and Med 3 groups increases from 5 to 6 to 7. The formulas make use of the group counts returned by the Data Analysis add-in to pick up the number of observations in each treatment group.

The differences between the treatment group means and the control group mean are shown in the range J8:J10. They are the numerators for the t-ratios, which appear in the range J4:J6. Each t-ratio is the result of dividing the difference between two group means by the associated denominator, as follows:

J4: =J8/I4

J5: =J9/I5

J6: =J10/I6

In sum, when your group sizes are unequal, traditional methods have you calculate different denominators for each of the t-ratios that contrast all group means but one (here, Med 1, Med 2, and Med 3) with another mean (here, Control). Then for each pair of means, calculate the mean difference and divide by the denominator for that pair.

You'll also want to compare the values of the t-ratios with the values in Dunnett's tables. In this case you would want to locate the values associated with 18 within-groups degrees of freedom (from cell C22 in the ANOVA table) and 4 groups. The intersection of those values in the table is 2.21 for an alpha level of 0.05 and 3.01 for an alpha level of 0.01 (see cells J12 and J13 in Figure 7.7). Therefore, only the difference between Med 2 and Control, with a t-ratio of 2.79, is beyond the cutoff for 5% of the Dunnett t distribution, and it does not exceed the cutoff for 1% of the distribution. You can reject the null for Med 2 versus Control at the 5% level of confidence but not at the 1% level. You cannot reject the null hypothesis for the other two contrasts at even the 5% level of confidence.

Notice that the F-ratio in the ANOVA table, 4.757 in cell E21, will appear in a central F distribution with 3 and 18 degrees of freedom only 1.3% of the time. (A central F distribution in the context of an ANOVA is one in which the estimate of the population variance due to the differences among group means is equal to the estimate of the population variance due to the average within-group variance.) So the ANOVA informs you that an F-ratio of 4.757 with 3 and 18 degrees of freedom is unlikely to occur by chance if the population means equal one another.

That likelihood, 1.3%, echoes the result of the contrast of the Med 2 group with the Control group. The t-ratio for that contrast, 2.79, exceeds the critical value for 5% of the Dunnett distribution but not the critical value for 1% of the distribution. So the objective of the multiple comparison procedure, to pinpoint the difference in means that the ANOVA's F-ratio tells you must exist, has been met.

Things go a lot more smoothly if you use LINEST() instead. See Figure 7.8.

Figure 7.8
LINEST() calculates the mean differences and t-ratio denominators for you.

J15 =F.DIST.RT(G5,3,H5)

	A	B	C	D	E	F	G	H	I	J	K
1	Treatment	Outcome	Med 1 Vector	Med 2 Vector	Med 3 Vector		=LINEST(B2:B23,C2:E23,,TRUE)				
2	Control	164	0	0	0		-1.32	29.92	18.05	146.75	
3	Control	141	0	0	0		10.40	10.71	11.13	8.29	
4	Control	144	0	0	0		0.44	16.59	#N/A	#N/A	
5	Control	138	0	0	0		4.76	18	#N/A	#N/A	
6	Med 1	153	1	0	0		3926.721	4952.6	#N/A	#N/A	
7	Med 1	191	1	0	0						
8	Med 1	192	1	0	0		t-ratio, Med 1 vs. Control			1.62	=I2/I3
9	Med 1	126	1	0	0		t-ratio, Med 2 vs. Control			2.79	=H2/H3
10	Med 1	162	1	0	0		t-ratio, Med 3 vs. Control			-0.13	=G2/G3
11	Med 2	165	0	1	0						
12	Med 2	168	0	1	0		Critical value, 0.05			2.21	
13	Med 2	175	0	1	0		Critical value, 0.01			3.01	
14	Med 2	189	0	1	0						
15	Med 2	182	0	1	0		Probability of F-ratio			0.013	
16	Med 2	181	0	1	0						
17	Med 3	160	0	0	1						
18	Med 3	132	0	0	1						
19	Med 3	123	0	0	1						
20	Med 3	155	0	0	1						
21	Med 3	149	0	0	1						
22	Med 3	147	0	0	1						
23	Med 3	152	0	0	1						

Figure 7.8 has the same underlying data as Figure 7.7: Four groups with a different number of subjects in each. (The group membership vectors were created just as shown in Figure 7.6, using VLOOKUP(), but to save space in the figure the formulas were converted to values and the key deleted.)

This LINEST() formula is array-entered in the range G2:J6:

```
=LINEST(B2:B23,C2:E23,,TRUE)
```

Compare the regression coefficients in cells G2, H2, and I2 with the mean differences shown in Figure 7.7 (J8:J10). Once again, just as in Figures 7.5 and 7.6, the regression coefficients are exactly equal to the mean differences between the groups that have 1's in the vectors and the group that has 0's throughout. So there's no need to calculate the mean differences explicitly.

The standard errors of the regression coefficients in Figure 7.8 also equal the denominators of the t-ratios in Figure 7.7 (in the range I4:I6). LINEST() automatically takes the differences in the group sizes into account. All there's left to do is divide the regression coefficients by their standards errors, as is done in the range J8:J10. The formulas in those cells are given as text in K8:K10. But don't forget, when you label each t-ratio with verbiage that states which two means are involved, that LINEST() returns the coefficients and their standard errors backwards: Med 3 versus Control in G2:G3, Med 2 versus Control in cell H2:H3, and Med 1 versus Control in I2:I3.

Figure 7.8 repeats in J12 and J13 the critical Dunnett values for 18 within-group degrees of freedom (picked up from cell H5 in the LINEST() results) and 4 groups at the 0.05 and 0.01 cutoffs. The outcome is, of course, the same: Your choice of whether to use regression or traditional ANOVA makes no difference to the outcome of the multiple comparison procedure.

Finally, as mentioned earlier, the LINEST() function does not return the probability of the F-ratio associated with the R^2 for the full regression. That figure is returned in cell J15 by this formula:

```
=F.DIST.RT(G5,3,H5)
```

Where G5 contains the F-ratio and H5 contains the within-group (or "residual") degrees of freedom. You have to supply the second argument (here, 3) yourself: It's the number of groups minus 1 (notice that it equals the number of vectors in LINEST()'s second argument, C2:E23) also known as the degrees of freedom between in an ANOVA or degrees of freedom regression in the context of LINEST().

Effect Coding

Another type of coding, called *effect coding*, contrasts each group mean following an ANOVA with the grand mean of all the observations.

> **NOTE** More precisely, effect coding contrasts each group mean with the mean of all the group means. When each group has the same number of observations, the grand mean of all the observations is equal to the mean of the group means. With unequal group sizes, the two are not equivalent. In either case, though, effect coding contrasts each group mean with the mean of the group means.

This aspect of effect coding—contrasting group means with the grand mean rather than with a specified group, as with dummy coding—is due to the use of –1 instead of 0 as the code for the group that gets the same code throughout the coded vectors. Because the contrasts are with the grand mean, each contrast represents the effect of being in a particular group.

Coding with –1 Instead of 0

Let's take a look at an example before getting into the particulars of effect coding. See Figure 7.9.

Figure 7.9
The coefficients in LINEST() equal each group's distance from the grand mean.

L5 =J2-SUM(G2:I2)

	A	B	C	D	E	F	G	H	I	J	K	L	M
1	Treat-ment	Out-come	Med 1 Vector	Med 2 Vector	Med 3 Vector		=LINEST(B2:B21,C2:E21,,TRUE)					Group Mean	
2	Med 1	153	1	0	0		-14.3	17.7	6.7	158.1		164.8	=J2+I2
3	Med 1	191	1	0	0		6.77	6.77	6.77	3.91		175.8	=J2+H2
4	Med 1	192	1	0	0		0.41	17.47	#N/A	#N/A		143.8	=J2+G2
5	Med 1	126	1	0	0		3.63	16	#N/A	#N/A		148.0	=J2-SUM(G2:I2)
6	Med 1	162	1	0	0		3323.4	4882.4	#N/A	#N/A			
7	Med 2	165	0	1	0								
8	Med 2	168	0	1	0		Anova: Single Factor						
9	Med 2	175	0	1	0								
10	Med 2	189	0	1	0		SUMMARY						
11	Med 2	182	0	1	0		*Groups*	*Count*	*Sum*	*Average*	*Variance*		
12	Med 3	150	0	0	1		Med 1	5	824	164.8	769.7		
13	Med 3	132	0	0	1		Med 2	5	879	175.8	97.7		
14	Med 3	123	0	0	1		Med 3	5	719	143.8	241.7		
15	Med 3	155	0	0	1		Control	5	740	148.0	111.5		
16	Med 3	159	0	0	1								
17	Control	164	-1	-1	-1								
18	Control	141	-1	-1	-1		ANOVA						
19	Control	144	-1	-1	-1		*Source of Variation*	*SS*	*df*	*MS*	*F*	*P-value*	*F crit*
20	Control	138	-1	-1	-1		Between Groups	3323.4	3	1107.8	3.63	0.04	3.238871517
21	Control	153	-1	-1	-1		Within Groups	4882.4	16	305.15			
22													
23							Total	8205.8	19				

Figure 7.9 has the same data set as Figure 7.6, except that the Control group has the value –1 throughout the three coded vectors, instead of 0 as in dummy coding. Some of the LINEST() results are therefore different than in Figure 7.6. The regression coefficients in Figure 7.9 differ from those in Figure 7.6, as do their standard errors. All the remaining values are the same: R^2, the standard error of estimate, the F-ratio, the residual degrees of freedom, and the regression and residual sums of squares—all those remain the same, just as they do with the third method of coding that this chapter considers, planned orthogonal contrasts.

In dummy coding, the constant returned by LINEST() is the mean of the group that's assigned 0's throughout the coded vectors—usually a control group. In effect coding, the constant is the grand mean. The constant is easy to find. It's the value in the first row (along with the regression coefficients) and in the rightmost column of the LINEST() results.

Because the constant equals the grand mean, it's easy to calculate the group means from the constant and the regression coefficients. Each coefficient, as I mentioned at the start of this section, represents the difference between the associated group's mean and the grand mean. So, to calculate the group means, add the constant to the regression coefficients. That's been done in Figure 7.9, in the range L2:L4. The formulas used in that range are given as text in M2:M4.

Notice that the three formulas add the constant (the grand mean) to a regression coefficient (a measure of the effect of being in that group, the distance of the group mean above or below the grand mean). The fourth formula in L5 is specific to the group assigned codes of −1, and it subtracts the other coefficients from the grand mean to calculate the mean of that group.

Also notice that the results of the formulas in L2:L5 equal the group means reported in the range J12:J15 by the Data Analysis add-in's ANOVA: Single Factor tool. It's also worth verifying that the F-ratio, the residual degrees of freedom, and the regression and residual sums of squares equal those reported by that tool in the range H20:K21.

Relationship to the General Linear Model

The general linear model is a useful way of conceptualizing the components of a value on an outcome variable. Its name makes it sound a lot more forbidding than it really is. Here's the general linear model in its simplest form:

$$Y_{ij} = \mu + \alpha_j + \varepsilon_{ij}$$

The formula uses Greek instead of Roman letters to emphasize that it's referring to the population from which observations are sampled, but it's equally useful to consider that it refers to a sample taken from that population:

$$Y_{ij} = \overline{Y} + a_j + e_{ij}$$

The idea is that each observation Y_{ij} can be considered as the sum of three components:

- The grand mean, μ
- The effect of treatment j, α_j
- The quantity ε_{ij} that represents the deviation of an individual score Y_{ij} from the combination of the grand mean and the jth treatment's effect.

Here it is in the context of a worksheet (see Figure 7.10):

Figure 7.10
Observations broken down in terms of the components of the general linear model.

H2 =SUMSQ(F2:F21)

	A	B	C	D	E	F	G	H
1	Treat-ment	Outcome		Grand Mean	Treatment Effect	Error		Sum of Squared Errors
2	Med 1	153		158.1	6.7	-11.8		4882.4
3	Med 1	191		158.1	6.7	26.2		
4	Med 1	192		158.1	6.7	27.2		
5	Med 1	126		158.1	6.7	-38.8		
6	Med 1	162		158.1	6.7	-2.8		
7	Med 2	165		158.1	17.7	-10.8		
8	Med 2	168		158.1	17.7	-7.8		
9	Med 2	175		158.1	17.7	-0.8		
10	Med 2	189		158.1	17.7	13.2		
11	Med 2	182		158.1	17.7	6.2		
12	Med 3	150		158.1	-14.3	6.2		
13	Med 3	132		158.1	-14.3	-11.8		
14	Med 3	123		158.1	-14.3	-20.8		
15	Med 3	155		158.1	-14.3	11.2		
16	Med 3	159		158.1	-14.3	15.2		
17	Control	164		158.1	-10.1	16.0		
18	Control	141		158.1	-10.1	-7.0		
19	Control	144		158.1	-10.1	-4.0		
20	Control	138		158.1	-10.1	-10.0		
21	Control	153		158.1	-10.1	5.0		

In Figure 7.10, each of the 20 observations in Figure 7.9 have been broken down into the three components of the general linear model: the grand mean in the range D2:D21, the effect of each treatment group in E2:E21, and the so-called "error" involved with each observation.

NOTE

The term *error* is used for some not especially good historical reasons, and it's made its way into other terms such as *mean square error* and even the symbol ∈. There's nothing erroneous about these values. *Residuals* is a perfectly descriptive term that isn't misleading, but statistical jargon tends to prefer *error*.

If you didn't expect that one or more treatments would have an effect on the subjects receiving that treatment, then your best estimate of the value of a particular observation would be the grand mean (in this case, that's 158.1).

But suppose you expected that the effect of a treatment would be to raise the observed values for the subjects receiving that treatment above, or lower them below, the grand mean. In that case your best estimate of a given observation would be the grand mean plus the effect, whether positive or negative, associated with that treatment. In the case of, say, the observation in row 5 of Figure 7.10, your expectation would be 158.1 + 6.7, or 164.8. If you give the matter a little thought, you'll see why that figure, 164.8, must be the mean outcome score for the Med 1 group.

Although the mean of its group is your best expectation for any one of its members, most—typically all—of the members of a group will have a score on the outcome variable different from the mean of the group. Those quantities (differences, deviations, residuals, errors, or whatever you prefer to call them) are shown in the range F2:F21 as the result of subtracting the grand mean and the group's treatment effect from the actual observation. For example, the value in cell F2 is returned by this formula:

=B2-D2-E2

The purpose of a regression equation is to minimize the sum of the squares of those errors. When that's done, the minimized result is called the Residual Sum of Squares in the context of regression, and the Sum of Squares Within in the context of ANOVA.

Note the sum of the squared errors in cell H2. It's returned by this formula:

=SUMSQ(F2:F21)

The SUMSQ() function squares the values in its argument and totals them. That's the same value as you'll find in Figure 7.9, cells H21, the Sum of Squares Within Groups from the ANOVA, and H6, the residual sum of squares from LINEST(). As Figure 7.10 shows, the sum of squares is based on the mean deviations from the grand mean, and on the individual deviations from the group means.

It's very simple to move from dummy coding to effect coding. Rather than assigning codes of 0 throughout the coding vectors to one particular group, you assign codes of –1 to one of the groups—not necessarily a control group—throughout the vectors. If you do that in the key range used by VLOOKUP(), you need to make that replacement in only as many key range cells as you have vectors. You can see in Figure 7.9 that this has been done in the range C17:E21, which contains –1's rather than 0's. I assigned the –1's to the control group not because it's necessarily desirable to do so, but to make comparisons with the dummy coding used in Figure 7.6.

Regression analysis with a single factor and effect coding handles unequal group sizes accurately, and so does traditional ANOVA. Figure 7.11 shows an analysis of a data set with unequal group sizes.

Figure 7.11
The ANOVA: Single Factor tool returns the group counts in the range H11:H14.

L5 =J2-SUM(G2:I2)

	A	B	C	D	E	F	G	H	I	J	K	L	M
1	Treat- ment	Out- come	Med 1 Vector	Med 2 Vector	Med 3 Vector		=LINEST(B2:B23,C2:E23,,TRUE)					Group Mean	
2	Med 1	153	1	0	0		-12.98	18.26	6.39	158.41		164.80	=J2+I2
3	Med 1	191	1	0	0		5.72	6.00	6.37	3.61		176.67	=J2+H2
4	Med 1	192	1	0	0		0.44	16.59	#N/A	#N/A		145.43	=J2+G2
5	Med 1	126	1	0	0		4.76	18	#N/A	#N/A		146.75	=J2-SUM(G2:I2)
6	Med 1	162	1	0	0		3926.7	4952.6	#N/A	#N/A			
7	Med 2	165	0	1	0								
8	Med 2	168	0	1	0		Anova: Single Factor						
9	Med 2	175	0	1	0		SUMMARY						
10	Med 2	189	0	1	0		*Groups*	*Count*	*Sum*	*Average*	*Variance*		
11	Med 2	182	0	1	0		Med 1	5	824	164.80	769.70		
12	Med 2	181	0	1	0		Med 2	6	1060	176.67	82.67		
13	Med 3	160	0	0	1		Med 3	7	1018	145.43	174.29		
14	Med 3	132	0	0	1		Control	4	587	146.75	138.25		
15	Med 3	123	0	0	1								
16	Med 3	155	0	0	1								
17	Med 3	149	0	0	1		ANOVA						
18	Med 3	147	0	0	1		*Source of Variation*	*SS*	*df*	*MS*	*F*	*P-value*	*F crit*
19	Med 3	152	0	0	1		Between Groups	3926.7	3	1308.907	4.757	0.013	3.16
20	Control	164	-1	-1	-1		Within Groups	4952.6	18	275.1443			
21	Control	141	-1	-1	-1								
22	Control	144	-1	-1	-1		Total	8879.3	21				
23	Control	138	-1	-1	-1								

7

Notice that the regression equation returns as the regression coefficients the effect of being in each of the treatment groups. In the range L2:L5, the grand mean (which is the constant in the regression equation) is added to each of the regression coefficients to return the actual mean for each group. Compare the results with the means returned by the Data Analysis add-in in the range J11:J14.

Notice that the grand mean is the average of the group means, 158.41, rather than the mean of the individual observations, 158.6. This situation is typical of designs in which the groups have different numbers of observations.

Both the traditional ANOVA approach and the regression approach manage the situation of unequal group sizes effectively. But if you have groups with very discrepant numbers of observations *and* very discrepant variances, you'll want to keep in mind the discussion from Chapter 6 regarding their combined effects on probability estimates: If your larger groups also have the larger variances, your apparent tests will tend to be conservative. If the larger groups have the smaller variances, your apparent tests will tend to be liberal.

Multiple Comparisons with Effect Coding

Dummy coding largely defines the comparisons of interest to you. The fact that you choose dummy coding as the method of populating the vectors in the data matrix implies that you want to compare one particular group mean, usually that of a control group, with the other group means in the data set. The Dunnett method of multiple comparisons is often the method of choice when you've used dummy coding.

A more flexible method of multiple comparisons is called the Scheffé method. It is a post hoc method, meaning that you can use it after you've seen the results of the overall analysis and that you need not plan ahead of time what comparisons you'll make. The Scheffé method also enables you to make complex contrasts, such as the mean of two groups versus the mean of three other groups.

There's a price to that flexibility, and it's in the statistical power of the Scheffé method. The Scheffé will fail to declare comparisons as statistically significant that other methods would. That's a problem and it's a good reason to consider other methods such as planned orthogonal contrast (discussed later in this chapter).

To use the Scheffé method, you need to set up a matrix that defines the contrasts you want to make. See Figure 7.12.

Consider Contrast A, in the range I2:I6 of Figure 7.12. Cell I3 contains a 1 and cell I4 contains a –1; the remaining cells in I2:I6 contain 0's. The values in the matrix are termed *contrast coefficients*. You multiply each contrast coefficient by the mean of the group it belongs to. Therefore, I2:I6 defines a contrast in which the mean of Med 2 (coefficient of –1) is subtracted from the mean of Med 1 (coefficient of 1), and the remaining group means do not enter the contrast.

Similarly, Contrast B, in J2:J6, also contains a 1 and a –1, but this time it's the difference between Med 3 and Med 4 that's to be tested.

Figure 7.12
The matrix of contrasts defines how much weight each group mean is given.

L17 =SUMPRODUCT(I2:I6,I15:I19)

	A	B	C	D	E	F	G	H	I	J	K	L	M
1	Out-come	Treat-ment	Vector 1	Vector 2	Vector 3	Vector 4		Contrast	A	B	C	D	
2	29	Control	-1	-1	-1	-1		Control	0	0	0	1	
3	39	Control	-1	-1	-1	-1		Med 1	1	0	0.5	-0.25	
4	34	Control	-1	-1	-1	-1		Med 2	-1	0	0.5	-0.25	
5	31	Control	-1	-1	-1	-1		Med 3	0	1	-0.5	-0.25	
6	38	Control	-1	-1	-1	-1		Med 4	0	-1	-0.5	-0.25	
7	38	Control	-1	-1	-1	-1							
8	41	Control	-1	-1	-1	-1		=LINEST(A2:A51,C2:F51,,TRUE)					
9	42	Control	-1	-1	-1	-1		-2.96	3.94	0.14	5.44	43.26	
10	42	Control	-1	-1	-1	-1		1.52	1.52	1.52	1.52	0.76	
11	33	Control	-1	-1	-1	-1		0.43	5.38	#N/A	#N/A	#N/A	
12	47	Med 1	1	0	0	0		8.39	45	#N/A	#N/A	#N/A	
13	44	Med 1	1	0	0	0		969.32	1300.30	#N/A	#N/A	#N/A	
14	38	Med 1	1	0	0	0							
15	54	Med 1	1	0	0	0		Control	36.7				Critical
16	53	Med 1	1	0	0	0		Med 1	48.7		Contrast	Value	Value
17	52	Med 1	1	0	0	0		Med 2	43.4		A, Med 1 vs Med 2	5.3	7.7
18	54	Med 1	1	0	0	0		Med 3	47.2		B, Med 3 vs Med 4	6.9	7.7
19	40	Med 1	1	0	0	0		Med 4	40.3		C, 1 & 2 vs 3 & 4	2.3	5.46
20	53	Med 1	1	0	0	0					D, Control vs Meds	-8.2	6.10

More complex contrasts are possible, of course. Contrast C compares the *average* of Med 1 and Med 2 with the average of Med 3 and Med 4, and Contrast D compares Control with the average of the four medication groups.

A regression analysis of the effect-coded data in A2:F51 appears in H9:L13. An F-test of the full regression (which is equivalent to a test of the deviation of the R^2 value in H11 from 0.0) could be managed with this formula:

=F.DIST.RT(H12,4,I12)

It returns 0.00004, and something like 4 in 100,000 replications of this experiment would return an F-ratio of 8.39 or greater if there were no differences between the population means. So you move on to a multiple comparisons procedure to try to pinpoint the differences that bring about so large an F-ratio.

You can pick up the group means by combining the constant returned by LINEST() in cell L9 (which, with effect coding, is the mean of the group means) with the individual regression coefficients. For example, the mean of the Med 1 group is returned in cell I16 with this formula:

=K9+$L$9

The formula for the mean of the group assigned –1's throughout the coding matrix is just a little more complicated. It is the grand mean minus the sum of the remaining regression coefficients. So, the formula for the control group in cell I15 is:

=$L$9-SUM(H9:K9)

With the five group means established in the range I15:I19, you can apply the contrasts you defined in the range I2:L6 by multiplying each group mean by the associated contrast coefficient for that contrast. Excel's SUMPRODUCT() function is convenient for that: It multiplies the corresponding elements in two arrays and returns the sum of the products. Therefore, this formula in cell L17:

```
=SUMPRODUCT(I2:I6,I15:I19)
```

has this effect:

```
=I2*I15 + I3*I16 + I4*I17 + I5*I18 + I6*I19
```

which results in the value 5.3. The formula in cell L18 moves one column to the right in the contrast matrix:

```
=SUMPRODUCT(J2:J6,I15:I19)
```

and so on through the fourth contrast.

The final step in the Scheffé method is to determine a critical value that the contrast values in L17:L20 must exceed to be regarded as statistically significant. Here's the formula, which looks a little forbidding in Excel syntax:

```
=SQRT((5−1)*F.INV(0.95,4,I12))*SQRT(($I$13/$I$12)
 *(J2^2/10+J3^2/10+J4^2/10+J5^2/10+J6^2/10))
```

Here it is using more conventional notation:

$$\sqrt{(k - 1)F_{df1,df2}}\sqrt{MSR\sum C_j^2/n_j}$$

where:

- k is the number of groups.
- $F_{df1,df2}$ is the value of the F distribution at the alpha level you select, such as 0.05 or 0.01. In the Excel version of the formula just given, I chose the .05 level, using 0.95 as the argument to the F.INV() function because it returns the F-ratio that has, in this case, 0.95 of the distribution to its left. I could have used, instead, F.INV.RT(0.05,4,I12) to return the same value.
- MSR is the mean square residual from the LINEST() results, obtained by dividing the residual sum of squares by the degrees of freedom for the residual.
- C is the contrast coefficient. Each contrast coefficient is squared and divided by n_j, the number of observations in the group. The results of the divisions are summed.

The critical value varies across the contrasts that have different coefficients. To complete the process, compare the value of each contrast with its critical value. If the absolute value of the contrast exceeds the critical value, then the contrast is considered significant at the level you chose for the F value in the formula for the critical value.

In Figure 7.12, the critical values are shown in the range M17:M20. Only one contrast has an absolute value that exceeds its associated critical value: Contrast D, which contrasts the mean of the Control group with the average of the means of the four remaining groups.

I mentioned at the start of this section that the Scheffé method is at once the least statistically powerful and the most flexible of the multiple comparison methods. You might want to compare the results reported here with the results of planned orthogonal contrasts, discussed in the next section. Planned orthogonal contrasts are at once the most statistically powerful and the least flexible of the multiple comparison methods. When we get to the multiple comparisons in the next section, you'll see that the same data set returns very different outcomes.

Orthogonal Coding

A third useful type of coding, besides dummy coding and effect coding, is *orthogonal coding*. You can use orthogonal coding in both planned and post hoc situations. I'll be discussing planned orthogonal coding (also termed planned orthogonal *contrasts*) here, because this approach is most useful when you already know something about how your variables work, and therefore are in a position to specify in advance which comparisons you will want to make.

Establishing the Contrasts

Orthogonal coding (I'll explain the term *orthogonal* shortly) depends on a matrix of values that define the contrasts that you want to make. Suppose that you plan an experiment with five groups: say, four treatments and a control. To define the contrasts that interest you, you set up a matrix such as the one shown in Figure 7.13.

Figure 7.13
The sums of products in G9:G14 satisfy the condition of orthogonality.

D11 =D2*D5

	A	B	C	D	E	F	G
1	Contrast	Control	Med 1	Med 2	Med 3	Med 4	
2	A	0	1	-1	0	0	
3	B	0	0	0	1	-1	
4	C	0	0.5	0.5	-0.5	-0.5	
5	D	1	-0.25	-0.25	-0.25	-0.25	
6							
7							
8	Contrast Pair			Products of Coefficients			Sum of Products
9	AB	0	0	0	0	0	0
10	AC	0	0.5	-0.5	0	0	0
11	AD	0	-0.25	0.25	0	0	0
12	BC	0	0	0	-0.5	0.5	0
13	BD	0	0	0	-0.25	0.25	0
14	CD	0	-0.125	-0.125	0.125	0.125	0

In orthogonal coding, just defining the contrasts isn't enough. Verifying that the contrasts are orthogonal to one another is also necessary. One fairly tedious way to verify that is also shown in Figure 7.13. The range B9:F14 contains the products of corresponding coefficients for each pair of contrasts defined in B2:F5. So row 10 tests Contrasts A and C, and the coefficients in row 2 and row 4 are multiplied to get the products in row 10. For example, the formula in cell C10 is:

=C2*C4

In row 11, testing Contrast A with Contrast D, cell D11 contains this formula:

```
=D2*D5
```

Finally, total up the cells in each row of the matrix of coefficient products. If the total is 0, those two contrasts are orthogonal to one another. This is done in the range G9:G14. All the totals in that range are 0, so each of the contrasts defined in B2:F5 are orthogonal to one another.

Planned Orthogonal Contrasts Via ANOVA

Figure 7.14 shows how the contrast coefficients are used in the context of an ANOVA. I'm inflicting this on you to give you a greater appreciation of how much easier the regression approach makes all this.

Figure 7.14
The calculation of the t-ratios involves the group means and counts, the mean square within and the contrast coefficients.

C23 =T.DIST.2T(ABS(B23),$J$15)

	A	B	C	D	E	F	G	H	I	J	K	L	M	N
1		Control	Med 1	Med 2	Med 3	Med 4		Anova: Single Factor						
2		48	66	59	64	58								
3		58	63	60	74	60		SUMMARY						
4		53	57	58	76	57		*Groups*	*Count*	*Sum*	*Average*	*Variance*		
5		50	73	63	69	69		Control	10	557	55.7	21.8		
6		57	72	61	59	52		Med 1	10	677	67.7	36.7		
7		57	71	74	63	62		Med 2	10	624	62.4	29.2		
8		60	73	70	62	59		Med 3	10	662	66.2	29.7		
9		61	59	62	68	55		Med 4	10	593	59.3	27.1		
10		61	72	59	64	66								
11		52	71	58	63	55								
12								ANOVA						
13	Contrast	Control	Med 1	Med 2	Med 3	Med 4		*Source of Variation*	*SS*	*df*	*MS*	*F*	*P-value*	*F crit*
14	A	0	1	-1	0	0		Between Groups	969.32	4	242.33	8.4	3.83E-05	2.58
15	B	0	0	0	1	-1		Within Groups	1300.3	45	28.89556			
16	C	0	0.5	0.5	-0.5	-0.5								
17	D	1	-0.25	-0.25	-0.25	-0.25		Total	2269.62	49				
18														
19	Contrast	t-ratio	Prob of t-ratio											
20	A	2.20	0.03263											
21	B	2.87	0.00623											
22	C	1.35	0.18280											
23	D	-4.31	0.00009											

Figure 7.14 shows a new data set, laid out for analysis by the ANOVA: Single Factor tool. That tool has been run on the data, and the results are shown in H1:N17. The matrix of contrast coefficients, which has already been tested for orthogonality, is in the range B14:F17. Each of these is needed to compute the t-ratios that test the significance of the difference established in each contrast.

The formulas to calculate the t-ratios are complex. Here's the formula for the first contrast, Contrast A, which tests the difference between the mean of the Med 1 group and the Med 2 group:

```
=SUMPRODUCT(B14:F14,TRANSPOSE($K$5:$K$9))/
 SQRT($K$15*SUM(B14:F14^2/TRANSPOSE($I$5:$I$9)))
```

7

The formula must be array-entered using Ctrl+Shift+Enter. Here it is in general form, using summation notation:

$$t = \sum C_j\overline{X}_j / \sqrt{MSE \sum C_j^2 / n_j}$$

where:

- C_j is the contrast coefficient for the jth mean.
- $\overline{X}_j$ is the jth sample mean.
- MSE is the mean square error from the ANOVA table. If you don't want to start by running an ANOVA, just take the average of the sample group variances. In this case, MSE is picked up from cell K15, calculated and reported by the Data Analysis tool.
- n_j is the number of observations in the jth sample.

The prior two formulas, in Excel and summation syntax, are a trifle more complicated than they need be. They allow for unequal sample sizes. As you'll see in the next section, unequal sample sizes generally—not always—result in nonorthogonal contrasts. If you have equal sample sizes, the formulas can treat the sample sizes as a constant and simplify as a result.

Returning to Figure 7.14, notice the t-ratios and associated probability levels in the range B20:C23. Each of the t-ratios is calculated using the Excel array formula just given, adjusted to pick up the contrast coefficients for different contrasts.

The probabilities are returned by the T.DIST.2T() function, the non-directional version of the t-test. The probability informs you how much of the area under the t-distribution with 45 degrees of freedom is to the left of, in the case of Contrast A, −2.20 and to the right of +2.20. If you had specified alpha as 0.01 prior to seeing the data, you could reject the null hypothesis of no population difference for Contrast B and Contrast D. The probabilities of the associated t-ratios occurring by chance in a central t distribution are lower than your alpha level. The probabilities for Contrast A and Contrast C are higher than alpha and you must retain the associated null hypotheses.

Planned Orthogonal Contrasts Using LINEST()

As far as I'm concerned, there's a lot of work—and opportunity to make mistakes—involved with planned orthogonal contrasts in the context of the traditional ANOVA. Figure 7.15 shows how much easier things are using regression, and in an Excel worksheet that means LINEST().

Using regression, you still need to come up with the orthogonal contrasts and their coefficients. But they're the same ones needed for the ANOVA approach. Figure 7.15 repeats them, transposed from Figure 7.14, in the range I1:M6.

Figure 7.15
With orthogonal coding, the regression coefficients and their standard errors do most of the work for you.

I18 | =T.DIST.2T(ABS(I17),$J$13)

	A	B	C	D	E	F	G	H	I	J	K	L	M
1	Out- come	Treat- ment	V1	V2	V3	V4			Contrast	A	B	C	D
2	29	Control	0	0	0	1			Control	0	0	0	1
3	39	Control	0	0	0	1			Med 1	1	0	0.5	-0.25
4	34	Control	0	0	0	1			Med 2	-1	0	0.5	-0.25
5	31	Control	0	0	0	1			Med 3	0	1	-0.5	-0.25
6	38	Control	0	0	0	1			Med 4	0	-1	-0.5	-0.25
7	38	Control	0	0	0	1							
8	41	Control	0	0	0	1			=LINEST(A2:A51,C2:F51,,TRUE)				
9	42	Control	0	0	0	1		Contrast:	D	C	B	A	
10	42	Control	0	0	0	1			-6.56	2.3	3.45	2.65	43.26
11	33	Control	0	0	0	1			1.52	1.70	1.20	1.20	0.76
12	47	Med 1	1	0	0.5	-0.25			0.43	5.38	#N/A	#N/A	#N/A
13	44	Med 1	1	0	0.5	-0.25			8.39	45	#N/A	#N/A	#N/A
14	38	Med 1	1	0	0.5	-0.25			969.32	1300.30	#N/A	#N/A	#N/A
15	54	Med 1	1	0	0.5	-0.25							
16	53	Med 1	1	0	0.5	-0.25		Contrast:	D	C	B	A	
17	52	Med 1	1	0	0.5	-0.25		t-ratios	-4.31	1.35	2.87	2.20	
18	54	Med 1	1	0	0.5	-0.25		Prob	0.00009	0.18280	0.00623	0.03263	
19	40	Med 1	1	0	0.5	-0.25							
20	53	Med 1	1	0	0.5	-0.25			V1	V2	V3	V4	
21	52	Med 1	1	0	0.5	-0.25		V1	1.00				
22	40	Med 2	-1	0	0.5	-0.25		V2	0.00	1.00			
23	41	Med 2	-1	0	0.5	-0.25		V3	0.00	0.00	1.00		
24	39	Med 2	-1	0	0.5	-0.25		V4	0.00	0.00	0.00	1.00	

The difference with orthogonal coding and regression, as distinct from the traditional ANOVA approach shown in Figure 7.14, is that you use the coefficients to populate the vectors, just as you do with dummy coding (1's and 0's) and effect coding (1's, 0's, and –1's). Each vector represents a contrast and the values in the vector are the contrast's coefficients, each associated with a different group.

So, in Figure 7.15, Vector 1 in Column C has 0's for the Control group, 1's for Med 1, –1's for Med 2, and—although you can't see them in the figure—0's for Med 3 and Med 4. Those are the values called for in Contrast A, in the range J2:J6. Similar comments apply to vectors 2 through 4. The vectors make the contrast coefficients a formal part of the analysis.

The regression approach also allows for a different slant on the notion of orthogonality. Notice the matrix of values in the range I21:L24. It's a correlation matrix showing the correlations between each pair of vectors in columns C through F. Notice that each vector has a 0.0 correlation with each of the other vectors. They are independent of one another. That's another way of saying that if you plotted them, their axes would be at right angles to one another (*orthogonal* means *right angled*).

NOTE

As an experiment, I suggest that you try adding at least one case to at least one of the groups in columns A through F of the worksheet for Figure 7.15—it's in the workbook for this chapter, which you can download from quepublishing.com/title/9780789756558. For example, to add a case to the control group, insert cells in A12:F12 and put 0's in C12:E12 and a 1 in F12. Then rebuild the correlation matrix starting in cell I21, either entering the CORREL( ) functions yourself or running the Data Analysis add-in's Correlation tool on the data in columns C through F. Notice that the correlations involving vectors where you have changed the group count no longer equal 0.0. They're no longer orthogonal.

This effect has implications for designs with two or more factors and unequal group frequencies. A distinction is made between situations in which the treatments might be causally related to the unequal frequencies—differential experimental mortality by treatment—and inequality in group counts due to causes unrelated to the treatments.

Planned orthogonal contrasts have the greatest amount of statistical power of any of the multiple comparison methods. That means that planned orthogonal contrasts are more likely to identify true population differences than the alternatives (such as Dunnett and Scheffé). However, they require that you be able to specify your hypotheses in the form of contrasts before the experiment, and that you are able to obtain equal group sizes. If you add even one observation to any of the groups, the correlations among the vectors will no longer be 0.0, you'll have lost the orthogonality, and you'll need to resort to (probably) planned nonorthogonal contrasts, which, other things equal, are less powerful.

It's easy to set up the vectors using the general VLOOKUP() approach described earlier in this chapter. For example, this formula is used to populate Vector 1:

```
=VLOOKUP($B2,$I$2:$M$6,2,0)
```

It's entered in cell C2 and can be copied and pasted into columns D through F (you'll need to adjust the third argument from 2 to 3, 4 and 5). Then make a multiple selection of C2:F2 and drag down through the end of the Outcome values.

With the vectors established, array-enter this LINEST() formula into a five-row by five-column range:

```
=LINEST(A2:A51,C2:F51,,TRUE)
```

You now have the regression coefficients and their standard errors. The t-ratios—the same ones that show up in the range B20:B23 of Figure 7.14—are calculated by dividing a regression coefficient by its standard error. So the t-ratio in cell L17 of Figure 7.15 is returned by this formula:

```
=L10/L11
```

The coefficients and standard errors come back from LINEST() in reverse of the order that you would like, so the t-ratios are in reverse order, too. However, if you compare them to the t-ratios in Figure 7.14, you'll find that their values are precisely the same.

You calculate the probabilities associated with the t-ratios just as in Figure 7.14, using the T.DIST() function that's appropriate to the sort of research hypothesis (directional or nondirectional) that you would specify at the outset.

Factorial Analysis

One of the reasons that the development of the analysis of variance represents such a major step forward in the science of data analysis is that it provides the ability to study the simultaneous effects of two or more factors on the outcome variable. Prior to the groundwork that Fisher did with ANOVA, researchers were limited to studying one variable at a time, usually just two levels of that factor at a time.

This situation meant that researchers could not investigate the *joint* effect of two or more factors. For example, it may be that men have a different attitude toward a politician when they are over 50 years of age than they do earlier in their lives. Furthermore, it may be that women's attitude toward that politician do not change as a function of their age. If we had to study the effects of sex and age separately, we wouldn't be able to determine that a joint effect—termed an *interaction* in statistical jargon—exists.

But we can accommodate more than just one factor in an ANOVA—or, of course, in a regression analysis. When you simultaneously analyze how two or more factors are related to an outcome variable, you're said to be using *factorial analysis*.

And when you can study and analyze the effects of more than just one variable at a time, you get more bang for your buck. The costs of running an experiment are often just trivially greater when you study additional variables than when you study only one.

It also happens that adding one or more factors to a single factor ANOVA can increase its statistical power. In a single-factor ANOVA, variation in the outcome variable that can't be attributed to the factor gets tossed into the mean square residual. It can happen that such variation might be associated with another factor (or, as you'll see in the next chapter, a covariate). Then that variation could be removed from the mean square error—which, when decreased, increases the value of F-ratios in the analysis, thus increasing the tests' statistical power.

Excel's Data Analysis add-in includes a tool that accommodates two factors at once, but it has drawbacks. In addition to a problem I've noted before, that the results do not come back as formulas but as static values, the ANOVA: Two-Factor with Replication tool requires that you arrange your data in a highly idiosyncratic fashion, and it cannot accommodate unequal group sizes, nor can it accommodate more than two factors. Covariates are out.

If you use regression instead, you don't have to live with those limits. To give you a basis for comparison, let's look at the results of the ANOVA: Two-Factor with Replication tool.

> **NOTE** Excel's Data Analysis add-in also offers a tool called ANOVA: Two-Factor Without Replication. It is a fairly old-fashioned approach to analyzing what's often termed a *repeated measures design*. This is the last you'll hear of it in this book.

The ANOVA tool used in Figure 7.16 is helpful in that it returns the average and variance of the outcome variable, as well as the count, for each group in the design. My own preference would be to use a pivot table to report these descriptive statistics, because that's a live analysis and the table returned by the ANOVA tool is, again, static values. With a pivot table I can add, delete, or edit observations and have the pivot table update itself. With static values I have to run the ANOVA tool over again.

The ANOVA table at the end of the results shows a couple of features that don't appear in the Data Analysis add-in's Single Factor version. Notice that rows 27 and 28 show a Sample and a Column source of variation. The Column source of variation refers to sex: Values for males are in column B and values for females are in column C. The Sample data source refers to whatever variable has values that occupy different rows. In Figure 7.16, values for Med 1 are in rows 2 through 6, Med 2 in rows 7 through 11, and Med 3 in rows 12 through 16.

Figure 7.16
The ANOVA: Two-Factor with Replication tool will not run if different groups have different numbers of observations.

	A	B	C	D	E	F	G	H	I	J	K
1		Male	Female		Anova: Two-Factor With Replication						
2	Med 1	12	17								
3		15	16		SUMMARY	Male	Female	Total			
4		19	13		*Med 1*						
5		16	19		Count	5	5	10			
6		11	18		Sum	73	83	156			
7	Med 2	16	22		Average	14.6	16.6	15.6			
8		15	20		Variance	10.3	5.3	8.0			
9		17	22		*Med 2*						
10		22	16		Count	5	5	10			
11		18	16		Sum	88	96	184			
12	Med 3	18	24		Average	17.6	19.2	18.4			
13		24	23		Variance	7.3	9.2	8.0			
14		21	21		*Med 3*						
15		16	21		Count	5	5	10			
16		23	22		Sum	102	111	213			
17					Average	20.4	22.2	21.3			
18					Variance	11.3	1.7	6.7			
19					*Total*						
20					Count	15	15				
21					Sum	263	290				
22					Average	17.5	19.3				
23					Variance	14.3	10.2				
24											
25					ANOVA						
26					*Source of Variation*	*SS*	*df*	*MS*	*F*	*P-value*	*F crit*
27					Sample	162.47	2	81.23	10.81	0.000	3.403
28					Columns	24.3	1	24.3	3.23	0.085	4.260
29					Interaction	0.2	2	0.1	0.01	0.987	3.403
30					Within	180.4	24	7.52			
31											
32					Total	367.37	29				

You'll want to draw your own conclusions regarding the convenience of the data layout (required, by the way, by the Data Analysis tool) and regarding the labeling of the factors in the ANOVA table.

The main point is that both factors, Sex (labeled *Columns* in cell E28) and Medication (labeled *Sample* in cell E27), exist as sources of variation in the ANOVA table. Males' averages differ from females' averages, and that constitutes a source of variation. The three kinds of medication also differ from one another's averages—another source of variation.

There is also a third source labeled *Interaction*, which refers to the joint effect of the Sex and Medication variables. At the interaction level, groups are considered to constitute combinations of levels of the main factors: For example, Males who get Med 2 constitute a group, as do Females who get Med 1. Differences due to the combined main effects—not just Male compared to Female, or Med 1 compared to Med 3—are collectively referred to as the *interaction* between, here, Sex and Treatment.

The ANOVA shown in Figure 7.16 evaluates the effect of Sex as not significant at the .05 level (see cell J28, which reports the probability of an F-ratio of 3.23 with 1 and 24 degrees of freedom as 8.5% when there is no difference in the populations). Similarly, there is no significant difference due to the interaction of Sex with Treatment. Differences between the means of the six design cells (two sexes times three treatments) are not great enough to reject the null hypothesis of no differences among the six groups. Only the Treatment main effect is statistically significant. If, from the outset, you intended to use planned orthogonal contrasts to test the differences between specific means, you could do so now and enjoy the statistical power available to you. In the absence of such planning, you could use the Scheffé procedure, hoping that you wouldn't lose too much statistical power as a penalty for having failed to plan your contrasts.

Factorial Analysis with Orthogonal Coding

However, there's no reason that you couldn't use orthogonal coefficients in the vectors. You wouldn't do so on a post hoc basis to increase statistical power, because that requires you to choose your comparisons before seeing the results. However, with equal group sizes you could still use orthogonal codes in the vectors to make some of the computations more convenient. Figure 7.17 shows the data from Figure 7.16 laid out as a list, with vectors that represent the Sex and the Treatment variables.

The data set in Figure 7.17 has one vector in column D to represent the Sex variable. Because that factor has only two levels, one vector is sufficient to represent it. The data set also has two vectors in columns E and F to represent the Treatment factor. That factor has three levels, so two vectors are needed. Finally, two vectors representing the interaction between Sex and Treatment occupy columns G and H.

The interaction vectors are easily populated by multiplying the main effect vectors. The vector in column G is the result of multiplying the Sex vector by the Treatment 1 vector. The vector in column H results from the product of the Sex vector and the Treatment 2 vector.

Figure 7.17
The two-factor problem from Figure 7.16 laid out for regression analysis.

	A	B	C	D	E	F	G	H
1	Sex	Treat-ment	Out-come	Sex Vector	Treatment Vector 1	Treatment Vector 2	Sex by Treatment 1	Sex by Treatment 2
2	Male	Med 1	12	1	1	1	1	1
3	Male	Med 1	15	1	1	1	1	1
4	Male	Med 1	19	1	1	1	1	1
5	Male	Med 1	16	1	1	1	1	1
6	Male	Med 1	11	1	1	1	1	1
7	Male	Med 2	16	1	-1	1	-1	1
8	Male	Med 2	15	1	-1	1	-1	1
9	Male	Med 2	17	1	-1	1	-1	1
10	Male	Med 2	22	1	-1	1	-1	1
11	Male	Med 2	18	1	-1	1	-1	1
12	Male	Med 3	18	1	0	-2	0	-2
13	Male	Med 3	24	1	0	-2	0	-2
14	Male	Med 3	21	1	0	-2	0	-2
15	Male	Med 3	16	1	0	-2	0	-2
16	Male	Med 3	23	1	0	-2	0	-2
17	Female	Med 1	17	-1	1	1	-1	-1
18	Female	Med 1	16	-1	1	1	-1	-1
19	Female	Med 1	13	-1	1	1	-1	-1
20	Female	Med 1	19	-1	1	1	-1	-1
21	Female	Med 1	18	-1	1	1	-1	-1
22	Female	Med 2	22	-1	-1	1	1	-1
23	Female	Med 2	20	-1	-1	1	1	-1

The choice of codes in the Sex and Treatment vectors is made so that all the vectors will be mutually orthogonal. That's a different reason from the one used in Figure 7.15, where the idea is to specify contrasts that are of particular theoretical interest—the means of particular groups, and the combinations of group means, that you hope will inform you about the way that independent variables work together and with the dependent variable to bring about the observed outcomes.

But in Figure 7.17, the codes are chosen simply to make the vectors mutually orthogonal because it makes the subsequent analysis easier. The most straightforward way to do this is as follows.

1. Supply the first vector with codes that will contrast the first level of the factor with the second level, and ignore other levels. In Figure 7.17, the first factor has only two levels, so it requires only one vector, and the two levels exhaust the factor's information. Therefore, give one level of Sex a code of 1 and the other level a code of –1.
2. Do the same for the first level of the second factor. In this case the second factor is Treatment, which has three levels and therefore two vectors. The first level, Med 1, gets a 1 in the first vector and the second level, Med 2, gets a –1. All other levels, in this case Med 3, get 0's. This conforms to what was done with the Sex vector in Step 1.
3. In the second (and subsequent) vectors for a given factor, enter codes that contrast the first two levels with the third level (or the first three with the fourth, or the first four with the fifth, and so on). That's done in the second Treatment variable by assigning the code 1 to both Med 1 and Med 2, and –2 to Med 3. This contrasts the first two levels from the third. If there were other levels shown in this vector they would be assigned 0's.

The interaction vectors are obtained by multiplication of the main effect vectors, as described in the preceding steps. Now, Figure 7.18 shows the analysis of this data set.

Figure 7.18
The orthogonal vectors all correlate 0.0 with one another.

P18 =SUM(K18:O18)

	I	J	K	L	M	N	O	P
1			*Sex Vector*	*Treatment Vector 1*	*Treatment Vector 2*	*Sex by Treatment 1*	*Sex by Treatment 2*	
2		Sex Vector	1.0					
3		Treatment Vector 1	0.0	1.0				
4		Treatment Vector 2	0.0	0.0	1.0			
5		Sex by Tratment 1	0.0	0.0	0.0	1.0		
6		Sex by Treatment 2	0.0	0.0	0.0	0.0	1.0	
7								
8			=LINEST(C2:C31,D2:H31,,TRUE)					
9		0.00	-0.10	-1.43	-1.40	-0.90	18.43	
10		0.35	0.61	0.35	0.61	0.50	0.50	
11		0.51	2.74	#N/A	#N/A	#N/A	#N/A	
12		4.97	24	#N/A	#N/A	#N/A	#N/A	
13		186.97	180.40	#N/A	#N/A	#N/A	#N/A	
14								
15		Significance of F for full regression			0.005			
16								
17			**Sex Vector**	**Treatment Vector 1**	**Treatment Vector 2**	**Sex by Treatment 1**	**Sex by Treatment 2**	**Total R Squared**
18		R squared with Outcome	0.0661	0.1067	0.3355	0.0005	0.0000	0.51

In Figure 7.18, the correlation matrix in the range K2:O6 shows that the correlations between each pair of vectors is 0.0.

TIP

The Correlation tool in the Data Analysis add-in is a convenient way to create a matrix such as the one in K2:06.

The fact that all the correlations between the vectors are 0.0 means that the vectors share no variance. Because they share no variance, it's impossible for the relationships of two vectors with the outcome variable to overlap. Any variance shared by the outcome variable and, say, the first Treatment vector is unique to that Treatment vector. *When all the vectors are mutually orthogonal, there is no ambiguity about where to assign variance shared with the outcome variable.*

In the range J9:O13 of Figure 7.18 you'll find the results returned by LINEST() for the data shown in Figure 7.17. Not that it matters for present purposes, but the statistical significance of the overall regression is shown in cell M15, again using the F.DIST.RT() function. More pertinent is that the R^2 for the regression, 0.51, is found in cell J11.

The range K18:O18 contains the R^2 values for each coded vector with the outcome variable. Excel provides a function, RSQ(), that returns the square of the correlation between two variables. So the formula in cell K18 is:

```
=RSQ($C$2:$C$31,D2:D31)
```

Cell P18 shows the sum of the five R^2 values. That sum, 0.51, is identical to the R^2 for the full regression equation that's returned by LINEST() in cell J11. We have now partitioned the R^2 for the full equation into five constituents.

Figure 7.19 ties the results of the regression analysis back to the two-factor ANOVA in Figure 7.16.

Figure 7.19
Compare the result of using sums of squares with the using proportions of variance.

D8 =C7+C8

	A	B	C	D	E	F	G	H
1								
2		Treatment Vector 1	Treatment Vector 2	Sex Vector	Sex by Treatment 1	Sex by Treatment 2	Total R Squared	
3	R^2 with Outcome	0.1067	0.3355	0.0661	0.0005	0.0000	0.51	
4								
5								
6			R^2	Factor Total R^2	df	MS	F	
7		Treatment Vector 1	0.1067					
8		Treatment Vector 2	0.3355	0.4422	2	0.2211	10.81	
9		Sex Vector	0.0661	0.0661	1	0.0661	3.23	
10		Sex by Treatment 1	0.0005					
11		Sex by Treatment 2	0.0000	0.0005	2	0.0003	0.01	
12		Within	0.4911	0.4911	24	0.0205		
13								
14		ANOVA						
15		*Source of Variation*		*SS*	*df*	*MS*	*F*	*P-value*
16		Sample		162.47	2	81.23	10.81	0.000
17		Columns		24.3	1	24.3	3.23	0.085
18		Interaction		0.2	2	0.1	0.01	0.987
19		Within		180.4	24	7.52		
20								
21		Total		367.37	29			

In Figure 7.19 I have brought forward the R^2 values for the coded vectors, and the total regression R^2, from the range K18:P18 in Figure 7.18. Recall that these R^2 values are actually measures of the proportion of total variance in the outcome variable associated with each vector. So the total amount of variance explained by the coded vectors is 51%, and therefore 49% of the total variance remains unexplained. That 49% of the variance is represented by the mean square residual (or mean square within, or mean square error) component—the divisor for the F-ratios.

The three main points to take from Figure 7.19 are discussed next.

Unique Proportions of Variance

The individual R^2 values for each vector can simply be summed to get the R^2 for the total regression equation. The simple sum is accurate because the vectors are mutually orthogonal. Each vector accounts for a unique proportion of the variance in the outcome. Therefore there is no double-counting of the variance, as there would be if the vectors were correlated. The total of the individual R^2 values equals the total R^2 for the full regression.

Proportions of Variance Equivalent to Sums of Squares

Compare the F-ratios from the analysis in the range C7:G12, derived from proportions of variance, with the F-ratios from the ANOVA in B15:H21, derived from the sums of squares. The F-ratios are identical, and the conclusions that you would draw from each analysis: that the sole significant difference is due to the treatments, and no difference emerges as a function of either sex or the interaction of sex with treatment.

The proportions of variance bear the same relationships to one another as do the sums of squares. That's not surprising. It's virtually by definition, because each proportion of variance is simply the sum of squares for that component divided by the total sum of squares—a constant. All the proportions of variance, including that associated with mean square residual, total to 1.0, so if you multiply an individual proportion of variance such as 0.4422 in cell D8, by the total sum of squares (367.37 in cell D21), you wind up with the sum of squares for that component (162.47 in cell D16). Generally, proportions of variance speak for themselves, while sums of squares don't. If you say that 44.22% of the variance in the outcome variable is due to the treatments, I immediately know how important the treatments are. If you say that the sum of squares due to treatments is 162.47, I suggest that you're not communicating with me.

Summing Component Effects

The traditional ANOVA shown in B15:H21 of Figure 7.19 does not provide inferential information for each comparison. The sums of squares, the mean squares and the F-ratios are for the full factor. Traditional methods cannot distinguish, for example, the effect of Med 1 versus Med 2 from the effect of Med 2 versus Med 3. That's what multiple comparisons are for.

However, we can get an R^2 for each vector in the regression analysis. For example, the R^2 values in cells C7 and C8 are 0.1067 and 0.3355. Those proportions of variance are attributable to whatever comparison is implied by the codes in their respective vectors. As coded in Figure 7.17, Treatment Vector 1 compares Med 1 with Med 2, and Treatment Vector 2 compares the average of Med 1 and Med 2 with Med 3. Notice that if you add the two proportions of variance together and multiply by the total sum of squares, 367.37, you get the sum of squares associated with the Treatment factor in cell D16, returned by the traditional ANOVA.

The individual vectors can be tested in the same way as the collective vectors (one for each main effect and one for the interaction). Figure 7.20 demonstrates that more fine-grained analysis.

The probabilities in the range G7:G11 of Figure 7.20 indicate the likelihoods of obtaining an F-ratio as large as those in F7:F11 if the differences in means that are defined by the vectors' codes were all 0.0 in the population. So, if you had specified an alpha of 0.05, you could reject the null hypothesis for the Treatment 1 vector (Med 1 versus Med 2) and the Treatment 2 vector (the average of Med 1 and Med 2 versus Med 3). But if you had selected an alpha of 0.01, you could reject the null hypothesis for only the comparison in Treatment Vector 2.

Figure 7.20
The question of what is being tested by a vector's F-ratio depends on how you have coded the vector.

F7 | f_x =E7/$E$12

	Treatment Vector 1	Treatment Vector 2	Sex Vector	Sex by Treatment 1	Sex by Treatment 2	Total R Squared
R^2 with Outcome	0.1067	0.3355	0.0661	0.0005	0.0000	0.51

	R^2	df	MS	F	Prob of F
Treatment Vector 1	0.1067	1	0.1067	5.22	0.032
Treatment Vector 2	0.3355	1	0.3355	16.40	0.000
Sex Vector	0.0661	1	0.0661	3.23	0.085
Sex by Treatment 1	0.0005	1	0.0005	0.03	0.872
Sex by Treatment 2	0.0000	1	0.0000	0.00	1.000
Within	0.4911	24	0.0205		

Factorial Analysis with Effect Coding

Chapter 3, in the section titled "Partial and Semipartial Correlations," discussed how the effect of a third variable can be statistically removed from the correlation between two other variables. The third variable's effect can be removed from both of the other two variables (partial correlation) or from just one of the other two (semipartial correlation). We'll make use of semipartial correlations—actually, the squares of the semipartial correlations—in this section. The technique also finds broad applicability in situations that involve unequal numbers of observations per group.

Figure 7.21 shows how the orthogonal coding from Figure 7.17, used in Figures 7.18 through 7.20, has been changed to effect coding.

Figure 7.21
As you'll see, effect coding results in vectors that are not wholly orthogonal.

Sex	Treat-ment	Out-come	Sex Vector	Treatment Vector 1	Treatment Vector 2	Sex by Treatment 1	Sex by Treatment 2
Male	Med 1	12	1	1	0	1	0
Male	Med 1	15	1	1	0	1	0
Male	Med 1	19	1	1	0	1	0
Male	Med 1	16	1	1	0	1	0
Male	Med 1	11	1	1	0	1	0
Male	Med 2	16	1	0	1	0	1
Male	Med 2	15	1	0	1	0	1
Male	Med 2	17	1	0	1	0	1
Male	Med 2	22	1	0	1	0	1
Male	Med 2	18	1	0	1	0	1
Male	Med 3	18	1	-1	-1	-1	-1
Male	Med 3	24	1	-1	-1	-1	-1
Male	Med 3	21	1	-1	-1	-1	-1
Male	Med 3	16	1	-1	-1	-1	-1
Male	Med 3	23	1	-1	-1	-1	-1
Female	Med 1	17	-1	1	0	-1	0
Female	Med 1	16	-1	1	0	-1	0
Female	Med 1	13	-1	1	0	-1	0

7

In the Sex vector, Males are assigned 1's and Females are assigned –1's. With a two-level factor such as Sex, orthogonal coding is identical to effect coding.

The first Treatment vector assigns 1's to Med 1, 0's to Med 2, and –1's to Med 3. So Treatment Vector 1 contrasts Med 1 with Med 3. Treatment Vector 2 assigns 0's to Med 1, 1's to Med 2 and (again) –1's to Med 3, resulting in a contrast of Med 2 with Med 3. Thus, although they both provide tests of the Treatment variable, the two Treatment vectors define different contrasts than are defined by the orthogonal coding used in Figure 7.17.

Figure 7.22 displays the results of effect coding on the outcome variable, which has the same values as in Figure 7.17.

Figure 7.22
Vectors that represent different levels of a given factor are correlated if you use effect coding.

M19 =RSQ($C$2:$C$31,F2:F31-TREND(F2:F31,$D2:E31))

	J	K	L	M	N	O	P
1		*Sex Vector*	*Treatment Vector 1*	*Treatment Vector 2*	*Sex by Treatment 1*	*Sex by Treatment 2*	
2	Sex Vector	1.0					
3	Treatment Vector 1	0.0	1.0				
4	Treatment Vector 2	0.0	0.5	1.0			
5	Sex by Treatment 1	0.0	0.0	0.0	1.0		
6	Sex by Treatment 2	0.0	0.0	0.0	0.5	1.0	
7							
8			=LINEST(C2:C31,D2:H31,,TRUE)				
9	0.10	-0.10	-0.03	-2.83	-0.90	18.43	
10	0.71	0.71	0.71	0.71	0.50	0.50	
11	0.51	2.74	#N/A	#N/A	#N/A	#N/A	
12	4.97	24	#N/A	#N/A	#N/A	#N/A	
13	186.97	180.40	#N/A	#N/A	#N/A	#N/A	
14							
15		Significance of F for full regression		0.005			
16							
17		**Sex Vector**	**Treatment Vector 1**	**Treatment Vector 2**	**Sex by Treatment 1**	**Sex by Treatment 2**	**Total R Squared**
18	R squared with Outcome	0.0661	0.4422	0.1145	0.0001	0.0001	0.62
19	Adjusted R squared with Outcome	0.0661	0.4422	0.0000	0.0001	0.0004	0.51

Notice that not all the off-diagonal entries in the correlation matrix, in the range K2:O6, are 0.0. Treatment Vector 1 has a 0.50 correlation with Treatment Vector 2, and the two vectors that represent the Sex by Treatment interaction also correlate at 0.50. This is typical of effect coding, although it becomes evident in the correlation matrix only when a main effect has at least three levels (as does Treatment in this example).

The result is that the vectors are not all mutually orthogonal, and therefore we cannot simply add up each variable's R^2 to get the R^2 for the full regression equation, as is done in Figures 7.18 through 7.20. Furthermore, because the R^2 of the vectors do not represent unique proportions of variance, we can't simply use those R^2 values to test the statistical significance of each vector.

Instead, it's necessary to use squared semipartial correlations to adjust the R^2 values so that they *are* orthogonal, representing unique proportions of the variance of the outcome variable.

In Figure 7.22, the LINEST() analysis in the range J9:O13 returns the same values as the LINEST() analysis with orthogonal coding in Figure 7.18, *except* for the regression coefficients, the constant, and their standard errors. In other words, the differences between orthogonal and effect coding make no difference to the equation's R^2, its standard error of estimate, the F-ratio, the degrees of freedom for the residual, or the regression and residual sums of squares. This is not limited to effect and orthogonal coding. Regardless of the method you apply—dummy coding, for example—it makes no difference to the statistics that pertain to the equation generally. The differences in coding methods show up when you start to look at variable-to-variable quantities, such as a vector's regression coefficient or its simple R^2 with the outcome variable.

Notice the table of R^2 values in rows 18 and 19 of Figure 7.22. The R^2 values in row 18 are raw, unadjusted proportions of variance. They do not represent unique proportions shared with the outcome variable. As evidence of that, the totals of the R^2 values in rows 18 and 19 are shown in cells P18 and P19. The value in cell P18, the total of the unadjusted R^2 values in row 18, is 0.62, well in excess of the R^2 for the full regression reported by LINEST() in cell J11.

Most of the R^2 values in row 19, by contrast, are actually squared semipartial correlations. Two that should catch your eye are those in L19 and M19. Because the two vectors for the Treatment variable are correlated, the proportions of variance attributed to them in row 18 via the unadjusted R^2 values double-count some of the variance shared with the outcome variable. It's that double-counting that inflates the total of the R^2 values to 0.62 from its legitimate value of 0.51.

What we want to do is remove the effect of the vectors to the left of the second Treatment vector from the second Treatment vector itself. You can see how that's done most easily by starting with the R^2 in cell K19, and then following the trail of bread crumbs through cells L19 and M19, as follows:

Cell K19: The formula is as follows:

```
=RSQ($C$2:$C$31,D2:D31).
```

The vector in column D, Sex, is the leftmost variable in LINEST()'s X-value arguments. (Space limits prevent the display of column D in Figure 7.22, but it's visible in Figure 7.21 and, of course, in the downloaded workbook for Chapter 7.) No variables precede the vector in column D and so there's nothing to partial out of the Sex vector: We just accept the raw R^2.

Cell L19: The formula is:

```
=RSQ($C$2:$C$31,E2:E31-TREND(E2:E31,$D2:D31))
```

The fragment TREND(E2:E31,$D2:D31) predicts the values in E2:E31 (the first Treatment vector) from the values in the Sex vector (D2:D31). You can see from the correlation matrix at the top of Figure 7.22 that the correlation between the Sex vector and the first Treatment vector is 0.0. In that case, the regression of Treatment 1 on Sex predicts the mean of the Treatment 1 vector. The vector has equal numbers of 1's, 0's and –1's, so its mean is 0.0. In short, the R^2 formula subtracts 0.0 from the codes in E2:E31 and we wind up with the same result in L19 as we do in L18.

Cell M19: The formula is:

```
=RSQ($C$2:$C$31,F2:F31-TREND(F2:F31,$D2:E31))
```

Here's where the squared semipartial kicks in. This fragment:

```
TREND(F2:F31,$D2:E31)
```

predicts the values for the second Treatment vector, F2:F31, based on its relationship to the vectors in column D *and* column E, via the TREND() function. When those predicted values are subtracted from the actual codes in column F, via this fragment:

```
F2:F31-TREND(F2:F31,$D2:E31)
```

you're left with residuals: the values for the second Treatment vector in F2:F31 that have their relationship with the Sex and the first Treatment vector removed. With those effects gone, what's left of the codes in F2:F31 is unique and unshared with either Sex or Treatment Vector 1. As it happens, the result of removing the effects of the Sex and the Treatment 1 vectors eliminates the original relationship between the Treatment 2 vector and the outcome variable, leaving both a correlation and an R^2 of 0.0. The double-counting of the shared variance is also eliminated and, when the adjusting formulas are extended through O19, the sum of the proportions of variance in K19:O19 equals the R^2 for the full equation in cell J11.

The structure of the formulas that calculate the squared semipartial correlations deserves a little attention because it can save you time and headaches. Here again is the formula used in cell L19:

```
=RSQ($C$2:$C$31,E2:E31-TREND(E2:E31,$D2:D31))
```

The address $C$2:$C$31 contains the outcome variable. It is a fixed reference (although it might as well be treated as a mixed reference, $C2:$C31, because we won't intentionally paste it outside row 19). As we copy and paste it to the right of column L, we want to continue to point the RSQ function at C2:C31, and anchoring the reference to column C accomplishes that.

The other point to note in the RSQ() formula is the mixed reference $D2:D31. Here's what the formula changes to as you copy and paste it, or drag and drop it, from L19 one column right into M19:

```
=RSQ($C$2:$C$31,F2:F31-TREND(F2:F31,$D2:E31))
```

Notice first that the references to E2:E31 in L19 have changed to F2:F31, in response to copying the formula one column right. We're now looking at the squared semipartial correlation between the outcome variable in column C and the second Treatment vector in column F.

But the TREND() fragment shows that we're adjusting the codes in F2:F31 for their relationship to the codes in columns D *and* E. By dragging the formula on column to the right:

- $C$2:$C$31 remains unchanged. That's where the outcome variable is located.
- E2:E31 changes to F2:F31. That's the address of the second Treatment vector.
- $D2:D31 changes to $D2:E31. That's the address of the preceding predictor variables. We want to remove from the second Treatment vector the variance it shares with the preceding predictors: the Sex vector and the first Treatment vector.

By the time the formula reaches O19:

- $C$2:$C$31 remains unchanged.
- E2:E31 changes to H2:H31.
- $D2:D31 changes to $D2:G31.

The techniques I've outlined in this section become even more important in Chapter 8, where we take up the analysis of covariance (ANCOVA). In ANCOVA you use variables that are measured on an interval or ratio scale as though they were factors measured on a nominal scale. The idea is not just to make successive R^2 values unique, as discussed in the present section, but to equate different groups of subjects as though they entered the experiment on a common footing—in effect, giving random assignment an assist.

Statistical Power, Type I and Type II Errors

In previous chapters I have mentioned a topic termed *statistical power* from time to time. Because it is a major reason to carry out factorial analyses as discussed in this chapter, and to carry out the analysis of covariance as discussed in Chapter 8, it's important to develop a more thorough understanding of what statistical power is and how to quantify it.

On a purely conceptual level, statistical power refers to a statistical test's ability to identify the difference between two or more group means as genuine, when in fact the difference *is* genuine at the population level. You might think of statistical power as the sensitivity of a test to the difference between groups.

Suppose you're responsible for bringing a collection of websites to the attention of consumers who are shopping online. Your goal is to increase the number of hits that your websites experience; any resulting revenue and profit are up to the people who choose which products to market and how much to charge for them.

You arrange with the owner of a popular site for web searches to display links to 16 of your sites, randomly selected from among those that your company controls. The other randomly selected 16 of your sites will, for a month, get no special promotion.

Your intent is to compare the average number of hourly hits for the sites whose links get prominent display with the average number of hourly hits for the remaining sites. You decide to make a directional hypothesis at the 0.05 alpha level: Only if the specially promoted sites have a higher average number of hits, and only if the difference between the two groups of sites is so large that it could come about by chance only once in 20 replications of this trial, will you reject the hypothesis that the added promotion makes no difference to the hourly average number of hits.

Your data come in a month later and you find that your control group—the sites that received no special promotion—have an average of 45 hits each hour, and the specially promoted sites have an average hourly hit rate of 55. The standard error of the mean is 5. Figure 7.23 displays the situation graphically.

Figure 7.23
Both power and alpha can be thought of as probabilities and depicted as areas under a curve.

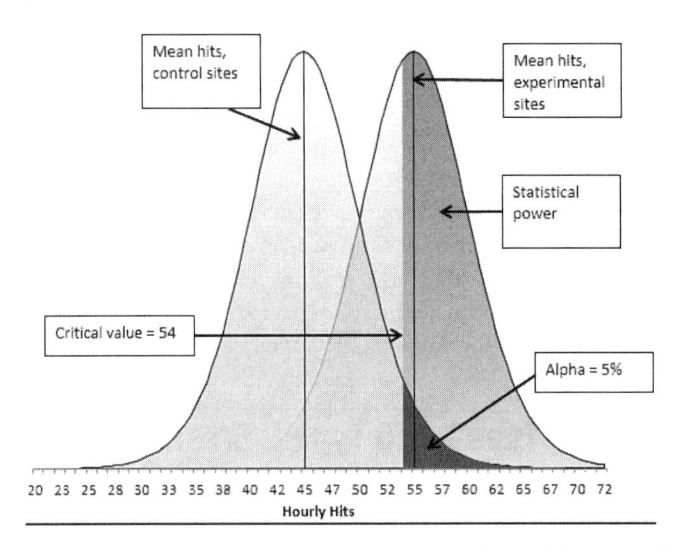

Assume that two populations exist: The first consists of websites like yours that get no special promotion. The second consists of websites that are promoted via links on another popular site, but that are otherwise equivalent to the first population. If you repeated your month-long study hundreds or perhaps thousands of times, you might get two distributions that look like the two curves in Figure 7.23.

The curve on the left represents the population of websites that get no special promotion. Over the course of a month, some of those sites—a very few—get as few as 25 hits per hour, and an equally small number get 62 hits per hour. The great majority of those sites average 45 hits per hour: the mode, mean and median of the curve on the left.

The curve on the right represents the specially promoted websites. They tend to get about 10 hits more per hour than the sites represented by the curve on the left. Their overall average is 55 hits per hour.

Now, most of this information is hidden from you. You don't have access to information about the full populations, just the results of the two samples you took—but that's enough.

Suppose that at the end of the month the two populations have the same mean, as would be the case if the extra promotion had no effect on the average hourly hits.

In that case, the difference in the average hit rate returned by your 16 experimental sites would have been due to nothing more than sampling error. That average of 55 hourly hits is among the averages in the right-hand tail of the curve on the left: the portion of the curve designated as *alpha*, shown in the chart in Figure 7.23 in a darker shade than the rest of the curve on the left.

Calculating Statistical Power

The boundary between alpha and the rest of the curve on the left is the critical value established by alpha. When you adopted 5% as your alpha level, with a directional hypothesis, you committed to the 5% of the right-hand tail of the curve. The critical value cuts off that 5%, and you can find that critical value using Excel's T.INV() function:

=T.INV(0.95,30)

That is, what is the value in the t distribution with 30 degrees of freedom that separates the lowest 95% of the values in the distribution from the top 5%? The result is 1.7. If you go up from the mean of the distribution by 1.7 standard errors, you account for the lowest 95% of the distribution. In this case the standard error is 5 (you learned that when you got the data on mean hourly hits), and 5 times 1.7 is 8.5. Add that to the mean of the curve on the left, and you get a critical value of 53.5.

In sum: The value of alpha is entirely under your control—it's *your* decision rule. You have made a directional hypothesis and you have set alpha to 0.05. Therefore, you have decided to reject the null hypothesis of no difference between the groups at the population level if, and only if, the experimental group's sample mean turns out to be at least 1.7 standard errors above the control group's mean.

Sometimes, the experimental group's mean will come from that right-hand tail of the left curve's distribution, just because of sampling error. Because the experimental group's mean, in that case, is at least 1.7 standard errors above the control group's mean, you'll reject the null hypothesis even though both populations have the same mean. That's Type I error, the probability of incorrectly rejecting a true null hypothesis.

Now suppose that in reality the populations are distributed as shown in Figure 7.23. If the sample experimental group has a mean at least 1.7 standard errors above the critical value of 54—which is 1.7 standard errors above the control group mean—then you'll *correctly* reject the null hypothesis of no difference at the population level.

Focus on the right curve in Figure 7.23. The area to the right of the critical value in that curve is the statistical power of your t-test. It is the probability that the experimental group mean comes from the curve on the right, in a reality where the two groups are distributed as shown at the population level.

Quantifying that probability is easy enough. Just take the difference between the critical value and the experimental group mean and divide by the standard error of 5:

=(54 – 55)/5

7

To get –0.2. That's a t-value. Evaluate it using the T.DIST() function:

=T.DIST(–0.2,15,TRUE)

using 15 as the degrees of freedom, because at this point we're working solely with the experimental group of 16 websites. The result is 0.422. That is, 42.2% of the area beneath the curve that represents the experimental group lies below the critical value of 54. Therefore 57.8% of the area under the curve lies to the right of the critical value, and the statistical power of the t-test is 57.8%. See Figure 7.24.

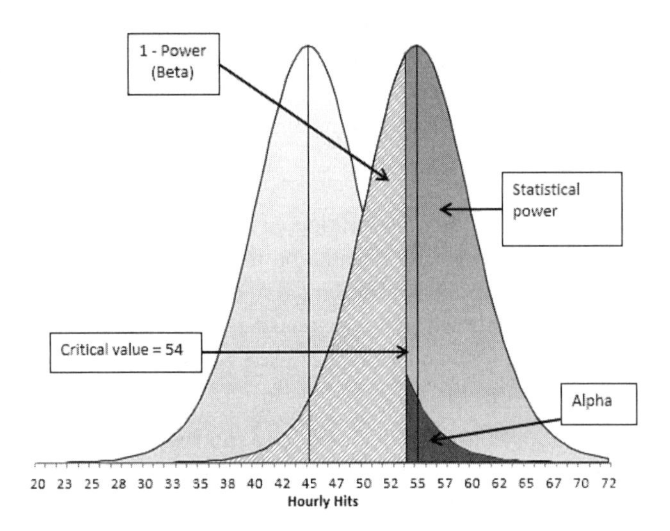

Figure 7.24
Type I error and alpha have counterparts in Type II error and beta.

In Figure 7.24 you can see the area that corresponds to statistical power in the curve on the right, to the right of the critical value. The remaining area under that curve is usually termed *beta*. It is alpha's counterpart.

If you incorrectly reject a true null hypothesis (for example, by deciding that two population means differ when in fact they don't), that's a Type I error and it has a probability of alpha. You decide the value of alpha, and your decision is typically based on the cost of making a Type I error, in the context of the benefits of correctly rejecting a false null hypothesis.

If you incorrectly reject a true alternative hypothesis (for example, by deciding that two population means are identical when in fact they differ), that's a Type II error and it has a probability of beta. The value of beta is not directly in your control. However, you can influence it, along with the statistical power of your test, as discussed in the next section.

Increasing Statistical Power

One excellent time to perform a power analysis is right after concluding a pilot study. At that point you often have the basic numbers on hand to calculate the power of a planned full study, and you're still in a position to make changes to the experimental design if the power study warrants. While a comparison of costs and benefits does not always argue for an increase in statistical power, it can warn you against pointless use of costly resources.

For example, if you can't get the estimated statistical power above 50%, you might decide that the study just isn't feasible—your odds of getting a reliable treatment effect are too low. Or it might turn out that increasing the sample size by 50% will result in an increase of only 5% in statistical power, so you're not getting enough bang for your buck.

You have available several methods of increasing statistical power. Some are purely theoretical, and have little chance of helping in real-world conditions. Others can make good sense.

One way is to reduce the size of the denominator of the test statistic. That denominator is typically a measure of the variability in the individual measures: a t-test, for example, might use either the standard error of the mean or the standard error of the difference between two means as the denominator of the t-ratio. An F-test uses the mean square residual (depending on the context, also known as mean square within or mean square error) as the denominator of the F-ratio.

When the denominator of a ratio decreases, the ratio itself increases. Other things equal, a larger t-ratio is more likely to be significant in a statistical sense than is a smaller t-ratio. One way to decrease the standard error or the mean square residual is to increase the sample size. Recall that the standard error of the mean divides the standard deviation by the square root of the sample size, and the mean square residual is the result of dividing the residual sum of squares by the residual degrees of freedom. In either case, increasing the sample size decreases the size of the t-ratio's or the F-ratio's denominator, which in turn increases the t-ratio or the F-ratio—improving the statistical power.

Another method of decreasing the size of the denominator is directly pertinent to factorial analysis, discussed in this chapter, and the analysis of covariance, discussed in Chapter 8. Both techniques add one or more predictors to the analysis: predictors that might have a substantial effect on the outcome variable. In that case, some of the variability in the individual measures can be attributed to the added factor or covariate and in that way kept out of the ratio's denominator.

So, adding a factor or covariate to the analysis might result in moving some of the variation out of the t-test's or the F-test's denominator and into the regression sum of squares (or the sum of squares between), thus increasing the size of the ratio and therefore its statistical power. Furthermore, and perhaps more importantly, adding the factor or the covariate could better illuminate the outcome of the study, particularly if two or more of the factors turn out to be involved in significant interactions.

You should also bear in mind three other ways to increase statistical power (neither of them directly related to the topics discussed in this chapter or in Chapter 8). One is to increase the treatment effect—the numerator of the t-ratio or the F-ratio, rather than its denominator. If you can increase the size of the treatment without also increasing the individual variation, your statistical test will be more powerful.

Consider making directional hypotheses ("one-tailed tests") instead of nondirectional hypotheses ("two-tailed tests"). One-tailed tests put all of alpha into one tail of the distribution. That moves the critical value toward the distribution's mean value. The closer the critical value is to the mean, the more likely you are to obtain an experimental result that exceeds the critical value—again, increasing the statistical power.

A related technique is to relax alpha. Notice in Figure 7.24 that if you increase (or *relax*) alpha from 0.05 to, say, 0.10, one result takes place in the distribution the right: the area representing statistical power increases as the critical value moves toward the mean of the curve on the left. By increasing the likelihood of making a Type I error, you reduce the likelihood of making a Type II error.

Coping with Unequal Cell Sizes

In Chapter 6 we looked at the combined effects of unequal group sizes and unequal variances on the nominal probability of a given t-ratio with a given degrees of freedom. You saw that the results can differ from the expected probabilities depending on whether the larger or the smaller group has the larger variance. You saw how Welch's correction can help compensate for unequal cell sizes.

Things are more complicated with more than just two groups (as in a t-test), particularly in factorial designs with two or more factors. Then, there are several—rather than just two—groups to compare as to both group size and variance.

In a design with at least two factors, and therefore at least four cells, several options exist, based primarily on the models comparison approach that is discussed at some length in Chapter 5. Unfortunately, these approaches do not have names that are generally accepted. Of the two discussed in this section, one is sometimes termed the *regression* approach and sometimes the *experimental design* approach; the other is sometimes termed the *sequential* approach and sometimes the *a priori ordering* approach. There are other terms in use. I use *experimental design* and *sequential* here.

The additional difficulty imposed by factorial designs when cell sizes are unequal concerns correlations between the vectors that define group membership: the 1's, 0's and –1's used in dummy, effect and orthogonal coding. Recall from earlier chapters that with equal cell sizes, the correlations between the vectors are largely 0.0. That feature means the sums of squares (and equivalently the variance) of the outcome variable can be assigned unambiguously to one vector or another.

But when the vectors are correlated, the unambiguous assignment of variability to a given vector becomes ambiguous. The vectors share variance with the outcome variable, of course. If they didn't there would be little point to retaining them in the analysis. But with unequal cell frequencies, the vectors share variance not only with the outcome variable but with one another, and in that case it's not possible to tell whether, say, 2% of the outcome variance belongs to Factor A, to Factor B, or to some sort of shared assignment such as 1.5% and 0.5%.

Despite the fact that I've cast this problem in terms of the vectors used in multiple regression analysis, the traditional ANOVA approaches are subject to the problem too.

But there's no single, generally applicable answer to the problem in the traditional framework either. The reliance there is typically on proportional (if unequal) cell frequencies and unweighted means analysis. Most current statistical packages use one of the approaches discussed here, or on one of their near relatives. (Not that it's representative of applications such as SAS or R, but Excel's Data Analysis add-in does not support designs with unequal cell frequencies in its 2-Factor ANOVA tool.)

Using the Regression Approach

This approach is also termed the *unique* approach because it treats each vector as though it were the last to enter the regression equation. If, say, a factor named Treatment is the last to enter the equation, all the other sources of regression variation are already in the equation and any variance shared by Treatment with the other vectors has already been assigned to them. Therefore, any remaining variance attributable to Treatment belongs to Treatment alone. It's unique to the Treatment vector.

Figure 7.25 shows an example of how this works.

Figure 7.25
This design has four cells with different numbers of observations.

H22 =SUM(H17:H20)

	A	B	C	D	E	F	G	H	I	J	K	L
1	Attitude	Sex	Affiliation	Sex by Affiliation		=LINEST(A2:A29,B2:B29,,TRUE)			=LINEST(A2:A29,B2:C29,,TRUE)			
2	2	-1	-1	1		0.1778	2.6222		-0.6753	0.2078	2.7273	
3	2	-1	-1	1		0.2120	0.2120		0.1602	0.1654	0.1671	
4	3	-1	-1	1		0.0263	1.0750		0.4310	0.8381	#N/A	
5	3	-1	-1	1		0.7033	26		9.4675	25	#N/A	
6	3	-1	-1	1		0.8127	30.0444		13.2987	17.5584	#N/A	
7	3	-1	-1	1								
8	3	-1	-1	1								
9	3	-1	-1	1		=LINEST(A2:A29,C2:C29,,TRUE)			=LINEST(A2:A29,B2:D29,,TRUE)			
10	1	-1	1	-1		-0.6667	2.6667		-0.5708	-0.8458	0.3042	2.7792
11	1	-1	1	-1		0.1618	0.1618		0.1256	0.1256	0.1256	0.1256
12	2	-1	1	-1		0.3951	0.8473		0.6943	0.6269	#N/A	#N/A
13	2	-1	1	-1		16.9796	26		18.1686	24	#N/A	#N/A
14	2	-1	1	-1		12.1905	18.6667		21.4238	9.4333	#N/A	#N/A
15	2	-1	1	-1								
16	3	-1	1	-1		Source	Formula	% of Var	DF	MS	F	p
17	3	-1	1	-1		Sex	=I4-F12	0.0359	1	0.03591	2.81951	0.1061
18	3	-1	1	-1		Affiliation	=I4-F4	0.4046	1	0.40464	31.76651	0.0000
19	3	-1	1	-1		Interaction	=I12-I4	0.2633	1	0.26331	20.67165	0.0001
20	4	1	-1	-1		Residual	=1-I12	0.3057	24	0.01274		
21	4	1	-1	-1								
22	5	1	-1	-1			Total	1.0096				
23	5	1	-1	-1								

The idea is to use the models comparison approach to isolate the variance explained uniquely by each variable. Generally, we want to assess the factors one by one, before moving on to their interactions. So the process with this design is to subtract the variance explained by each main effect from the variance explained by all the main effects.

For example, in Figure 7.25, the range F10:G14 returns LINEST() results for Attitude, the outcome variable, regressed onto Affiliation, one of the two factors. Affiliation explains 39.51% of the variability in Attitude when Affiliation is the only vector entered: see cell F12.

Similarly, the range I2:K6 returns LINEST() results for the regression of Attitude on Sex *and* Affiliation. Cell I4 shows that together, Sex and Affiliation account for 43.1% of the variance in Attitude.

Therefore, with this data set, we can conclude that 43.10% – 39.51%, or 3.59%, of the variability in the outcome measure is specifically and uniquely attributable to the Sex factor: the proportion attributable to the two main effects less that attributable to Affiliation. This finding, 3.59%, differs from the result obtained from the LINEST() analysis in F2:G6, where cell F4 tells us that Sex accounts for 2.63% of the variance in the outcome measure. The difference between 3.59% and 2.63% is due to the fact that the unequal cell frequencies induce correlations between the vectors, introducing ambiguity into how the variance is allocated to the vectors.

NOTE

It's worth noting that if the cell frequencies were equal, the proportions of variance attributable to each factor would be the same whether a factor was isolated by means of the models comparison approach or by including only the one factor in LINEST(). In terms of the example in Figure 7.25, if each of the four design cells had the same number of observations, the same proportion of variance would appear in both cells F4 and H17. The coded vectors would be orthogonal to one another, and it would make no difference whether a factor were the first or last to enter the regression equation: It would always account for the same proportion of variance.

The proportion of variance due to Affiliation is calculated in the same fashion. The proportion returned by LINEST() for the outcome measure regressed onto Sex, in cell F4, is subtracted from (once again) the proportion for both main effects, in cell I4. The result of that subtraction is 40.46%, a bit more than the 39.51% returned by the single factor LINEST() in cell F12.

After the two main effects, Sex and Affiliation, are assessed individually by subtracting their proportions of variance from the proportion accounted for by both, the analysis moves on to the interaction of the main effects. That's managed by subtracting the proportion of variance for the main effects, returned by LINEST() in cell I4, from the total proportion explained by the main effects and the interaction, returned by LINEST() in cell I12.

We can test the statistical significance of each main effect and the interaction in an ANOVA table, substituting proportions of total variance for sums of squares. That's done in the range F17:L20. Just divide the proportion of variance associated with each source of variation by its degrees of freedom to get a stand in for the mean square. Divide the mean square for each main effect and for the interaction by the mean square residual to obtain the F-ratio for each factor and for the interaction. The F-ratios are tested as usual with Excel's F.DIST.RT() function.

This analysis can be duplicated for sums of squares instead of proportions of variance, simply by multiplying each proportion by the sum of the squared deviations of the outcome variable from the grand mean.

Notice, by the way, in cell H22 that the proportions of variance do not total to precisely 100.00%, although the total is quite close. With unequal cell frequencies, even when managed by this unique variance approach, the total of the proportions of variance is not necessarily equal to exactly 100%. If you were working with sums of squares rather than proportions of variance, the factors' sums of squares do not necessarily add up precisely to the total sum of squares. Again, this is due to the adjustment of each factor and interaction vector for its correlation with the other vectors.

The sequential approach to dealing with unequal cell frequencies and correlated vectors, discussed in the next section, usually leads to a somewhat different outcome.

Sequential Variance Assignment

Bear in mind that the technique discussed in this section is just one of several methods for dealing with unequal cell frequencies in factorial designs. The unique assignment technique, described in the preceding section, is another such method. It differs from the sequential method in that it adjusts each factor's contribution to the regression sum of squares for that of the other factor or factors. In the sequential method, factors that are entered earlier are *not* adjusted for factors entered later.

Figure 7.26 shows how the sequential method works.

Figure 7.26
Consider using the sequential approach when one factor might have a causal effect on another.

H22 =SUM(H17:H20)

	A	B	C	D	E	F	G	H	I	J	K	L
1	Attitude	Sex	Affiliation	Sex by Affiliation		=LINEST(A2:A29,B2:B29,,TRUE)			=LINEST(A2:A29,B2:C29,,TRUE)			
2	2	-1	-1	1		0.1770	2.0222		-0.6753	0.2078	2.7273	
3	2	-1	-1	1		0.2120	0.2120		0.1602	0.1654	0.1671	
4	3	-1	-1	1		0.0263	1.0750		0.4310	0.8381	#N/A	
5	3	-1	-1	1		0.7033	26		9.4675	25	#N/A	
6	3	-1	-1	1		0.8127	30.0444		13.2987	17.5584	#N/A	
7	3	-1	-1	1								
8	3	-1	-1	1								
9	3	-1	-1	1		=LINEST(A2:A29,C2:C29,,TRUE)			=LINEST(A2:A29,B2:D29,,TRUE)			
10	1	-1	1	-1		-0.6667	2.6667		-0.5708	-0.8458	0.3042	2.7792
11	1	-1	1	-1		0.1618	0.1618		0.1256	0.1256	0.1256	0.1256
12	2	-1	1	-1		0.3951	0.8473		0.6943	0.6269	#N/A	#N/A
13	2	-1	1	-1		16.9796	26		18.1686	24	#N/A	#N/A
14	2	-1	1	-1		12.1905	18.6667		21.4238	9.4333	#N/A	#N/A
15	2	-1	1	-1								
16	3	-1	1	-1		Source	Formula	% of Var	DF	MS	F	p
17	3	-1	1	-1		Sex	=F4	0.0263	1	0.02634	2.06764	0.1634
18	3	-1	1	-1		Affiliation	=I4-F4	0.4046	1	0.40464	31.76651	0.0000
19	3	-1	1	-1		Interaction	=I12-I4	0.2633	1	0.26331	20.67165	0.0001
20	4	1	-1	-1		Residual	=1-I12	0.3057	24	0.01274		
21	4	1	-1	-1								
22	5	1	-1	-1			Total	1.0000				

Figure 7.26 includes only two changes from Figure 7.25, but they can turn out to be important. In the sequential analysis shown in Figure 7.26, the variability associated with the Sex variable is unadjusted for its correlation with the Affiliation variable, whereas that adjustment occurs in Figure 7.25. Compare cell H17 in the two figures.

Here's the rationale for the difference. In this data set, the subjects are categorized according to their sex and their party affiliation. It is known that, nationally, women show a moderate preference for registering as Democrats rather than as Republicans, and the reverse is true for men. Therefore, any random sample of registered voters will have unequal cell frequencies (unless the researcher takes steps to ensure equal group sizes, a dubious practice at best when the variables are not directly under experimental control). And with those unequal cell frequencies come the correlations between the coded vectors that we're trying to deal with.

In this and similar cases, however, there's a good argument for assigning all the variance shared by Sex and Affiliation to the Sex variable. The reasoning is that a person's sex might influence his or her political preference (mediated, no doubt, by social and cultural variables that are sensitive to a person's sex). The reverse idea, that a person's sex is influenced by his or her choice of political party, is absurd.

Therefore, it's arguable that variance shared by Sex and Affiliation is due to Sex and not to Affiliation. In turn, that argues for allowing the Sex variable to retain all the variance that it can claim in a single factor analysis, and not to adjust the variance attributed to Sex according to its correlation with Affiliation.

In that case you can use the entire proportion of variance attributable to Sex in a single factor analysis as its proportion in the full analysis. That's what has been done in Figure 7.26, where the formula in cell H17 is:

```
=F4
```

instead of this:

```
=I4-F12
```

in cell H17 of Figure 7.25. In that figure, the variance attributable to the Sex factor is adjusted by subtracting the variance attributable to the Affiliation factor (cell F12) from the variance attributable to both main effects (cell I4). But in Figure 7.26, the variance attributable to Sex in a single-factor analysis in cell F4 is used in cell H17, unadjusted for variance it shares with Affiliation. Again, this is because in the researcher's judgment any variance shared by the two factors belongs to the Sex variable as, to some degree, causing variability in the Affiliation factor.

The adjustment, or the lack thereof, makes no practical difference in this case: The variance attributable to Sex is so small that it will not approach statistical significance whether it is adjusted or not. But with a different data set, the decision to adjust one variable for another as in the unique variance approach, or to retain the first factor's full variance, as in the sequential approach, could easily make a meaningful difference.

Notice, by the way, in Figure 7.26 that the coded vectors have in effect been made orthogonal to one another. The total of the proportions of variance in the range H17:H20 now comes to 1.000, as shown in cell H22. That demonstrates that the overlap in variability has been removed by the decision *not* to adjust Sex's variance for its correlation with Affiliation.

Why not follow the sequential approach in all cases? Because you need a good, sound reason to treat one factor as causal with respect to other factors. In this case, the fact of causality is underscored by the patterns in the full population, and the logic of the situation argues for the directionality of the cause: that Sex causes Affiliation rather than the other way around.

Nevertheless, this is a case in which the subjects assign themselves to groups by being of one sex or the other and by deciding which political party to belong to. If the researcher were to selectively discard subjects in order to achieve equal group sizes, without regard to causality, he would be artificially imposing orthogonality on the variables. So doing alters the reality of the situation, and you therefore need to be able to show that causality exists and what its direction is.

This consideration does not tend to arise in true experimental designs, where the researcher is in a position to randomly assign subjects to treatments and conditions. Broccoli plants are not in a position to decide whether they prefer organic or inorganic fertilizer, or whether they flourish in sun or in shade.

Let's move on to Chapter 8 now, and look further into the effect of adding a covariate to your design.

The Analysis of Covariance

8

IN THIS CHAPTER

Contrasting the Results 297

Structuring a Conventional ANCOVA 308

Structuring an ANCOVA Using Regression 315

Checking for a Common Regression Line 316

Testing the Adjusted Means: Planned Orthogonal Coding in ANCOVA 321

ANCOVA and Multiple Comparisons Using the Regression Approach 328

Multiple Comparisons via Planned Nonorthogonal Contrasts 330

Multiple Comparisons with Post Hoc Nonorthogonal Contrasts 332

In previous chapters I've noted that it's possible to use both a factor, measured originally on a nominal scale, and a *covariate*, measured on an interval scale, as predictors in the same regression equation. For example, suppose that you assign members of a group of 20 people who want to lose weight to one of two groups: one group of 10 is assigned supervised exercises and a particular diet, and the other group of 10 is assigned the exercises only. Your aim is to determine whether the exercise and diet group loses more weight than the exercise-only group, and, in that case, the amount of the difference.

Those are pretty small samples and even with random assignment to groups, you shouldn't be surprised to find a meaningful difference between the groups' mean weights at the outset of the study. You might get the weight of each subject before beginning the treatments and take the pretest to posttest weight loss as your dependent variable. But that opens you up to regression toward the mean (see Chapter 3, "Simple Regression"). The 20 people who want to lose weight are probably heavier than the adult mean weight, and at the end of your study you won't know how much of the weight loss is due to regression toward the mean and how much to the effects of the treatments. (This concern is typical any time that subjects are selected due to their deviation from the mean of some variable.)

This is just the sort of situation that the analysis of covariance, or ANCOVA, is suited for. With ANCOVA, you can use both the pre-treatment weight *and* the treatment factor as predictors in a regression equation. In so doing, you separate the effect of pre-treatment weight from the analysis of the treatment effects. As you'll see in this chapter,

ANCOVA enables you to adjust the pre-treatment weights so that each treatment group starts on the same footing. The post-treatment weights are adjusted accordingly. Your analysis of the adjusted post-treatment weights acts as though the treatment groups had started with the same average weight.

That feature of ANCOVA is termed its *bias reduction* function. ANCOVA is intended to reduce the bias in the comparison of group means when the groups do not start out on an equivalent basis. That basis is measured using a variable, the covariate, that's quantitatively related to the outcome variable.

The other principal function of ANCOVA is to increase statistical power. Recall from Chapter 7 that the F-test is one way to assess the reliability of a regression equation. The F-test calculates an F-ratio with this formula:

$$F = MS_{Regression}/MS_{Residual}$$

where MS stands for mean square. You calculate $MS_{Residual}$ with this formula:

$$MS_{Residual} = SS_{Residual}/df_{Residual}$$

When you add a covariate to a regression equation, some portion of the total sum of squares gets included in the $SS_{Regression}$ instead of in the $SS_{Residual}$. That reduces the $SS_{Residual}$ and, in turn, the $MS_{Residual}$. With a smaller $MS_{Residual}$ the F-ratio increases and it becomes more likely that you will reject a null hypothesis of no differences between group means. In effect, you reclassify variability in the outcome measure. In an ANOVA, without the covariate involved, the variability goes to the error term. In an ANCOVA, some variability that had been in the error term is assigned instead to the covariate, where it becomes part of the $SS_{Regression}$.

Although both these functions—bias reduction and increasing statistical power—can be useful, it's probably true that more meaningful improvements to the analysis come from increasing power than from adjustments that reduce bias. We'll take several looks at both in the course of this chapter.

Before moving to the first example, I want to clarify the meaning of the term *analysis of covariance*. Because of its similarity to *analysis of variance*, it's all too easy to conclude that you replace the variance with covariance. That is, because an ANOVA tests differences between means by partitioning the variance, it's reasonable to assume that an ANCOVA does the same thing with covariance.

That's not the case. Both ANOVA and ANCOVA aim at an F-test, which is based on the ratio of two estimates of the overall variance. ANCOVA does indeed analyze covariance, but not by forming the ratio of two covariances. ANCOVA analyzes covariance by quantifying the relationship of a covariate (such as pre-treatment weight) to an outcome variable (such as post-treatment weight). Quantifying that covariance typically enables you to reallocate some of the variance in the outcome variable from the residual variance to the effect of the covariate, thus increasing the F-test's statistical power. But the F-test is still the ratio of two variances.

Contrasting the Results

Let's take a look at a data set analyzed first by means of ANOVA and then by means of ANCOVA. See Figure 8.1.

Figure 8.1
In this case ANOVA returns a result that indicates the regression equation's results are not reliable.

	A	B	C	D	E	F	G	H	I	J	K	L
1	Treatment	Cognitive Outcome	Treatment Vector		=LINEST(B2:B21,C2:C21,,TRUE)				Strength	Puzzles		
2	Strength	52	1		2.9	75.3			52	61		
3	Strength	61	1		3.12	3.12			61	54		
4	Strength	83	1		0.05	13.95			83	63		
5	Strength	69	1		0.86	18			69	66		
6	Strength	79	1		168.20	3504.00			79	74		
7	Strength	77	1						77	63		
8	Strength	80	1		0.93	t-ratio			80	79		
9	Strength	85	1		0.36	Prob of t			85	80		
10	Strength	101	1						101	91		
11	Strength	95	1						95	93		
12	Puzzles	61	-1			Anova: Single Factor						
13	Puzzles	54	-1			SUMMARY						
14	Puzzles	63	-1			*Groups*	*Count*	*Sum*	*Average*	*Variance*		
15	Puzzles	66	-1			Strength	10	782	78.2	215.956		
16	Puzzles	74	-1			Puzzles	10	724	72.4	173.378		
17	Puzzles	63	-1									
18	Puzzles	79	-1			ANOVA						
19	Puzzles	80	-1			*Source of Variation*	*SS*	*df*	*MS*	*F*	*P-value*	*F crit*
20	Puzzles	91	-1			Between Groups	168.2	1	168.20	0.86	0.36	4.41
21	Puzzles	93	-1			Within Groups	3504	18	194.67			
22						Total	3672	19				

Suppose you're evaluating the results of a small pilot study of 20 people thought to be at risk of developing dementia. The study compares the relative effects of physical exercise designed to increase muscle strength with the effects of cognitive exercises such as crossword puzzles. The subjects have been randomly assigned to the two groups in order to equate them at the start of treatment. After three months, a measure of cognitive function has been taken, with the results, labeled Cognitive Outcome, shown in the range B2:B21 in Figure 8.1.

The data is laid out in A2:C21 for analysis by means of LINEST(), using effect coding. The LINEST() function itself is in E2:F6, with the regression coefficient associated with the treatment vector in cell E2. Cell E8 contains the t-ratio of the regression coefficient to its standard error, and cell E9 tests the reliability of that t-ratio. Cell E9 tells us that we can expect a difference between the two group means as large as this one 36% of the time, when the group means in the population are equal. Few would regard this finding as "statistically significant."

Just for easy comparison, Figure 8.1 also lays out the data for analysis by a conventional ANOVA. To comply with the requirements of the Data Analysis add-in's Single Factor ANOVA tool, I've repeated the outcome measures in the range I2:J11. The tool's results appear in the range F12:L22.

Notice that although I tested the difference between the group means using a *t*-ratio in cell E8, the probability of that t-ratio in cell E9 is identical to the probability of the ANOVA's *F*-ratio in cell K20. Also notice that the F-ratio reported by LINEST() in cell E5 is identical to the F-ratio returned by the ANOVA tool in cell J20.

And notice that the t-ratio in cell E8 is the square root of the F-ratios in cells E5 and J20. Recall from Chapter 4 that this is always true when exactly two group means are involved in the analysis.

Figure 8.2 shows what could happen when you add a covariate to the analysis. I've added each subject's age in years to the data set.

Figure 8.2
The ANCOVA returns a very different picture than the one shown in Figure 8.1.

F4: {=LINEST(B2:B21,C2:D21,,TRUE)}

	A	B	C	D
1	Treatment	Cognitive Outcome	Age	Treatment Vector
2	Strength	52	48	1
3	Strength	61	58	1
4	Strength	83	61	1
5	Strength	69	62	1
6	Strength	79	62	1
7	Strength	77	65	1
8	Strength	80	68	1
9	Strength	85	73	1
10	Strength	101	70	1
11	Strength	95	72	1
12	Puzzles	61	59	-1
13	Puzzles	54	62	-1
14	Puzzles	63	67	-1
15	Puzzles	66	67	-1
16	Puzzles	74	68	-1
17	Puzzles	63	59	-1
18	Puzzles	79	75	-1
19	Puzzles	80	75	-1
20	Puzzles	91	77	-1
21	Puzzles	93	78	-1

=LINEST(B2:B21,C2:D21,,TRUE)

F	G	H
6.94	1.68	-36.17
1.55	0.21	13.96
0.80	6.56	#N/A
34.21	17	#N/A
2941.41	730.79	#N/A

4.48	t-ratio
0.0003	Prob of t

0.75	Correlation, Age with Outcome

=LINEST(B2:B21,D2:D21,,TRUE)

F	G
2.90	75.30
3.12	3.12
0.05	13.95
0.86	18
168.20	3504.00

It's beside the point, but notice the arrangement of the data in the range B2:D21 of Figure 8.2. The syntax of the LINEST() function forces you to keep the predictor variables—here, Age and the Treatment vector—adjacent. If for some reason you had placed the Age variable in column M instead of column C, you would need to use something that Excel finds unacceptable such as this:

=LINEST(B2:B21,D2:D21,M2:M21,,TRUE)

instead of this:

=LINEST(B2:B21,C2:D21,,TRUE)

The LINEST() function expects to find a TRUE, a FALSE or nothing at all as its third argument and it will complain if instead you present it with a worksheet address.

NOTE Those of you who are Defined Names mavens will doubtless wonder what happens if you define a name that refers to nonadjacent ranges and use that, in place of the actual range addresses, in the argument list. Nice try, but LINEST( ) returns a #REF! error in response.

Back to the main point. The LINEST() results in F2:H6 indicate a much stronger outcome than shown in Figure 8.1. The values for the Cognitive Outcome and for the Treatment vector are the same, and the only change is the addition of the Age covariate, but the difference is considerable.

The R^2 for the regression equation that uses the covariate is 0.80—see cell F4. In the range F15:G19, where I have repeated the ANOVA without the covariate, the R^2 is only 0.05—see cell F17. So the regression equation with the covariate predicts the Cognitive Outcome variable much more accurately than does the Treatment vector alone.

From a more inferential perspective, the F-ratio in cell F5, for the ANCOVA, could occur by chance less than 0.01% of the time with 2 and 17 degrees of freedom when the R^2 is 0.0 in the population. By contrast, the F-ratio in cell F18, for the ANOVA, could occur by chance 36.49% of the time with 1 and 18 degrees of freedom when the R^2 is 0.0 in the population.

To get those probabilities, <0.01% and 36.49%, use the F.DIST.RT() worksheet function. For the ANCOVA analysis in F2:H6, use these arguments:

```
=F.DIST.RT(F5,2,G5)
```

and for the ANOVA in F15:G19, use these arguments:

```
=F.DIST.RT(F18,1,G18)
```

> **NOTE** The numerator of the F-ratio for the ANCOVA has two degrees of freedom: one for the Treatment vector and one for the covariate. The numerator of the F-ratio for the ANOVA has one degree of freedom only, for the Treatment vector.

However, we don't yet know whether to ascribe the more powerful regression equation from the ANCOVA to the covariate Age, or to the adjustment of the group means due to ANCOVA's reduction of bias, or both. The way to do that is via the models comparison approach, which you'll find in Figure 8.3.

Figure 8.3 shows the same data set as appears in Figures 8.1 and 8.2. It also shows the ANCOVA analysis obtained from LINEST() in the range F2:H6, just as in Figures 8.1 and 8.2. Another LINEST() analysis appears in F10:G14. It assesses the relationship between the covariate Age and the Cognitive Outcome variable. Any effect attributable to the treatments is left out of the analysis in F10:G14.

So we have two analyses of the outcome variable:

- A less restricted model in F2:H6, which analyzes the combined relationships of Age and Treatment with the outcome variable
- A more restricted model in F10:G14, which analyzes the relationship of Age only with the outcome

Figure 8.3
The models comparison approach generally subtracts a more restricted model from a less restricted model.

F10 {=LINEST(B2:B21,C2:C21,,TRUE)}

	A	B	C	D	E	F	G	H	I	J	K
1	Treatment	Cognitive Outcome	Age	Treatment Vector		=LINEST(B2:B21,C2:D21,,TRUE)					
2	Strength	52	48	1		6.94	1.68	-36.17			
3	Strength	61	58	1		1.55	0.21	13.96			
4	Strength	83	61	1		0.80	6.56	#N/A			
5	Strength	69	62	1		34.21	17	#N/A			
6	Strength	79	62	1		2941.41	730.79	#N/A			
7	Strength	77	65	1							
8	Strength	80	68	1							
9	Strength	85	73	1	=LINEST(B2:B21,C2:C21,,TRUE)						
10	Strength	101	70	1		1.38	-16.04				
11	Strength	95	72	1		0.28	18.95				
12	Puzzles	61	59	-1		0.57	9.40				
13	Puzzles	54	62	-1		23.53	18				
14	Puzzles	63	67	-1		2080.55	1591.65				
15	Puzzles	66	67	-1							
16	Puzzles	74	68	-1							
17	Puzzles	63	59	-1			Delta R²	df	MS	F	Prob of F
18	Puzzles	79	75	-1		Regression	0.23	1	0.234	20.03	0.0003
19	Puzzles	80	75	-1		Residual	0.20	17	0.012		
20	Puzzles	91	77	-1							
21	Puzzles	93	78	-1			0.20	=G6/(F6+G6)			
22							0.20	=1-F4			

We can begin the comparison of the two models in cell G18, which contains the difference in the overall R^2 values of the two regression equations:

=F4 – F12

That formula returns 0.23 in cell G18. The model that includes both Age and Treatment as predictors accounts for 23% more variability in the outcome variable than does the model that includes Age alone as a predictor. Because the only difference between the two models is the presence of Treatment as a predictor in the unrestricted regression equation, we can attribute that 23% to the difference between the treatments as measured by the outcome variable. That seems sizable, but it might not be big enough to rely on, given a sample of only 20 subjects. The remaining cells in the range G18:K19 address that issue.

Together, those cells constitute an analysis of variance of the difference between the unrestricted and the restricted models. For now, we work with proportions of variance (shortly, I'll reconstruct this analysis using the more conventional sums of squares). We have the difference between the two models' proportion of the regression sum of squares in cell G18.

The residual proportion of the variance is in cell G19, and is based on the proportion of variance explained in the full, unrestricted model. That proportion of explained variance, 0.80, is found in cell F4. To get the proportion of *un*explained variance, just subtract the explained proportion from 1: (1 – 0.80) or 0.20. Again, that's the residual proportion of the variance from the full model, and it's stored in cell G19.

NOTE

Getting the proportion of unexplained variance by subtracting R^2 from 1.0 is quick and easy. Another approach that's a little more explicit is to divide the residual sum of squares by the total sum of squares. In Figure 8.3 you could arrange that by this formula: =G6/(F6+G6). Both formulas are shown in G21:G22 so that you can compare them.

The degrees of freedom come next. For the proportion of variance explained by the regression, it's the regression degrees of freedom for the full model less the regression degrees of freedom for the restricted model. This calculation parallels that of the delta in the two values of R^2. In this example, the regression sum of squares for the full model has two degrees of freedom: one for the covariate and one for the Treatment vector. The regression sum of squares for the restricted model has one degree of freedom, for the covariate. The difference between the two, 2 – 1, is 1. So the difference in the proportions of explained variance has 1 degree of freedom.

The residual proportion of the explained variance, 0.20, has as many degrees of freedom as the sum of squares on which it's based. Although we calculate that proportion by subtracting the full model's R^2 from 1.0, that's a shortcut. The residual sum of squares divided by the total sum of squares for the full model also returns the proportion of unexplained variance, 0.20. The associated degrees of freedom is 17 (from cell G5, the residual degrees of freedom from the full model), so that's the degrees of freedom we use for the residual in the analysis shown in G18:K19.

The values labeled "MS" in cells I18:I19 are the proportions of variance in G18:G19 divided by the degrees of freedom in H18:H19. The results are not truly Mean Squares, but proportions of variance divided by degrees of freedom. However, because we don't have a term for that calculation, and because of the close conceptual relationship to actual mean squares, I've retained the terminology in cell I17.

We can now calculate an F-ratio that speaks to the question of the reliability of the difference between the unrestricted and restricted models. That's done, as usual, by dividing the "mean square" for the residual into the "mean square" for the regression. The result, in cell J18, is 20.03. Also, this formula:

```
=F.DIST.RT(J18,H18,H19)
```

returns 0.0003, indicating that we would get an F-ratio of 20.03 about 3 times in 10,000 if there were no difference between the unrestricted and the restricted models. In other words, 0.0003 is the F-ratio's probability if there were no difference between the treatments in the population after the covariate is accounted for.

I'll go into that clause "after the covariate is accounted for" in more detail shortly—it has to do with the adjustment of the treatment group means. First, though, it's useful to look at the models comparison shown in Figure 8.3 using sums of squares instead of proportions of variance. See Figure 8.4.

Figure 8.4
Using sums of squares leads to the same conclusions, with a little more work.

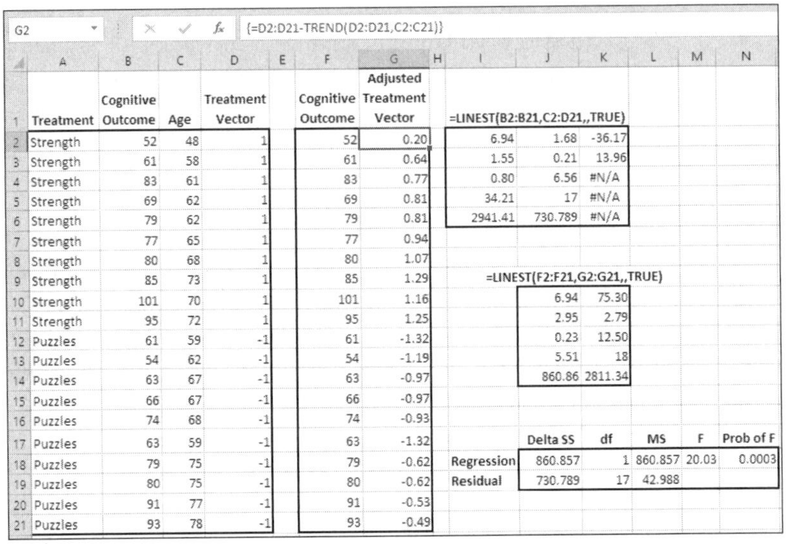

G2 {=D2:D21-TREND(D2:D21,C2:C21)}

	A	B	C	D	F	G
1	Treatment	Cognitive Outcome	Age	Treatment Vector	Cognitive Outcome	Adjusted Treatment Vector
2	Strength	52	48	1	52	0.20
3	Strength	61	58	1	61	0.64
4	Strength	83	61	1	83	0.77
5	Strength	69	62	1	69	0.81
6	Strength	79	62	1	79	0.81
7	Strength	77	65	1	77	0.94
8	Strength	80	68	1	80	1.07
9	Strength	85	73	1	85	1.29
10	Strength	101	70	1	101	1.16
11	Strength	95	72	1	95	1.25
12	Puzzles	61	59	-1	61	-1.32
13	Puzzles	54	62	-1	54	-1.19
14	Puzzles	63	67	-1	63	-0.97
15	Puzzles	66	67	-1	66	-0.97
16	Puzzles	74	68	-1	74	-0.93
17	Puzzles	63	59	-1	63	-1.32
18	Puzzles	79	75	-1	79	-0.62
19	Puzzles	80	75	-1	80	-0.62
20	Puzzles	91	77	-1	91	-0.53
21	Puzzles	93	78	-1	93	-0.49

=LINEST(B2:B21,C2:D21,,TRUE)

I	J	K
6.94	1.68	-36.17
1.55	0.21	13.96
0.80	6.56	#N/A
34.21	17	#N/A
2941.41	730.789	#N/A

=LINEST(F2:F21,G2:G21,,TRUE)

J	K
6.94	75.30
2.95	2.79
0.23	12.50
5.51	18
860.86	2811.34

	Delta SS	df	MS	F	Prob of F
Regression	860.857	1	860.857	20.03	0.0003
Residual	730.789	17	42.988		

Figure 8.4 removes the effect of the Age covariate from the Treatment vector. It does so via a different route than the one taken in Figure 8.3, which subtracted the R^2 due to the covariate from the full R^2 in the unrestricted model.

Figure 8.4 brings semipartial correlation to bear on the relationship between Treatment and Cognitive Outcome. Chapter 3 discusses the notion of semipartial correlation. If X and Y are correlated variables, you can get a simple semipartial correlation between X and Y by first removing from X the portion that can be predicted from Z.

The present case removes from the Treatment vector the portion of its values that can be predicted by the Age covariate. The range G2:G21 contains the key to this approach. That range contains this array formula:

```
=D2:D21–TREND(D2:D21,C2:C21)
```

Start with this fragment of the formula:

```
TREND(D2:D21,C2:C21)
```

The TREND() function is also discussed at length in Chapter 3. Here, it's used to predict values of the Treatment vector from the relationship of the actual Treatment vector values in D2:D21 with the values of the Age covariate in C2:C21. The result is a set of values that represent the portion of the Treatment vector that can be predicted from the Age covariate.

Then, that set of values is subtracted from the actual values in the Treatment vector using this array formula in G2:G21:

```
=D2:D21–TREND(D2:D21,C2:C21)
```

The result is the set of residual values that are left after subtracting predicted Treatment values from actual Treatment values.

The prediction-and-subtraction carried out by the array formula is just another way of expressing the adjustment in Figure 8.3. There, the process subtracts the R^2 of Age with Cognitive Outcome from the R^2 of Age *and* the Treatment vector with Cognitive Outcome. The result, 0.23 in cell G18 of Figure 8.3, is the proportion of variance shared by the Treatment vector with Cognitive Outcome.

By removing the portion of the Treatment vector's values that can be predicted by the Age covariate, you remove the portion of the R^2 that can be jointly shared by Age, the Treatment vector, and Cognitive Outcome. Notice that the removal is from the Treatment vector only, not from both the Treatment vector and the Cognitive Outcome variable. The analysis follows the logic of *semi*-partial correlation, rather than partial correlation, which would remove Age from both Treatment and Cognitive Outcome. Removal of common variance from the outcome variable would alter its distributional characteristics. This chapter examines how the group means on the outcome variable are altered by ANCOVA in later sections.

So, with the variability shared with Age removed from the Treatment vector, it's interesting to analyze Cognitive Outcome in terms of the Treatment vector's residuals.

That's done in the range J10:K14 of Figure 8.4, which contains the results of this LINEST() array formula:

```
=LINEST(F2:F21,G2:G21,,TRUE)
```

which of course calculates and analyzes the regression equation that predicts Cognitive Outcome, repeated in F2:F21, from the Treatment vector's residuals, in G2:G21. Bear in mind that the variance shared with Age has been removed from the equation by calculating the residual Treatment vector values in G2:G21.

Now we're in a position to construct another conventional ANOVA of the sort shown in the range G18:K19 of Figure 8.3, using sums of squares instead of proportions of variance. The ANOVA table in the range J18:N19 of Figure 8.4 is built as follows:

- The value in cell J18, 860.857, is the sum of squares for the regression of Cognitive Outcome on the *residuals* of the Treatment vector, after removing the portion of that vector that you can predict from Age. It is conceptually and computationally equivalent to the 0.23 proportion of total variability for the regression, in cell G18 of Figure 8.3. The metric in Figure 8.3 is proportion of total variance; the metric in Figure 8.4 is the sum of squares.
- The value in cell J19, 730.789, is the residual or error variance for the unrestricted model. The unrestricted LINEST() analysis is repeated in the range I2:K6 in Figure 8.4, and we pick up the value used in cell J19 from that analysis, in cell J6. It is conceptually and computationally equivalent to the 0.20 proportion of total variability for the residual of the unrestricted model, in cell G19 of Figure 8.3. The formulas in cells G21 and G22 of Figure 8.3 demonstrate this equivalence and show how you can reach the same value by subtracting R^2 from 1.0, and by expressing the residual sum of squares as a proportion of the total sum of squares.

- Both the regression and the residual degrees of freedom in Figure 8.4 are obtained as shown in Figure 8.3. The overall regression has 2 degrees of freedom, for the covariate and for the Treatment vector. The regression of the Cognitive Outcome on Age has 1 degree of freedom. So the difference between the two regression sums of squares has 2 − 1 = 1 degree of freedom, shown in cell K18. The residual sum of squares from the unrestricted regression in I2:K6 has 17 degrees of freedom, so we carry that along with the residual sum of squares into the ANOVA table, cell K19.
- The Mean Squares in L18 and L19 are as usual calculated by dividing a source's degrees of freedom into its sum of squares.
- Lastly, the F-ratio is calculated by dividing the Mean Square for the regression by the Mean Square for the residual. The probability associated with the F-ratio, given its degrees of freedom, is calculated as shown in Figure 8.3, via F.DIST.RT(M18,K18,K19).

It's important to see that the two approaches followed in Figures 8.3 and 8.4 result in identical outcomes. That they do so is demonstrated most clearly by the F-ratio: in both cases, 20.03. The approaches differ in the way that they obtain their sums of squares.

In Figure 8.3, the sums of squares in cells G18 and G19 aren't sums of squares at all but proportions of the total sum of squares. The proportion of the sum of squares for the regression is obtained by subtracting the R^2 for the outcome regressed on the covariate from the R^2 for the outcome regressed on both the covariate and the treatment vector.

Subtracting the R^2 for the covariate only from the R^2 for the covariate *and* the treatment leaves only the R^2 uniquely associated with the treatment. Keep in mind that regressing the outcome on the Treatment vector leaves in the equation only the variability that's shared by Treatment and the covariate. When the Treatment vector has a non-zero correlation with the covariate, the normal state of affairs in ANCOVA, the Treatment vector and the covariate *must* share variance, and the only way to quantify the variance uniquely associated with Treatment is to subtract the R^2 for the covariate alone from the R^2 for the model with both Treatment and the covariate. So doing removes all the variance shared by the outcome and the covariate, as well as the variance shared by the Treatment vector and the covariate.

In sum, we've shown the following thus far:

Figure 8.1 shows that the difference between the treatment group means on the outcome variable, when the covariate is not taken into account at all, is not what most people would regard as statistically significant.

Figure 8.2 shows that adding the covariate to the equation results in an F-ratio for the full equation that most people *would* regard as statistically significant. But the analysis has not yet disaggregated the effect of the treatment on the outcome variable from the effect of the covariate, so we're not yet in a position to assess the difference between the treatments as measured by the outcome variable.

Figures 8.3 and 8.4 perform that disaggregation by way of models comparison, removing the effect of the covariate via subtraction from the full model. Figure 8.3 manages that by subtracting one proportion of variance from another. Figure 8.4 manages it by regressing the treatment vector onto the covariate and subtracting the predicted treatment values from the actual values. The result is a set of residual treatment values onto which the outcome variable is regressed.

You can use either approach to models comparison. I prefer to use proportions of variance and the general approach taken in Figure 8.3 because I'm comfortable thinking in those terms, because I know what's going on behind the scenes, and because it's a little quicker. The approach shown in Figure 8.4 shows more explicitly what's happening when the variance shared by the covariate and the treatment vector is removed from the model. Clearly, because both approaches take you to the same endpoint, you can use either and should use the one that you feel more at home with.

ANCOVA Charted

Before discussing the problem of how group means are adjusted—that is, how ANCOVA carries out its bias reduction function—it will help to take a look at a chart of how the function takes place. It's easier to understand the specifics when they're viewed in context.

Figure 8.5 looks complex at first but if you'll follow the discussion in this section I think you'll find that it clarifies nicely.

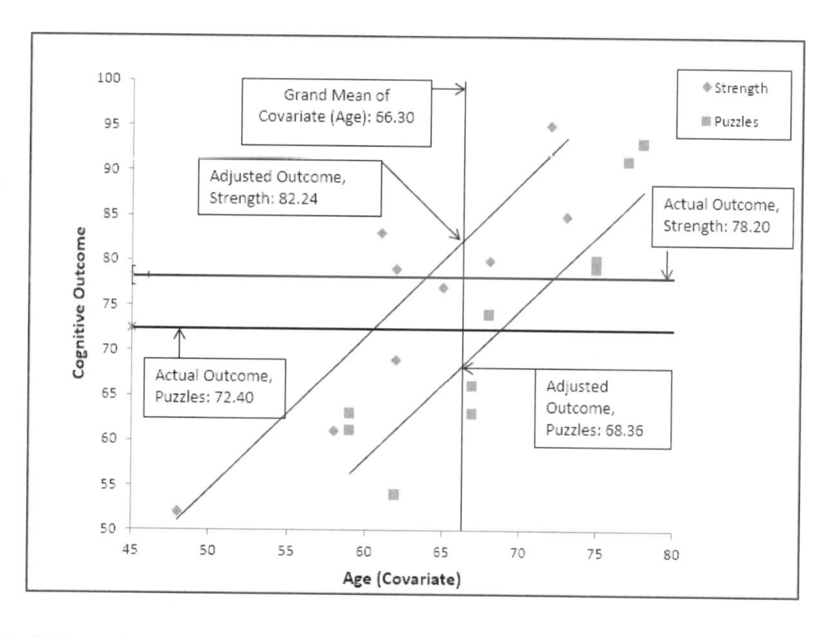

Figure 8.5
The chart projects where the two group means would show up on the outcome variable if they started out with the same mean on the covariate.

The idea behind bias reduction in ANCOVA is that if treatment groups start off differing on the covariate, they probably start off differing on whatever will be used as the outcome measure. And if the covariate is correlated with the outcome measure, then we can use our knowledge of the groups' means on the covariate to predict their values on the outcome measure.

Here's the basic equation:

$$\overline{Y}'_j = \overline{Y}_j - b(\overline{X}_j - \overline{X})$$

In words, for group *j*, the adjusted mean of the outcome measure ($\overline{Y}'_j$) equals the actual, observed mean of the outcome measure ($\overline{Y}_j$) minus the regression coefficient for the covariate (b) times the difference between the mean of group *j* on the covariate ($\overline{X}_j$) and the grand mean on the covariate ($\overline{X}$).

With that equation in mind, consider these possibilities:

- The groups each have the same mean on the covariate $\overline{X}_j$. If so, the difference between each group mean on the covariate and the grand mean on the covariate is 0.0. In that case, no adjustment of the group means on the outcome measure, ($\overline{Y}'_j$), takes place.
- The covariate is uncorrelated with the outcome measure. Then the regression coefficient for the covariate, b, equals 0.0 and again no adjustment of the outcome measure takes place.
- Suppose that the regression coefficient for the covariate, b, is positive. Suppose further that the mean on the covariate for the jth group is below the grand mean of the covariate. Then the quantity $b(\overline{X}_j - \overline{X})$ is negative. Because that negative value is subtracted from the observed mean of group *j* on the outcome variable ($\overline{Y}_j$), the result is to adjust the mean of group *j* upwards—compensating that group for the fact that it started out below the grand mean of the outcome variable.
- The reverse of that effect occurs if $\overline{X}_j$ is greater than the covariate's grand mean (and b is positive). In that case, $b(\overline{X}_j - \overline{X})$ is positive and the mean of group *j* on the outcome is adjusted down.

Of course, the prior two effects are reversed if the regression coefficient for the covariate b is negative.

Now, how does this equation for adjusting the group means pertain to what's going on in Figure 8.5? Start with the actual mean on the outcome variable for the Strength group, the subjects from Figures 8.1 through 8.4 whose treatment consists of physical exercise to improve muscle strength. Their mean on the outcome variable is 78.2 (see cell I15 in Figure 8.1). The upper horizontal line in Figure 8.5 represents the Strength group's actual mean value on the outcome. Notice that it intersects the Y-axis at 78.2.

The upper diagonal line in Figure 8.5 is the regression line for the outcome variable regressed onto the covariate. (This actually the *pooled* regression line, a distinction that this chapter will cover shortly.) Its slope, or b, or the regression coefficient for the covariate, is 1.681. It's the familiar rise over the run: For each unit of increase in the covariate on the X-axis, there's a concomitant increase of 1.681 on the Y-axis.

How does this work out in terms of the formula for adjusting the group means on the outcome variable? For the Strength group:

- The observed mean on the outcome variable is 78.2.
- The regression coefficient is 1.681.

- The observed mean on the covariate is 63.9. That's the point on the x-axis where the trendline for the Strength group crosses the horizontal line representing that group's average outcome.
- The grand mean on the covariate is 66.3.

So the adjusted mean for the Strength group on the outcome variable is:

$$78.2 - 1.681(63.9 - 66.3) = 82.24$$

Returning to Figure 8.5, run your eye along the regression line for the Strength group (the upper diagonal line). That regression line intersects the vertical line for the grand mean on the covariate at 82.24.

The same is true for the Puzzles group, the sample of subjects who are assigned mental puzzles to solve instead of physical exercises to complete. For the Puzzles group:

- The observed mean on the outcome variable is 72.4.
- The regression coefficient is 1.681.
- The observed mean on the covariate is 68.7. That's the point on the x-axis where the trendline for the Puzzles group crosses the horizontal line representing that group's average outcome.
- The grand mean on the covariate is 66.3.

So the adjusted mean for the Strength group on the outcome variable is:

$$72.4 - 1.681(68.7 - 66.3) = 68.36$$

Notice that the regression line for the Puzzles group—the lower diagonal line—intersects the vertical line for the covariate's grand mean at 68.36.

This analysis has adjusted the two group means on the outcome variable to estimates of what they would have been if the two groups had started out with the same mean age. Overall, age correlates with performance on the cognitive outcome in the 20 sampled subjects at 0.75, a substantial correlation. With a difference of 4.8 years in mean age, it's not surprising to find a substantial adjustment in the group means on the outcome measure. (You'll find the correlation and mean ages on the workbook for Chapter 8, on the worksheet corresponding to Figure 8.5.)

The Strength group is below the grand mean on the covariate, so with a positive b we expect to adjust the outcome measure up. The reverse is true for the Puzzles group: Their mean age is above the grand mean on the covariate, so we expect their mean value on the outcome to be adjusted down.

Therefore, because the Strength group has the higher observed mean on the outcome measure, the effect of the ANCOVA is to push the adjusted means on the outcome measure even farther apart. We saw earlier (Figure 8.2, for example) that adding the covariate to the analysis moves much of the residual sum of squares into the regression sum of squares.

That dramatically improves on the statistical power of the ANOVA. The increase in the difference between the adjusted means on the outcome measure makes the ANCOVA even more powerful.

You shouldn't expect this strong an effect of every ANCOVA. I deliberately structured the data to bring about the results discussed here in order to make the effect dramatic and easier to visualize on the chart in Figure 8.5. Still, differences that you see between your own data sets and the one used so far in this chapter are likely to be of degree, not of kind.

It's best to use ANCOVA in the context of random assignment to groups. If you take that approach with even a minimally reasonable sample size, you can expect the randomization to bring about group means on the covariate that are pretty close to one another. With a strong correlation between the covariate and the outcome measure, you're likely to get more benefit from a reduction in the residual variance than you will from the bias reduction due to adjusted group means.

Structuring a Conventional ANCOVA

This section walks you through the conventional method of preparing a conventional ANCOVA. I have two purposes in mind. One, the more minor of the two, is to provide a way to compare the results of the conventional approach with a regression approach that relies heavily on LINEST(). The more important purpose is to demonstrate how much easier it is to use regression, and how much easier it is to perceive what's going on in the analysis if you use regression instead of the conventional paint-by-numbers approach.

Analysis Without the Covariate

Let's start with the analysis in Figure 8.6.

Figure 8.6 contains a conventional ANOVA, similar to the one shown in the range F12:L22 of Figure 8.1. The two principal differences are:

- The factor in Figure 8.6 has three levels instead of two (as in Figure 8.1).
- The F-ratio in Figure 8.6 is highly unlikely under the null hypothesis of no differences between the group means. In contrast, the F-ratio in Figure 8.1 could come about by chance in 36% of repeated experiments when the null hypothesis is true.

For comparison, Figure 8.7 contains the same data set, laid out for analysis using LINEST() instead of the ANOVA tool in the Data Analysis add-in.

Notice that Figure 8.7 uses effect coding to represent group membership: Group 1 gets a 1 on the first vector and a 0 on the second; Group 2 gets a 0 on the first vector and a 1 on the second; Group 3 gets a –1 in both vectors. Choosing effect coding instead of some other method such as dummy coding or orthogonal coding makes a difference to the values of the regression coefficients, the intercept, and their standard errors, but makes no difference at all to the full equation's statistics such as the R^2, the standard error of estimate and the F-ratio.

Figure 8.6
In this example, the group means are significantly different *without* the use of a covariate.

	A	B	C	D	E	F	G	H	I	J	K
1	Group 1	Group 2	Group 3		Anova: Single Factor						
2	77	80	83								
3	80	79	85		SUMMARY						
4	80	82	82		*Groups*	*Count*	*Sum*	*Average*	*Variance*		
5	81	81	84		Group 1	6	473	78.83	2.97		
6	77	83	83		Group 2	6	488	81.33	2.67		
7	78	83	82		Group 3	6	499	83.17	1.37		
8											
9											
10					ANOVA						
11					*Source of Variation*	*SS*	*df*	*MS*	*F*	*P-value*	*F crit*
12					Between Groups	56.78	2	28.39	12.17	0.00	3.68
13					Within Groups	35	15	2.33			
14											
15					Total	91.78	17				

Figure 8.7
With three factor levels you need two vectors to code the group membership correctly.

I9 — *fx* =F.DIST.RT(F9,G9,H9)

	A	B	C	D	E	F	G	H	I	J
1	Group	Y	Group Vector 1	Group Vector 2		=LINEST(B2:B19,C2:D19,,TRUE)				
2	Group 1	77	1	0		0.22	-2.28	81.11		
3	Group 1	80	1	0		0.51	0.51	0.36		
4	Group 1	80	1	0		0.62	1.53	#N/A		
5	Group 1	81	1	0		12.17	15	#N/A		
6	Group 1	77	1	0		56.78	35.00	#N/A		
7	Group 1	78	1	0						
8	Group 2	80	0	1		F	DF1	DF2	Prob of F	
9	Group 2	79	0	1		12.17	2	15	0.00072	
10	Group 2	82	0	1						
11	Group 2	81	0	1			Grand Mean:	81.11	=AVERAGE(B2:B19)	
12	Group 2	83	0	1			Group 1 Mean:	78.83	=H11+G2	
13	Group 2	83	0	1			Group 2 Mean:	81.33	=H11+F2	
14	Group 3	83	-1	-1			Group 3 Mean:	83.17	=H11-(F2+G2)	
15	Group 3	85	-1	-1						
16	Group 3	82	-1	-1						
17	Group 3	84	-1	-1						
18	Group 3	83	-1	-1						
19	Group 3	82	-1	-1						

Just as a reminder, effect coding does bring about a result that proves handy from time to time: The regression coefficients express the difference between the associated group's mean and the grand mean. Thus, in Figure 8.7, the effect of being in Group 1 is to lower the mean of the group to 2.28 units below the grand mean (see cell G2). The effect of being in Group 2 is to raise that group's mean to 0.22 units above the grand mean (see cell F2), and the effect of being in Group 3 is to raise that group's mean to 2.06 units above the grand mean, returned by –(G2+F2).

Compare the F-ratio shown in Figure 8.6, cell I12, with that shown in Figure 8.7, cells F5 and F9. They are identical at 12.17. Regardless of the method of calculating the statistics in the analysis, the result is the same—a highly significant difference somewhere in the group means.

Analysis with the Covariate

Starting with Figure 8.8, the analysis, which now includes a covariate, becomes considerably more tedious and complicated. I include it here mainly to provide a sense of how much more straightforward it is to prepare an ANCOVA using regression methods (see the next section) than to do so using classic sums of squares and products of deviations. I certainly don't suggest that you submit to the regimen of the calculations used in Figure 8.8. But I've always found that some understanding of each approach helps to clarify the other.

Figure 8.8
Most of all this manipulation is to get a pooled or "common" regression coefficient.

I13 =F13/G8

	A	B	C	D	E	F	G	H	I	J
1	Group	Outcome	Covariate							
2	Group 1	77	18		Total R^2, covariate	0.8123	=RSQ(B2:B19,C2:C19)			
3	Group 1	80	23							
4	Group 1	80	18			SS Y	SS X	R^2 XY	SS'	
5	Group 1	81	22		Gp 1	14.83	22.83	0.31	10.31	
6	Group 1	77	19		Gp 2	13.33	26.83	0.78	2.98	
7	Group 1	78	21		Gp 3	6.83	18.83	0.80	1.35	
8	Group 2	80	23		Gp 1 + Gp 2 + Gp 3	35.00	68.50	0.57	14.63	
9	Group 2	79	20							
10	Group 2	82	24		Sum xy, Gp 1	10.17				
11	Group 2	81	21		Sum xy, Gp 2	16.67				
12	Group 2	83	25		Sum xy, Gp 3	10.17				
13	Group 2	83	26		Sum xy	37.00		Pooled beta:	0.5401	
14	Group 3	83	26							
15	Group 3	85	28		Source	Adjusted SS	df	Adjusted MS	F	Prob of F
16	Group 3	82	24		Between	2.21	2	1.11	1.03	0.3817
17	Group 3	84	27		Within	15.01	14	1.07		
18	Group 3	83	27		Total	17.23	16			
19	Group 3	82	23							
20					Source	Adjusted SS			Adjusted means	
21	Mean, Gp 1	78.83	20.17		Between	=F18-F17			80.39	
22	Mean, Gp 2	81.33	23.17		Within	=F8*(1-H8)			81.27	
23	Mean, Gp 3	83.17	25.83		Total	=DEVSQ(B2:B19)*(1-F2)			81.67	
24	Grand mean	81.11	23.06							

Bear in mind that the analysis in Figure 8.8 follows the approach presented in most classic texts on beginning-to-intermediate statistical analysis that were prepared before access to regression analysis became widely and cheaply available on personal computers. The classic approach relies on the accumulation of squared deviations and the products of deviations from the means of the outcome variable and the covariate.

The aim is to quantify and remove the effect of the covariate on the regression analysis, and *then* assess the effects of the treatments. To do so, it's necessary to calculate a statistic generally termed the *pooled beta* or the *common regression coefficient* between the covariate and the outcome measure.

The conventional approach begins with the figures needed to calculate beta *within* each level of the factor. The results are then summed and the sums are used to calculate the pooled beta. One of the formulas, used in simple linear regression, for beta is:

$$b = \Sigma xy/\Sigma x^2$$

where:

- b is the beta or regression coefficient.
- Σxy is the sum of the product of the deviations of each X (the predictor or covariate) from their mean and the deviations of each Y (the predicted or outcome measure) from their mean. The use of lowercase letters x and y instead of the capitals X and Y conventionally indicates deviations from a mean.
- Σx^2 is the sum of the squared deviations of the values of the covariate (or the predictor) from their mean.

However, the situation is more complicated in ANCOVA. We can't use a b calculated in exactly this way because, for example, a value of the covariate from one level of the factor would be deviated from the grand mean of the covariate rather than from the level's mean on the covariate. If the levels have different covariate means, as they almost certainly will, the result is a distortion of the pooled beta. The same argument applies to both the outcome measure and the covariate.

So, instead, we use this formula for a design with three factor levels:

$$b = (\Sigma xy_1 + \Sigma xy_2 + \Sigma xy_3)/(\Sigma x_1^2 + \Sigma x_2^2 + \Sigma x_3^2)$$

In words, get the sums of the products of the deviations of the covariate and the outcome variable from their own means in factor level 1, in factor level 2, and in factor level 3, and sum those sums. Do the same for the sums of the squared deviations of the covariate. Divide the former by the latter.

The result is the pooled beta or common regression coefficient for the covariate. Notice that for this data set the value of the pooled beta is .5401 (see cell I13). If you use LINEST() on the Outcome and the Covariate in this way:

```
=LINEST(B2:B19,C2:C19,,TRUE)
```

you'll find that you get a regression coefficient for the covariate of 0.67228. Obtaining a regression coefficient in this way, ignoring the differences between the groups, with this data, inflates the value of the pooled regression coefficient.

Getting the group-by-group statistics to calculate the pooled beta is the reason for the calculations in the ranges F5:I8 and F10:F13 of Figure 8.8. The calculations are discussed in the following sections.

Outcome Measure Sums of Squares, F5:F8

This range is labeled SS Y in Figure 8.8 because ANCOVA conventionally labels the outcome variable as Y (and the covariate as X). The formulas in F5:F7 are the sums of squared deviations in each of the three groups. So, the formula in F5 is:

=DEVSQ(B2:B7)

and similarly in cells F6 and F7, referencing the cells for Group 2 and Group 3 accordingly. Cell F8 totals the group-by-group sums of squares for the outcome measure in F5:F7.

Covariate Sums of Squares, G5:G8

This range, labeled SS X, performs exactly the same calculations as F5:F8, but on the covariate instead of the outcome measure.

R^2 for Covariate and Outcome, H5:H8

The range H5:H7 uses Excel's RSQ() function to return the R^2 value for the covariate and the outcome measure within each of the three groups or factor levels. For example, the formula in cell H5 is:

=RSQ(B2:B7,C2:C7)

The cell H8 takes a tack to calculating an R^2 that's similar to the one used by the pooled beta. The formula using sigma notation is:

$$R^2 = (\Sigma xy_1 + \Sigma xy_2 + \Sigma xy_3)/((\Sigma x_1^2 + \Sigma x_2^2 + \Sigma x_3^2)(\Sigma y_1^2 + \Sigma y_2^2 + \Sigma y_3^2))^2$$

The sum of the products of the deviations of the covariate and the outcome measure from their means is divided by the total sums of squares for the outcome measure times the total sums of squares of the covariate, and the denominator is squared. The result is one measure of the R^2 between the outcome measure and the covariate. It is used later in the ANCOVA to remove the effect of the covariate from the analysis so as to focus on the differences between the treatments.

Adjusted Regression Sum of Squares or SS', I5:I8

The range I5:I7 uses the within-cell R^2 values calculated in H5:H7 to remove the effect of the covariate from the within-cell sum of squares for the outcome variable. For example, the formula in cell I5 is:

=F5*(1–H5)

The value in cell F5 is the sum of squares of the outcome variable in Group 1, the first level of the factor. By subtracting the R^2 in cell H5 from 1.0, we get the proportion of variance in the outcome variable that is *not* shared by the covariate. So, the result of the formula in I5 is the sum of squares of the outcome variable in Group 1 that is unique to variability in the outcome measure, and not due to the relationship with the covariate.

Cells I6 and I7 do the same for Group 2 and Group 3, and cell I8 totals the results in I6:I7 to get a measure of the variability in the outcome measure across all three groups unaffected by the covariate.

Within-Cell Product of Deviations, F10:F13

In cells F10, F11, and F12 the ANCOVA calculates the sum of the products of the deviations of the covariate and the outcome measure from their respective group means. For example, the formula in cell F10 is:

```
=SUMPRODUCT(B2:B7–B21,C2:C7–C21)
```

Excel's SUMPRODUCT() function totals the products of the members of two arrays. As used in cell F10, the two arrays are:

- The deviations of the values in B2:B7 from their mean in cell B21.
- The deviations of the values in C2:C7 from their mean in cell C21.

So, SUMPRODUCT() multiplies the deviation of cell B2 from the mean in B21 times the deviation of cell C2 from the mean in cell C21. The product of the deviations is added to products calculated similarly for B3:B7 and C3:C7.

The same calculations are performed for Group 2 and Group 3 in cells F11 and F12. Then, the total of the SUMPRODUCT() results is taken in cell F13.

The result in cell F13 is used to calculate the pooled R^2 in cell H8 and the pooled beta in cell I13.

After all the work just described, you can calculate the pooled beta very quickly. Divide the quantity just calculated for cell F13 by the pooled sum of squares for the covariate in cell G8. The result, 0.5401, appears in cell I13.

We have now removed from the outcome variable the effects of its relationship with the covariate. We're ready to put together an ANOVA that tests the mean differences on the outcome variable with the covariate partialled out. Construct the ANOVA table in the range E16:J18 as follows (the formulas for the adjusted sums of squares are also given in the range F21:F23):

- The residual adjusted sum of squares in cell F17 after removing the covariate is obtained by this formula:

  ```
  =F8*(1–H8)
  ```

  Cell F8 contains the pooled sum of squares of the outcome measure. Multiply that by the proportion of variance in the outcome variable that is not shared with the covariate.
- The total adjusted sum of squares in cell F18, again after removing the covariate, is:

  ```
  =DEVSQ(B2:B19)*(1–F2)
  ```

where DEVSQ(B2:B19) is the sum of squares of the outcome variable and (1-F2) is the overall proportion of variance in the outcome that is not shared with the covariate.

- The adjusted sum of squares due to regression in cell F16 is obtained by subtracting the residual sum of squares in cell F17 from the total sum of squares in cell F18. You prefer not to calculate a critical value in this fashion. You would much prefer to calculate it independently, because doing it by subtraction can lead to the sort of error that for several years plagued LINEST()'s calculation of the regression sum of squares when the constant was forced to zero. However, it's the only method available to the conventional ANCOVA (the regression approach, discussed next, handles matters differently).
- The degrees of freedom for the regression is 2. The full equation uses 3 degrees of freedom for the regression, one for each predictor variable—one for the covariate and 2 for the coded vectors. However, we lose one of those three when we remove the covariate from the analysis. The residual sum of squares here has 14 degrees of freedom: the number of subjects less the number of factor levels less the number of covariates.
- The adjusted mean squares are calculated in the usual fashion by dividing the adjusted sums of squares by the degrees of freedom. The F-ratio is calculated by forming the ratio of the adjusted mean square for the regression to the adjusted mean square for the residual.

The F-ratio in cell I16 is just barely larger than 1.0 and is not statistically significant by most people's lights. Therefore, the differences between the adjusted factor level means are not large enough to produce an F-ratio that you can rely on to replicate.

Notice that this result is the opposite of the result shown in Figures 8.4 and 8.5. There, the adjustment process resulted in adjusted means that were farther apart than they were pre-adjustment. In the present example, the adjusted means are closer together than they were pre-adjustment. Apparently the random assignment did not in this case do a good job of equating the groups on the covariate, and the groups were not equivalent as measured by the covariate. That can happen with relatively small sample sizes and random assignment.

But the bias removal feature of the analysis of covariance equates the groups on the covariate. After that's done, it becomes clear that the different factor levels do not exert a sufficiently strong effect on the group means for the outcome measure to result in a sufficiently strong F-ratio.

And what are those adjusted means? They show up in Figure 8.8 in the range I21:I23. The formula for cell I21 is:

```
=B21–$I$13*(C21–$C$24)
```

The adjustment starts with the observed mean on the outcome variable for Group 1. Then it uses the pooled beta found in cell I13 to predict an outcome variable value based on the covariate. The value of the covariate is actually the difference between the group's observed mean for the covariate and the grand mean of the covariate. That difference, multiplied by the pooled beta, is the amount to adjust the observed value on the outcome variable.

If you compare all three adjusted means with all three observed means in B21:B23, you'll notice that the adjusted means are closer to one another than are the observed means. Hence the nonsignificant F-ratio for the adjusted means when the F-ratio shown in Figures 8.6 and 8.7 are highly significant—due not to the treatments, but to the pre-existing differences on the covariate.

Structuring an ANCOVA Using Regression

With the tedium of the conventional approach to ANCOVA behind us, let's have a look at doing the same analysis using LINEST() and proportions of variance. It's *much* smoother.

Figure 8.9
Using regression to calculate an ANCOVA, there's a lot less to do.

K15 =H15-$I$3*(I15-$I$18)

	A: Group	B: Out-come	C: Covari-ate	D: Group Vector 1	E: Group Vector 2
2	Group 1	77	18	1	0
3	Group 1	80	23	1	0
4	Group 1	80	18	1	0
5	Group 1	81	22	1	0
6	Group 1	77	19	1	0
7	Group 1	78	21	1	0
8	Group 2	80	23	0	1
9	Group 2	79	20	0	1
10	Group 2	82	24	0	1
11	Group 2	81	21	0	1
12	Group 2	83	25	0	1
13	Group 2	83	26	0	1
14	Group 3	83	26	-1	-1
15	Group 3	85	28	-1	-1
16	Group 3	82	24	-1	-1
17	Group 3	84	27	-1	-1
18	Group 3	83	27	-1	-1
19	Group 3	82	23	-1	-1

=LINEST(B2:B19,C2:E19,,TRUE)

G	H	I	J
0.16	-0.72	0.5401	68.66
0.35	0.50	0.13	2.90
0.84	1.04	#N/A	#N/A
23.86	14	#N/A	#N/A
76.76	15.01	#N/A	#N/A

=LINEST(B2:B19,C2:C19,,TRUE)

L	M
0.67	65.61
0.08	1.88
0.81	1.04
69.23	16
74.55	17.23

Source	SS Formula	Unique SS	df	MS	F	Prob of F
Between	=G7-L7	2.21	2	1.11	1.03	0.3817
Within	=H7	15.01	14	1.07		
Total	=M7	17.23	16			

	Outcome	Covariate	Adjusted Outcome
Mean, Grp 1	78.83	20.17	80.39
Mean, Grp 2	81.33	23.17	81.27
Mean, Grp 3	83.17	25.83	81.67
Grand Mean	81.11	23.06	

The only change to the data set in Figure 8.9, compared to that in Figure 8.8, is the addition of two vectors in columns D and E, to account for group membership.

We can replace all the work involved in calculating deviation products and R^2 values and so forth within each factor level, as is done in Figure 8.8, with two instances of LINEST() in Figure 8.9. The first, in the range G3:J7, regresses the outcome measure onto the covariate and the two vectors that define group membership. It uses this array formula:

=LINEST(B2:B19,C2:E19,,TRUE)

The second instance of LINEST(), in the range L3:M7, regresses the outcome measure onto the covariate only, with this array formula:

=LINEST(B2:B19,C2:C19,,TRUE)

With those two array formulas in place, we can subtract certain values in the covariate-only analysis from those in the covariate-plus-factor analysis, leaving only the variability unique to the factor. This is just another example of the models-comparison approach.

We can move directly from the LINEST() analyses to the ANOVA table, as follows (the formulas used to obtain the sum of squares in I10:I12 are found in H10:H12):

The sum of squares unique to the treatment vectors appears in cell I10. It's the difference between the regression sum of squares for the full model in G3:J7, and the regression sum of squares for the covariate-only model in L3:M7. Compare this sum of squares with the one shown in cell F16 of Figure 8.8.

The residual (or "within") sum of squares in cell I11 is the same as the residual sum of squares for the full model, in cell H7. Compare this residual sum of squares with the one shown in cell F17 of Figure 8.8.

The total sum of squares in cell I12 is the same as the residual sum of squares from the covariate-only analysis. The residual sum of squares in the covariate-only analysis is what's left over after the sum of squares due to the regression on the covariate has been accounted for. Because the ANOVA is intended to work on values from which the relationship with the covariate has been removed, the residual from the covariate-only analysis is the correct total sum of squares in the ANOVA.

Notice that the total sum of squares in I12 equals (within error of rounding induced by the cell format) the sum of the regression and the residual sum of squares. This is so even though none of the three values is calculated by backing into it from the other two values. They are arrived at independently—and the total in cell I12 equals that in cell F18 of Figure 8.8.

Finally, the adjusted means on the outcome variable for the three factor levels appear in K15:K17, and are identical to the adjusted means shown in Figure 8.8. To get the adjusted means, we need the observed means on the outcome variable (see H15:H17), the covariate means and their grand mean (see I15:I18), and the pooled beta. You'll find the pooled beta in cell I3. You get it automatically from the regression coefficient for the covariate in the full model.

Checking for a Common Regression Line

There's at least one further test needed to cross the t's in an analysis of covariance. You should check to see whether the within-group regression lines for the covariate and the outcome measure are, in statistical jargon, "homogeneous."

You don't worry if the regression lines aren't absolutely parallel to one another, as they would be if the within-cell regression coefficients were equal to one another. It typically happens that a little sampling error can make within-cell regression coefficients vary from one another, even when the same coefficients in the population are equal.

If the regression coefficients are wildly discrepant, you'll be adjusting the means using a pooled beta that may bear no resemblance at all to one or more of the actual within-cell regression coefficients. In that case, the adjustments you apply to the observed mean outcomes might be random, and the entire ANCOVA might be misleading.

However, given that the groups share a common regression line for the covariate and the outcome variable, random assignment to groups in combination with respectable sample sizes usually result in homogeneous regression coefficients. Even then, testing for differences is a good idea, if only to set aside one possible source of inaccuracy in the ANCOVA.

If the regression coefficients do in fact differ across the levels of the factor, then an interaction between the factor and the covariate exists. In that case, the difference will tend to show up in the slopes of regression lines charted separately for each level. See Figure 8.10.

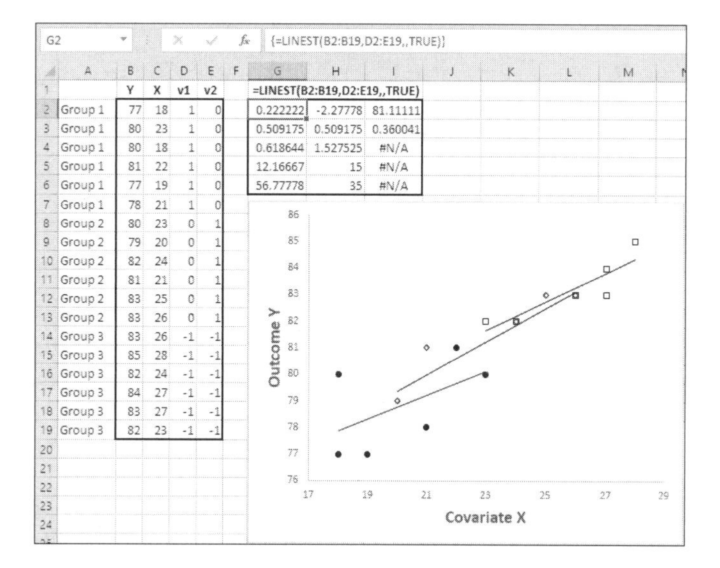

G2 {=LINEST(B2:B19,D2:E19,,TRUE)}

	A	B	C	D	E	F	G	H	I
1		Y	X	v1	v2		=LINEST(B2:B19,D2:E19,,TRUE)		
2	Group 1	77	18	1	0		0.222222	-2.27778	81.11111
3	Group 1	80	23	1	0		0.509175	0.509175	0.360041
4	Group 1	80	18	1	0		0.618644	1.527525	#N/A
5	Group 1	81	22	1	0		12.16667	15	#N/A
6	Group 1	77	19	1	0		56.77778	35	#N/A
7	Group 1	78	21	1	0				
8	Group 2	80	23	0	1				
9	Group 2	79	20	0	1				
10	Group 2	82	24	0	1				
11	Group 2	81	21	0	1				
12	Group 2	83	25	0	1				
13	Group 2	83	26	0	1				
14	Group 3	83	26	-1	-1				
15	Group 3	85	28	-1	-1				
16	Group 3	82	24	-1	-1				
17	Group 3	84	27	-1	-1				
18	Group 3	83	27	-1	-1				
19	Group 3	82	23	-1	-1				

Figure 8.10
A scatter chart of the outcome by the covariate often helps you to visualize the relationship at different factor levels.

The chart in Figure 8.10 makes two points clear:

- The three regression lines, one for each level of the factor, are very nearly parallel. The regression coefficients are therefore quite close to one another, any difference could be chalked up to sampling error, and the use of a common regression coefficient would be in order. Although we follow up this subjective analysis with one that's more objective, it's helpful to chart the separate regressions to get a better idea of what's going on in the data set.
- The groups clearly have discrepant means on the covariate and when, as here, the correlation between the covariate and the outcome is strong, you can expect considerable adjustments on the outcome measure.

> **NOTE** To chart the regression lines separately for each level of the factor, begin by charting the outcome by the covariate for the data in the factor's first level. With the chart active, click the Design tab and then click Select Data in the Data group. Click the Add button and supply the range addresses for the outcome and the covariate in the factor's second level. Repeat for as many factor levels as you have. You'll wind up with as many data series as you have levels, and each series will be represented by data markers with different shapes and colors.

As I noted, charting the separate regression coefficients can offer some insight into the data set, but it's not an objective test. To test more objectively for homogeneity of the regression coefficients, we once again use the models comparison approach. See Figure 8.11.

Figure 8.11
It's typical to get a nonsignificant F-ratio in a test for homogeneity of regression coefficients with random assignment to treatments.

J21 =J16*(H13+I13)

	A	B	C	D	E	F	G	H	I	J	K	L	M	N
1			Group	Group	Covariate	Covariate		=LINEST(A3:A20,B3:F20,,TRUE)						
2	Out- come	Covari- ate	Vector 1	Vector 2	by Group Vector 1	by Group Vector 2		0.09	-0.09	-1.73	1.18	0.54	68.67	
3	77	18	1	0	18	0		0.18	0.19	4.26	4.16	0.13	3.16	
4	80	23	1	0	23	0		0.841	1.10	#N/A	#N/A	#N/A	#N/A	
5	80	18	1	0	18	0		12.65	12	#N/A	#N/A	#N/A	#N/A	
6	81	22	1	0	22	0		77.14	14.63	#N/A	#N/A	#N/A	#N/A	
7	77	19	1	0	19	0								
8	78	21	1	0	21	0		=LINEST(A3:A20,B3:D20,,TRUE)						
9	80	23	0	1	0	23		0.16	-0.72	0.54	68.66			
10	79	20	0	1	0	20		0.35	0.50	0.13	2.90			
11	82	24	0	1	0	24		0.836	1.04	#N/A	#N/A			
12	81	21	0	1	0	21		23.86	14	#N/A	#N/A			
13	83	25	0	1	0	25		76.76	15.01	#N/A	#N/A			
14	83	26	0	1	0	26								
15	83	26	-1	-1	-26	-26		Source	R^2 Delta	R^2 Delta	df	MS Delta	F	Prob of F
16	85	28	-1	-1	-28	-28		Gp by X	=H4-H11	0.004	2	0.0021	0.156	0.698
17	82	24	-1	-1	-24	-24		Within	=1-H4	0.159	12	0.0133		
18	84	27	-1	-1	-27	-27		Total	=J16+J17	0.164				
19	83	27	-1	-1	-27	-27								
20	82	23	-1	-1	-23	-23			Source	SS Delta	df	MS Delta	F	Prob of F
21									Gp by X	0.38	2	0.19	0.156	0.698
22									Within	14.63	12	1.22		
23									Total	15.01				

In a test for homogeneous regression coefficients, we want to determine whether knowledge of the interaction between the factor and the covariate adds materially to the regression equation's R^2. To make that determination we create (in this example) two additional design vectors. Mechanically, you create the additional vector just as was done in Chapter 7, "Using Regression to Test Differences Between Group Means" (see Figure 7.17). Each interaction vector contains the products of a vector that represents a factor level with a vector that represents a covariate (as in this example) or a level of a different factor (as in Chapter 7).

In the present example, the factor has three levels and therefore two effect-coded vectors in columns C and D of Figure 8.11. There is one covariate, and so we need two additional vectors: one to represent the interaction between the first factor vector and the covariate, and one to represent the interaction between the second factor vector and the covariate. The first interaction vector, in column E, consists of the results of multiplying the values in column B by the corresponding values in column C. The second interaction vector, in column F, contains the products of the values in column B with corresponding values in column D.

With the two additional vectors established in columns E and F, it's time to continue with the models comparison. First, establish a LINEST() analysis of the full model (in Figure 8.11, the range H2:M6 contains the full model) using this array formula:

```
=LINEST(A3:A20,B3:F20,,TRUE)
```

Then establish another LINEST() analysis of the restricted model, one without the two interaction vectors, using this array formula:

```
=LINEST(A3:A20,B3:D20,,TRUE)
```

In Figure 8.11, the restricted model is in the range H9:K13.

The differences between the two sets of results are due solely to the presence of the two interaction vectors, columns E and F, in the full model. Notice the two values for R^2 associated with the two different models. The R^2 for the full model is 0.8406 (see cell H4) and for the restricted model it's 0.8364 (cell H11). Therefore, including the vectors that represent the interaction between the factor and the covariate accounts for an additional 0.8406 – 0.8364 = 0.0042 or 0.42% of the variability shared by the predictor variables and the outcome variable.

That difference, 0.0042, appears in cell J16 of Figure 8.11. It is the increment in the regression sum of squares, expressed as a percentage of total sum of squares, due to including the interaction vectors. (The formula to return that result is shown as text in cell I16.)

The remainder of the range J16:M18 completes an ANOVA that tests the statistical significance of that increment. The residual sum of squares, again expressed as a percentage of total sum of squares, is that amount that is not predicted via regression in the full model: 1 – H4, in cell J17. (Again, see cell I17 for the formula as text.) Note that a more explicit, if slightly more complicated way to express this quantity is as a ratio of the regression to the total sum of squares in the full model, I6 / (H6 + I6).

The degrees of freedom for the regression, in cell K16, is calculated by subtracting the number of predictors for the restricted model from the predictors in the full model: 5 – 3 = 2. For the residual, the degrees of freedom is the same as the residual degrees of freedom in the full model, or 12 (see cell I5).

The mean squares are calculated by dividing the proportions of sums of squares by the associated degrees of freedom, and the F-ratio is just the regression MS divided by the residual MS. In this case, the result is 0.156, much less than 1.0 and therefore not significant.

If you prefer to work with sums of squares, you can simply multiply the values labeled R^2 Delta in J16:J18 by the total sum of squares. This is the approach used in the range J21:N23 of Figure 8.11. For example, the value 0.38 in cell J21 is returned by this formula:

```
=J16*(H13+I13)
```

although you could get the result even more directly by subtracting the regression sum of squares for the restricted model from the regression sum of squares for the full model, as follows:

```
=H6 – H13
```

However you choose to complete the analysis, the interaction of the covariate with the factor in this data set is not significant in a statistical sense, and we can remove the covariate by factor vectors from the design in subsequent analyses (and there are some of those to come).

What if it turns out that you can't conclude that the regression lines differ only because of sampling error? The basic idea is to accept that the interaction exists and adopt a model that allows for and quantifies the interaction. Several such approaches exist, all beyond the scope of this book. They include the Johnson-Neyman technique, which involves the establishment of regions of significance, and (more recently) multilevel regression, which can involve *stratifying* the covariate to create a new factor within which observations are nested. Normally, it's a mistake to discard the information carried by a covariate, which is what you do when you convert a variable measured on an interval scale to a factor, but conditions exist in which it's both the best way to move forward and superior to the analysis of covariance.

Summarizing the Analysis

Figure 8.12 repeats some of the information in Figure 8.9, but tests the differences between the means via a comparison of the R^2 values.

Figure 8.12
Adjustment of the means using the covariate turns apparently significant mean differences into probable sampling error.

K20 =H20-$I$2*(I20-$I$23)

	A	B	C	D	E	F	G	H	I	J	K	L	M	N
1		Out-	Covari-	Group	Group		=LINEST(B3:B20,C3:E20,,TRUE)					=LINEST(B3:B20,C3:C20,,TRUE)		
2		come	ate	Vector 1	Vector 2		0.1622	-0.7174	0.5401	68.6577		0.6723	65.6113	
3	Group 1	77	18	1	0		0.3455	0.4998	0.1251	2.8952		0.0808	1.8789	
4	Group 1	80	23	1	0		0.8364	1.0356	#N/A	#N/A		0.8123	1.0377	
5	Group 1	80	18	1	0		23.8587	14	#N/A	#N/A		69.2291	16	
6	Group 1	81	22	1	0		76.7632	15.0146	#N/A	#N/A		74.5484	17.2294	
7	Group 1	77	19	1	0									
8	Group 1	78	21	1	0		Source	R^2 Delta	df	MS Delta	F	Prob of F		
9	Group 2	80	23	0	1		Treatments	0.0241	2	0.0121	1.0326	0.382		
10	Group 2	79	20	0	1		Residual	0.1636	14	0.0117				
11	Group 2	82	24	0	1									
12	Group 2	81	21	0	1		=LINEST(B3:B20,D3:E20,,TRUE)							
13	Group 2	83	25	0	1		0.2222	-2.2778	81.1111					
14	Group 2	83	26	0	1		0.5092	0.5092	0.3600					
15	Group 3	83	26	-1	-1		0.6186	1.5275	#N/A					
16	Group 3	85	28	-1	-1		12.1667	15	#N/A					
17	Group 3	82	24	-1	-1		56.7778	35.0000	#N/A					
18	Group 3	84	27	-1	-1									
19	Group 3	83	27	-1	-1			Outcome	Covariate		Adjusted Outcome			
20	Group 3	82	23	-1	-1		Mean, Grp 1	78.83	20.17		80.39			
21							Mean, Grp 2	81.33	23.17		81.27			
22							Mean, Grp 3	83.17	25.83		81.67			
23							Grand Mean	81.11	23.06					

The interaction between the covariate and the factor has been judged non-significant and so the interaction vectors have been deleted. The LINEST() results for the full model, involving the factor and the covariate main effects only, are in the range G2:J6, and the model employing the covariate only appears in the range L2:M6. The results are identical to those shown in Figure 8.9 (the apparent discrepancies are due to cell formatting).

However, the models comparison in Figure 8.12, shown in the range H9:L10, is performed using the difference in R^2 values in H9:H10, rather than the difference in the regression sums of squares, as is done in the range I10:I11 of Figure 8.9. I use R^2 in the models comparison shown in Figure 8.12 partly as a reminder that it's possible to make the comparison in terms of percentage of shared variance, but principally to stress how much more descriptive than sums of squares the R^2 measure is. In Figure 8.9, all you know is that the increment in the sum of squares "between"—that is, the increment in the regression sum of squares—is 2.21, and that the increment is not significant in a statistical sense.

In Figure 8.12, by contrast, you can see that the increment in R^2 is a mere 2.41% of the shared variance. Not only is that figure by itself more descriptive than a sum of squares, but it's more directly interpretable in the context of the 81.23% of shared variance attributable to the covariate (see cell L4 in Figure 8.12).

Finally, notice the LINEST() results in the range G13:I17 of Figure 8.12, which ignore the covariate. The results indicate a significant difference somewhere among the group means, whereas the ANCOVA, which evaluates the effect of the covariate, indicates no significant difference. In this case the adjustment of the group means causes what at first looked like evidence of a significant difference somewhere to vanish. This outcome is unusual in designs that involve random assignment to groups, because that assignment *tends* to equate the groups on the covariate—and the closer the groups' covariate means, the less that adjustment can occur.

Normally, at this point you would report that no significant difference exists among the adjusted group means and write your summary. But to illustrate how you might proceed if the models comparison resulted in a significant outcome, the next section discusses how to conduct multiple comparisons following a significant F-ratio for the models comparison.

Testing the Adjusted Means: Planned Orthogonal Coding in ANCOVA

Again, you would not necessarily proceed with a multiple comparison procedure if the ANCOVA resulted in no significant difference between adjusted means. It's usually pointless to go looking for the source of a significant difference when the omnibus F-test tells you that none exists. Nor would you run a multiple comparisons procedure if you had two groups only. In that case, the specific means responsible for a significant omnibus F-ratio must be the two accounted for in your design, and you can learn little new from a multiple comparisons procedure.

But the example discussed in the prior section involved three factor levels, each of which has both a raw and an adjusted mean. If the ANCOVA had reported, via the F-ratio for the models comparison, that a significant difference between adjusted means existed somewhere, then you would want to identify the factor levels involved in the significant difference. After all, Group 1 might differ from neither Group 2 nor Group 3, but the adjusted means of Groups 2 and 3 might well differ significantly.

Furthermore, if you have used orthogonal coding, you're not generally interested in the overall, omnibus F-ratio. You have already specified the comparisons that you're interested in, via the codes you selected for the coded vectors. So you would tend to be interested only in the significance of the individual vectors, which define the contrasts that you selected at the outset.

We'll start with a look at orthogonal comparisons in ANCOVA, and see how the outcomes compare to both a planned but nonorthogonal method, and with the post hoc Scheffé method.

Figure 8.13 shows how to test specific mean differences using orthogonal coding. See Chapter 7 for a discussion of the general topic of multiple comparisons as well as the use of planned orthogonal coding specifically—which requires that you specify the comparisons of interest *before* you actually conduct the analysis.

Figure 8.13
Using planned orthogonal coding in the factor vectors means that the t-ratios from LINEST( ) are directly interpretable.

E10 — fx: Gp 1 + Gp 2 + Gp 3 + Gp 4

	A	B	C	D	E	F	G	H	I	J	K	L
1		Out-	Covari-									
2	Group	come	ate									
3	Group 1	21	84		Total R^2, covariate/outcome	0.9445	=RSQ(B3:B34,C3:C34)					
4	Group 1	42	114									
5	Group 1	21	66			SS Y	SS X	R^2 XY	SS Y '			
6	Group 1	21	94		Gp 1	759.50	2118.00	0.77	174.62			
7	Group 1	7	64		Gp 2	294.00	403.50	0.69	89.86			
8	Group 1	14	70		Gp 3	287.88	699.50	0.64	103.88			
9	Group 1	14	66		Gp 4	588.00	1598.00	0.96	20.85			
10	Group 1	14	78		Gp 1 + Gp 2 + Gp 3 + Gp 4	1929.38	4819.00	0.79	389.22			
11	Group 2	28	94									
12	Group 2	35	98		Sum xy, Gp 1	1113.00						
13	Group 2	28	84		Sum xy, Gp 2	287.00						
14	Group 2	21	82		Sum xy, Gp 3	358.75						
15	Group 2	14	76		Sum xy, Gp 4	952.00						
16	Group 2	21	86		Sum xy	2710.75		Pooled beta:	0.5625			
17	Group 2	28	96									
18	Group 2	21	90		Source	Adjusted SS	df	Adjusted MS	F	Prob of F		
19	Group 3	49	122		Between	247.57	3	82.52	5.51	0.0044		
20	Group 3	56	130		Within	404.54	27	14.98				
21	Group 3	49	128		Total	652.11	30					
22	Group 3	42	112									
23	Group 3	35	104			Outcome	Covariate	Adjusted means		Comparison	t-ratio	Prob of t
24	Group 3	49	116		Mean, Gp 1	19.25	79.50	36.41		Gp 1 vs Gp 2	-0.164	0.871
25	Group 3	42	106		Mean, Gp 2	24.50	88.25	36.73		(Gp 1 + Gp 2)/2 vs Gp 3	-2.689	0.012
26	Group 3	49	108		Mean, Gp 3	46.38	115.75	43.14		(Gp 1 + Gp 2 + Gp 3)/3		
27	Group 4	49	130		Mean, Gp 4	63.00	156.50	36.84		vs Gp 4	0.504	0.618
28	Group 4	56	148		Grand mean	38.28	110.00					

Up to the point where you start assessing the mean differences, the approach taken in Figure 8.13 is the same as the one used in Figure 8.8. Both Figure 8.8 and 8.13 illustrate the traditional ANCOVA technique that relies not on regression and coding, but instead on calculating various statistics within each level of the factor.

You can still use the approach to multiple comparisons via planned orthogonal contrasts with the traditional ANCOVA. But I would not understand anyone's reason to do so. The traditional approach is tedious and error-prone (I'll walk you through it anyway, so that if you want you can compare the results of the traditional ANCOVA with those of the regression approach using data from some other source). Using the regression approach, all you need to do is add the coding vectors to the raw data, call on LINEST(), and get your group means. Finish up with a minimum of adding and multiplying and you're done.

The traditional ANCOVA approach offers little in the way of insight into the meaning of the process. There's not much to be said along the way apart from just stating what takes place, so we'll take it on a step-by-step basis. The first steps, through the completion of the ANCOVA table, are the same regardless of your intentions for subsequent multiple comparisons. These steps are keyed to Figure 8.13. I am assuming that you have already tested the possibility that the different groups have different regression lines, and that you have concluded they are homogeneous. (The process is discussed earlier in this chapter, in the section "Checking for a Common Regression Line.")

1. Calculate the total R^2 between the covariate and the outcome variable in cell F3, using this formula:

   =RSQ(B3:B34,C3:C34)

2. For each group (there are four groups in this example) calculate the sum of squares of both the outcome measure and the covariate using the DEVSQ() function. For example, cell F6 contains this formula:

   =DEVSQ(B3:B10)

   to return the sum of squares of the outcome measure in the first group. Cell G6 uses this formula:

   =DEVSQ(C3:C10)

   to get the sum of squares of the covariate in the first group.

3. In cell H6, get the R^2 between the covariate and the outcome measure in the first group with this formula:

   =RSQ(B3:B10,C3:C10)

4. Repeat steps 3 and 4 for each of the remaining groups. The results go into the range F7:H9.

5. We need to remove the effect of the covariate from the sum of squares for the outcome variable and thus arrive at the adjusted sum of squares in the range I6:I9. The way to do that is to multiply the original sum of squares for the outcome variable for each group, in the range F6:F9, by 1 minus the R^2 for the covariate and the outcome variable, also for that group. The pertinent R^2 value for Group 1 is found in cell H6. Therefore the formula in cell I6 is:

   =F6*(1–H6)

   We multiply the observed sum of squares for the first group, in cell F6, by the proportion of the outcome variable in the first group that *cannot* be predicted from the covariate.

6. Copy the formula in cell I6 into I7:I9.

7. Enter this formula in cell F10:

   =F6+F7+F8+F9

   Or use the SUM() function. Copy the formula in cell F10 into G10 and I10, but *skip* cell H10.

8. Enter this formula in cell H10:

=(F16/SQRT(F10*G10))^2

Assuming at this point that cell F16 is still empty, the formula in cell H10 will return 0.0. But H10 will show the pooled R^2 once the deviation products have been calculated.

Let's pause here to recapitulate what has gone on in steps 2 through 7. The idea is to accumulate the sums of squares of the outcome measure (Y), of the covariate (X), and the adjusted sum of squares of Y. But in the case of the sums of squares of the Y and the X variables, the deviations are centered on the mean of the group to which each observation belongs. In this way, the variability can be calculated *within* each group and thus remain independent of differences between group means. The appropriate calculations are carried out in the range F6:G9.

You can see the difference between an overall statistic and its pooled version in cells F3 and H10. Cell F3 contains the overall R^2, 0.9445, between the outcome measure and the covariate, calculated on all 32 observations without attending to group membership. The *grand mean* is subtracted from each observation to get the deviations. But the pooled R^2, 0.79, in cell H10, is based on deviations obtained by subtracting the *group mean* from each observation in that group. Those deviations are squared and accumulated in cells F10 and G10, and the eventual pooled R^2 is a measure of the variance shared by the outcome measure and the covariate *within each group*.

Because the basis for the calculation—that is, within groups—of the sums of squares for Y is the same as the calculation of the R^2 for Y and X, we can use this sort of formula to get a sum of squares for Y that is adjusted to remove the effect of the covariate on the outcome measure:

=F6*(1–H6)

Although the sums of squares in the ranges F6:F9 and G6:G9 are additive, the R^2 values in H6:H9 are not. To get the pooled R^2 in cell H10 it's necessary to calculate it using these quantities:

8A. The sum of the products of the Y and the X deviations in F16

8B. The result of step 8A divided by the square root of the product of the sums of squares in cells F10 and G10

8C. The result of step 8B, squared

Or, more succinctly:

=(F16/SQRT(F10*G10))^2

This is all part of the two processes, discussed at the beginning of this chapter, of removing bias in the observed outcome means and increasing the power of the eventual F-test. For example, the total of the within-group sums of squares for the covariate, found in cell G10, is used in the calculation of the pooled beta in I16, which is then used to adjust the group means in the range H24:H27. The pooled R^2 in cell H10 and the total of the within-groups adjusted sums of squares for Y, in cell F10, are both used to calculate the adjusted sum of squares within, in cell F20.

TIP The Trace Precedents and the Trace Dependents tools in the Formula Auditing group on the Ribbon's Formulas tab are useful for tracking this sort of flow.

To continue with the list of tasks necessary to complete the conventional ANCOVA, we need the summed products of the deviations within groups. Continuing with step 9:

9. The most straightforward way to get the summed product of two sets of values (in this case, the values are deviations from the mean) in Excel is to use the SUMPRODUCT() function. In cell F12 of Figure 8.13, it's used in this way:

   =SUMPRODUCT(B3:B10–F24,C3:C10–G24)

   The expression B3:B10-F24 returns the deviations from the mean of the outcome measure for Group 1. Similarly, the expression C3:C10-G24 returns the deviations from the mean of the covariate for Group 1. The SUMPRODUCT() function multiplies the individual members of the two arrays together and sums the products. The result is the sum of the deviation products for Group 1.

10. Formulas identical to the one used in cell F12, except for the addresses of the underlying records, are entered in F13:F15. The total of the within groups deviation products appears in cell F16.

11. The sum of the within-group deviation products in cell F16 is divided by the sum of within-group squares for the covariate, in cell G10. The result, in cell I16, is the pooled beta calculated on a within-group basis. Note the difference between the overall beta to predict Y from X, in cell F3, and the pooled within-group beta, in cell I16. The difference is due to deviating the observations from the grand mean, in cell F3, instead of from the individual group means, in cell I16.

12. To get the analysis of variance based on adjustments for the relationship between the covariate and the outcome measure, we start by obtaining the sum of squares within. The full analysis is shown in Figure 8.13 in the range E18:J21, and the adjusted sum of squares within is calculated in cell F20 using this formula:

    =F10*(1–H10)

    That is, the total unadjusted sum of squares for the outcome measure in cell F10 times the percentage of that variability that is *not* predicted by the relationship between the outcome measure and the covariate. Both of these factors are calculated using deviations of individual observations from the group means, not from the grand mean. This is because we're calculating the adjusted sum of squares within, and we want to keep the variation due to differences between the adjusted group means out of the calculation.

13. Unlike the adjusted sum of squares within, the basis for the total adjusted sum of squares for the outcome measure, in cell F21 of Figure 8.13, *is* the deviation from the grand mean. The reason is that in cell F21 we're after the total adjusted sum of squares, which includes both variability due to differences between group means and variability due to individual differences within groups. Those two sources must sum to the total adjusted sum of squares, so we return to the sum of squared deviations of all observations from the grand mean. We multiply that by 1.0 minus the proportion of variability shared by the covariate and the outcome in order to remove the effect of the covariate on the outcome. The formula for the total adjusted sum of squares in cell F21 is:

   =DEVSQ(B3:B34)*(1–F3)

   which calculates the original sum of squares of the values in the range B3:B34 and makes the adjustment based on the shared variance quantified in cell F3.

14. The adjusted sum of squares between groups is found by subtraction, finding the difference between the total sum of squares and the sum of squares within, via this formula in cell F19:

   =F21–F20

NOTE It's possible to derive the adjusted sum of squares between groups directly, rather than indirectly via subtraction, but doing so using the conventional calculations outlined in this section results in dependency between the numerator and the denominator of the F-ratio, and the mathematical basis for the distribution of the F-ratio requires that they be independent. This issue does not arise using the regression approach to making orthogonal comparisons, discussed in the next section.

15. The degrees of freedom in cells G19 and G20 are calculated in the usual way. The degrees of freedom for the adjusted sum of squares between is the number of groups, *k*, minus 1, or k−1 = 4−1 = 3. The DF for the sum of squares within is N−k−1 = 32−4−1 = 27.

16. The adjusted mean squares are found by dividing the adjusted sums of squares by their degrees of freedom. The F-ratio is found by dividing the mean square between by the mean square within.

17. The probability of getting an F-ratio of 5.51, with 3 and 27 degrees of freedom, assuming that all group means or combinations of means are the same in the full population, is given by this formula in cell J19:

   =F.DIST.RT(I19,G19,G20)

   The formula returns 0.0044. If that meets your criterion for "sufficiently unusual"—that is, if it's smaller than the alpha level you specified at the outset—then you can continue with your orthogonal contrasts to assess the planned comparisons.

That's done in the next few steps. But with planned orthogonal contrasts, this F-ratio is superfluous and you can proceed directly to the next step regardless of the value returned by the F.DIST.RT() function.

18. The observed means for the outcome measure and the covariate, for each group and for all groups, are shown in the range F24:G28. For example, the formula for the mean of the outcome measure in Group 1 is in cell F24 by way of this formula:

    =AVERAGE(B3:B10)

19. The adjusted means appear in the range H24:H27. The formulas for the adjusted means follow this pattern, which adjusts the mean of Group 1:

    =F24–$I$16*(G24–$G$28)

    It can be copied and pasted into H25:H27 to adjust the means of the outcome measure in the remaining three groups. The formula subtracts from the observed mean (F24) the pooled beta ($I$16) times the difference between Group 1's mean on the covariate minus the grand mean on the covariate. That is, the formula estimates the difference on the outcome measure that can be predicted by using the pooled beta in conjunction with the difference between Group 1's covariate mean and the grand covariate mean.

20. The next-to-last step is to calculate and assess the orthogonal comparisons, which are expressed as t-ratios. The first of three in this example is in cell K24, which uses this formula:

    =(H24–H25)/SQRT($H$20*(2/8+(G24–G25)^2/$G$10))

    The formula gets the difference between the adjusted means of Group 1 and Group 2, in cells H24 and H25. It divides that difference by the square root of:

    - The adjusted mean square within ($H$20)
    - Times the number of groups in the contrast (2)
    - Divided by the number of subjects per group (8) times the number of groups in the contrast minus 1 (1)
    - Plus the difference between the two groups' means on the covariate (G24 – G25), squared
    - Divided by the total (pooled) sum of squares of the covariate ($G$10).

    The remaining two t-ratios are calculated similarly, but the addresses of the adjusted group means must be altered to represent each planned contrast properly. Note that the number of groups may change the t-ratio's denominator, and therefore the number of subjects (for example, 2/8 in the first comparison becomes 3/16 in the second and 4/24 in the third).

21. The final step is to assess the probability of each t-ratio assuming the means in the comparison are equal. The formula in cell L24 is:

    =T.DIST.2T(ABS(K24),$G$20)

    It can be copied and pasted into L25 and L27. In this example I have not specified a directional hypothesis and so use the T.DIST.2T() function. Because that function requires a positive t-ratio, the formula uses the ABS() function to return the ratio's absolute value. The function's second argument, $G$20, returns the degrees of freedom for the "within" source of variation in the ANOVA of the covariate-adjusted values.

ANCOVA and Multiple Comparisons Using the Regression Approach

I inflicted that entire set of steps on you in the prior section for three reasons. One is that I believe the more ways you read about doing something even remotely complicated, the easier it is to get your head around the complexities. The second reason is that tracing the traditional approach helps underscore the process of deviating individual values from group means instead of from the grand mean. The third reason is that seeing how cumbersome the traditional approach is makes it easier to appreciate the speed and ease of the regression approach.

Figure 8.14 shows the entire regression process.

Figure 8.14
Using LINEST( ) and the regression approach, most of the work is done by the regression coefficients and their standard errors.

I19 =J3/J4

	B	C	D	E	F	G	H	I	J	K	L
1	Out-	Covari-	Group	Group	Group		=LINEST(B3:B34,C3:F34,,TRUE)				
2	come	ate	Vector 1	Vector 2	Vector 3		Vector 3	Vector 2	Vector 1	Cov	Const
3	21	84	1	0.5	1		0.479	-4.380	-0.164	0.5625	-23.595
4	42	114	1	0.5	1		0.950	1.629	0.998	0.0558	6.172
5	21	66	1	0.5	1		0.966	3.871	#N/A	#N/A	#N/A
6	21	94	1	0.5	1		189.446	27	#N/A	#N/A	#N/A
7	7	64	1	0.5	1		11353.926	404.543	#N/A	#N/A	#N/A
8	14	70	1	0.5	1						
9	14	66	1	0.5	1						
10	14	78	1	0.5	1					Adjusted	
11	28	94	-1	0.5	1			Outcome	Covariate	means	
12	35	98	-1	0.5	1		Mean, Gp 1	19.25	79.5	36.41	
13	28	84	-1	0.5	1		Mean, Gp 2	24.50	88.25	36.73	
14	21	82	-1	0.5	1		Mean, Gp 3	46.375	115.75	43.14	
15	14	76	-1	0.5	1		Mean, Gp 4	63.00	156.50	36.84	
16	21	86	-1	0.5	1		Grand mean	38.28	110.00		
17	28	96	-1	0.5	1						
18	21	90	-1	0.5	1		Comparison	t-ratio	Prob of t		
19	49	122	0	-1	1		Gp 1 vs Gp 2	-0.164	0.871		
20	56	130	0	-1	1		(Gp 1 + Gp 2)/2 vs Gp 3	-2.689	0.012		
21	49	128	0	-1	1		(Gp 1 + Gp 2 + Gp 3)/3				
22	42	112	0	-1	1		vs Gp 4	0.504	0.618		

Again, this assumes that you have already tested for homogeneity of regression coefficient and have found that the overall regression of the outcome measure on the covariate is worth keeping in the equation, but no persuasive evidence that the different groups have different regression lines.

Instead of 21 steps as in the prior section, there are just five steps, as follows:

1. Array-enter the LINEST() function in the range H3:L7.
2. In the range I12:J16, calculate the observed means for the outcome measure and the covariate as shown in the prior section. For example, the formula for Group 1's mean of the outcome measure in cell I12 is:

   =AVERAGE(B3:B10)

3. Calculate the adjusted group means on the outcome measure in the range K12:K15. The formula used in cell K12 is:

   =I12–$K$3*(J12–$J$16)

   With one difference, the formula is precisely the same as shown in the prior section. The difference is that the pooled beta in cell $K$3, used to adjust the difference between the Group 1 mean for the covariate in J12 and its grand mean in $J$16, is returned directly by LINEST(). It's unnecessary to jump through 11 hoops to get that pooled beta, as was done in Figure 8.13.

   You can copy and paste the formula in K12 into K13:K15.
4. The t-ratios in the range I19:I21 are just as easy to construct as the adjusted means in K12:K15. They are the ratios of the group vectors' regression coefficients to their standard errors and are returned directly by LINEST(). Recall that the orthogonal comparisons are built into the values for each vector, so no special range selection is needed to distinguish the t-ratio's numerators, or their group and observation counts in the denominators, as was done in Figure 8.13. The formula in cell I19 is:

   =J3/J4

   and the formulas in I20 and I22 are equally simple.
5. Test the probability of observing the t-ratios in I19, I20, and I22. The formula in cell J19 is:

   =T.DIST.2T(ABS(I19),$I$6)

   The only meaningful difference from the instances of T.DIST.2T() in Figure 8.13 is that you can pick up the degrees of freedom directly from cell $I$6 in the LINEST() results.

Be sure to note that the comparisons, the t-ratios and the probability of each t-ratio assuming no difference in the population are the same in Figure 8.13, J24:L27, as in Figure 8.14, H19:J22.

Multiple Comparisons via Planned Nonorthogonal Contrasts

Planned orthogonal contrasts are generally the most statistically powerful of the multiple comparison techniques, but as usual there's a tradeoff. If you opt for planned orthogonal contrasts, you must plan them beforehand. You agree not to wait until the data comes in to decide which interesting contrasts to make. Furthermore, only a subset of the possible contrasts are mutually orthogonal—generally, that's the number of levels in the factor, minus 1, or equivalently the factor's degrees of freedom.

In return for exercising your foresight in planning which contrasts to arrange, and your self-control in restricting the number of contrasts to the factor's degrees of freedom, you get a more powerful test. That test is more likely to meet your alpha level: the probability value you choose as a criterion for deciding that a difference in group means is too unlikely to be due to just sampling error.

Suppose you're in the planning stages of a test to determine what mix of screen layouts is most effective in driving new business on a website. Some preliminary estimates indicate that the most interesting contrasts will be between Options A and B, and between Options C and D. But a new option has just surfaced that you find intriguing, and you would like to contrast its results with the results of each of the existing options. Unfortunately, the contrasts of Option E with each of the other four would leave you with contrasts that are not orthogonal. (See Chapter 7 for instructions on setting up a matrix that enables you to assess which of available contrasts are orthogonal.)

You would like to accommodate those additional comparisons with Option E along with the two comparisons that you had originally contemplated, and you can do so in at least two general ways. One is to wait until all the data is in before you decide which comparisons to make. The best known way to follow that approach is the Scheffé method, and I'll cover it in the context of ANCOVA in the next section. You can also stick with planned contrasts but abandon the condition of orthogonality. In that way you could plan your comparisons in advance but give up the added statistical power afforded by the use of orthogonal comparisons only. Planned *non*orthogonal contrasts are less statistically powerful than planned orthogonal contrasts, but more powerful than the post facto Scheffé method. This section describes planned nonorthogonal contrasts.

Figure 8.15 shows one way to construct planned nonorthogonal contrasts. To make the comparison with planned orthogonal contrasts (and with the Scheffé method) more straightforward, I have used the same values for the outcome variable and for the covariate in Figure 8.15 as are used in Figure 8.14.

Figure 8.15
This data set uses effect coding instead of orthogonal coding in the coded vectors.

I25 =ABS(K18-K19)/SQRT(($I$7/$I$6)*(2/8)*(1+($H$14/(3*$I$14))))

	A	B	C	D	E	F	G	H	I	J	K	L
1		Out-	Covari-	Group	Group	Group		=LINEST(B3:B34,C3:F34,,TRUE)				
2	Group	come	ate	Vector 1	Vector 2	Vector 3		Vector 3	Vector 2	Vector 1	Cov	Const
3	Group 1	21	84	1	0	0		-1.438	4.859	-1.875	0.5625	-23.595
4	Group 1	42	114	1	0	0		2.851	1.228	2.073	0.0558	6.172
5	Group 1	21	66	1	0	0		0.966	3.871	#N/A	#N/A	#N/A
6	Group 1	21	94	1	0	0		189.446	27	#N/A	#N/A	#N/A
7	Group 1	7	64	1	0	0		11353.926	404.543	#N/A	#N/A	#N/A
8	Group 1	14	70	1	0	0						
9	Group 1	14	66	1	0	0		=LINEST(C3:C34,D3:F34,,TRUE)				
10	Group 1	14	78	1	0	0		46.500	5.750	-30.500	110.000	
11	Group 2	28	94	-1	-1	-1		4.017	4.017	4.017	2.319	
12	Group 2	35	98	-1	-1	-1		0.857	13.119	#N/A	#N/A	
13	Group 2	28	84	-1	-1	-1		55.758	28	#N/A	#N/A	
14	Group 2	21	82	-1	-1	-1		28789	4819	#N/A	#N/A	
15	Group 2	14	76	-1	-1	-1						
16	Group 2	21	86	-1	-1	-1					Adjusted	
17	Group 2	28	96	-1	-1	-1			Outcome	Covariate	means	
18	Group 2	21	90	-1	-1	-1		Mean, Gp 1	19.25	79.5	36.41	
19	Group 3	49	122	0	1	0		Mean, Gp 2	24.50	88.25	36.73	
20	Group 3	56	130	0	1	0		Mean, Gp 3	46.375	115.75	43.14	
21	Group 3	49	128	0	1	0		Mean, Gp 4	63.00	156.50	36.84	
22	Group 3	42	112	0	1	0		Grand mean	38.28	110.00		
23	Group 3	35	104	0	1	0						
24	Group 3	49	116	0	1	0		Comparison	t-ratio	Prob of t		
25	Group 3	42	106	0	1	0		Gp 1 vs Gp 2	0.098	0.923		
26	Group 3	49	108	0	1	0		(Gp 1 + Gp 2)/2 vs Gp 3	2.266	0.031		
27	Group 4	49	130	0	0	1		(Gp 1 + Gp 2 + Gp 3)/3				
28	Group 4	56	148	0	0	1		vs Gp 4	0.702	0.489		

I changed to effect coding in Figure 8.15. Compare the LINEST() results in Figure 8.15 with those in Figure 8.14. They differ as to the regression coefficients for the coded vectors and their standard errors, but the values of statistics such as the regression coefficient for the covariate, the standard error of estimate and the F-ratio—that is, all the values in the third through the fifth row of the LINEST() results—are unchanged. We'll use the regression coefficient for the covariate, the sums of squares for the regression and the residual, as well as the residual degrees of freedom, in the multiple comparisons, but not the regression coefficients or their standard errors. Therefore, for the purposes of your multiple comparisons, planned but nonorthogonal. You can use any coding method that accurately distinguishes between the groups.

Figure 8.15 also includes a second LINEST() analysis in the range H10:K14. Its purpose is to return the sums of squares for the covariate regressed onto the coded vectors, for use in calculating the t-ratios in cells I25, I26, and I28.

Compare the range H25:J28 in Figure 8.15 with the range H19:J22 in Figure 8.14. The means chosen for comparison are the same—Group 1 with Group 2, the average of the first two groups with the third, and the average of the first three groups with the fourth—but the results differ. The corresponding t-ratios in the two figures differ, with consequences for the probability associated with each t-ratio (under the assumption of no difference in the population between the means being compared).

Here's the formula for the t-ratios, in symbolic form:

$$t = \frac{\overline{Y}_{1,adj} - \overline{Y}_{2,adj}}{\sqrt{MSR(1/n_1 + 1/n_2)}\sqrt{1 + \left(\frac{SS_{reg(x)}}{kSS_{resid(x)}}\right)}}$$

The numerator subtracts the covariate-adjusted mean of one group from the adjusted mean of the other group. In the denominator:

- MSR is the mean square residual for the regression of the outcome measure on the covariate and the coded vectors.
- n_1 and n_2 are the number of observations—that is, the number of subjects—in each group. If a "group" is the average of two (or more) actual groups, n is the number of observations in the combination of the two.
- SSreg(x) and SSresid(x) are the sum of squares regression and the sum of squares residual from the regression of the covariate on the coded vectors.
- k is the number of coded vectors.

Here's how the formula translates to Excel syntax, from cell I25 in Figure 8.15:

```
=ABS(K18–K19)/SQRT(($I$7/$I$6)*(2/8)*(1+($H$14/(3*$I$14))))
```

- The references to the cells K18 and K19 establish the difference between the means of Group 1 and Group 2.
- The mean square residual comes from the ratio of the residual sum of squares in cell I7 to the residual degrees of freedom in cell I6.
- The fraction 2/8 is the sum of 1/8 and 1/8: the reciprocals of the two group sizes.
- The references to cells H14 and I14 bring into the formula the regression and the residual sums of squares for the regression of the covariate on the coded vectors.
- 3 represents k, the number of coded vectors.

Notice that the tests in H25:J26 of Figure 8.15 are less powerful than their orthogonal counterparts in Figure 8.14.

Multiple Comparisons with Post Hoc Nonorthogonal Contrasts

The third approach to multiple comparisons following an ANCOVA is generally known as the Scheffé method. It is the least powerful of the various multiple comparison methods: Comparisons that would be regarded as statistically significant, with other appropriate methods, are often regarded as not significant under the Scheffé method.

With the Scheffé you're permitted to decide on your comparisons *after* the data has come in and you've had a chance to evaluate it, so planning your comparisons in advance is not required. You're not limited to orthogonal contrasts, so you can call for as many comparisons as you want, including the more complex comparisons that involve averaging the means of two or more groups.

The price you pay for all that flexibility is the relatively low statistical power of the Scheffé test. However, for pilot studies in which you don't yet have good information about what you should expect of the data, the Scheffé can be a good temporary choice. You might decide to save the more powerful techniques for subsequent studies, which are often more carefully targeted than early pilot research and to which you can bring the information from pilot investigations.

Figure 8.16 shows the Scheffé method used on the data from Figures 8.14 and 8.15.

Figure 8.16
Again, the choice of coding method has no effect on such summary statistics as the R^2 and the sums of squares.

J25 =SQRT(3*F.INV.RT(0.05,3,I6))*SQRT((I7/I6)*(2/8))*SQRT(1+(H14/(3*I14)))

	C	D	E	F	G	H	I	J	K	L	M
6	94	1	0	0		189.446	27	#N/A	#N/A	#N/A	
7	64	1	0	0		11353.926	404.543	#N/A	#N/A	#N/A	
8	70	1	0	0							
9	66	1	0	0		=LINEST(C3:C34,D3:F34,,TRUE)					
10	78	1	0	0		5.750	-21.750	-30.500	110.000		
11	94	0	1	0		4.017	4.017	4.017	2.319		
12	98	0	1	0		0.857	13.119	#N/A	#N/A		
13	84	0	1	0		55.758	28	#N/A	#N/A		
14	82	0	1	0		28789	4819	#N/A	#N/A		
15	76	0	1	0							
16	86	0	1	0					Adjusted		
17	96	0	1	0			Outcome	Covariate	means		
18	90	0	1	0		Mean, Gp 1	19.25	79.5	36.41		
19	122	0	0	1		Mean, Gp 2	24.50	88.25	36.73		
20	130	0	0	1		Mean, Gp 3	46.375	115.75	43.14		
21	128	0	0	1		Mean, Gp 4	63.00	156.50	36.84		
22	112	0	0	1		Grand mean	38.28	110.00			
23	104	0	0	1							
24	116	0	0	1		Comparison	|D|	S	Result		
25	106	0	0	1		Gp 1 vs Gp 2	0.328	9.98	N/S at .05		
26	108	0	0	1		(Gp 1 + Gp 2)/2 vs Gp 3	6.570	8.64	N/S at .05		
27	130	-1	-1	-1		(Gp 1 + Gp 2 + Gp 3)/3			N/S at .05		
28	148	-1	-1	-1		vs Gp 4	1.917	8.12	N/S at .05		
29	160	-1	-1	-1		Gp 1 vs Gp 3	6.734	9.98	N/S at .05		

Just to demonstrate that the coding method for the factor levels has no effect on LINEST()'s summary statistics, I have changed from orthogonal coding in Figure 8.14 to effect coding in Figures 8.15 and 8.16. As long as your coding method accurately assigns the different observations to the correct group, the coding method (for example, effect, dummy, or orthogonal coding) makes no difference to ANCOVA's adjustment of the group means. It also makes no difference to other statistics, such as the sum of squares regression and residual, important to the multiple comparison process.

The range H10:K14 in Figure 8.16 contains the results of using LINEST() to regress the *covariate* on the group vectors. This analysis is needed in order to calculate the nonorthogonal comparisons.

The only other meaningful differences in Figure 8.16 from Figure 8.14's planned orthogonal contrasts and Figure 8.15's planned *non*orthogonal contrasts are in the range H24:J29 of Figure 8.16, which uses a different approach to evaluating comparisons than the corresponding ranges in earlier figures.

The usual approach to displaying the results of the Scheffé method is to establish two values for each contrast, D and S. The D value is the difference implied by the contrast.

So if a contrast compares Group 1 with Group 2, D is the difference between the covariance-adjusted means of the two groups. Because this method of multiple comparisons uses the F-ratio instead of the t-ratio, we use the absolute value of D, usually symbolized as |D|.

> **NOTE** The F-ratio, if calculated correctly, is always positive, so a negative value for D has no meaning in this method, whereas a negative value for a comparison is perfectly meaningful when compared to a t distribution.

So, in Figure 8.16, the formula in cell I25 compares the adjusted mean for Group 1 with that for Group 2:

```
=ABS(K18–K19)
```

The formula returns the absolute value of the difference between the adjusted mean of Group 1 and the adjusted mean of Group 2.

The second value for each contrast is referred to as *S*. If the absolute value of D exceeds the value of S for a given comparison, then the contrast is deemed "statistically significant" at the alpha level that you have already established. The alpha level in use in Figure 8.16 is 0.05 (you'll see how that comes about shortly). So in Figure 8.16, none of the comparisons is deemed significant. Recall that in Figure 8.14, using the more powerful method of planned orthogonal contrasts, the comparison of the average of Groups 1 and 2 with the mean of Group 3 was significant at the 0.05 level, and very nearly so at the 0.01 level (see cell J20 in Figure 8.14). The contrast is not deemed significant at the 0.05 level in Figure 8.16 using the Scheffé method—see H26:K26 in Figure 8.16.

Unfortunately, the formula for S is relatively complicated. Here it is in symbolic form:

$$S = \sqrt{kF_\alpha}\sqrt{MS_{resid}\sum C_j^2/n_j}\sqrt{1 + \left(\frac{SS_{reg(x)}}{kSS_{resid(x)}}\right)}$$

There's a lot of symbol shock in that formula. Here's what the symbols represent, after which we'll look at the Excel form:

- S is the value to be compared with |D|.
- k is the number of coded vectors that represent the factor levels.
- F_α is the value of the F distribution at your chosen level of alpha, with k and (N–k–2) degrees of freedom, where N is the total number of subjects in the regression.
- $MS_{residual}$ is the mean square residual from the ANCOVA, available from the LINEST() results.
- C_j^2 is the square of the coefficient of the jth contrast. If a contrast subtracts the mean of Group 2 from the mean of Group 1, then the Group 1 contrast coefficient is 1 and the Group 2 coefficient is –1.

- n_j is the number of subjects in the *j*th group.
- $SS_{reg}(x)$ is the sum of squares regression when the *covariate* is regressed on the group vectors.
- $SS_{resid}(x)$ is the sum of square residual when the covariate is regressed on the group vectors.

There's not a lot of insight to derive from studying that formula, aside from the size of the F value. The larger an F value, the less probable its occurrence in a central F distribution. Therefore, as you move the probability from, say, 0.05 to .01, you might (depending on its degrees of freedom) move the F value from 3.1 to 4.5. So doing in the formula for S makes the value returned by the full formula larger, and therefore more difficult for |D| to exceed.

Here's how the formula works out in Excel syntax, for cell J25 in Figure 8.16:

```
=SQRT(3*F.INV.RT(0.05,3,I6))*SQRT((I7/I6)*(2/8))*SQRT(1+(H14/(3*I14)))
```

The formula contains three square roots, so it's convenient to take them one by one. This fragment:

```
SQRT(3*F.INV.RT(0.05,3,I6))
```

uses the F.INV.RT() worksheet function to return the value in the F distribution that separates the rightmost (the RT tag) 5% (first argument) from the remainder of an F distribution. The F distribution has 3 (second argument) and whatever value is in cell I6 (third argument) degrees of freedom. In this case, the value in cell I6 is the residual degrees of freedom in the instance of LINEST() that regresses the outcome measure on the covariate and the coded group vectors.

Note that the F value is multiplied by 3, the number of levels of the factor minus 1. The result is of course the degrees of freedom for the factor, and therefore the number of coded vectors for the factor.

The second fragment:

```
SQRT((I7/I6)*(2/8))
```

picks up the mean square residual from the full LINEST() in H3:L7. It does so by dividing the sum of squares residual in cell I7 by the degrees of freedom residual in cell I6.

The fragment also evaluates the squares of the contrast coefficients for the comparison of, in this example, Group 1 with Group 2. The coefficients are 1 and –1, so their squares total to 2. Both groups have 8 observations, so we wind up with 2/8.

The third fragment:

SQRT(1+(H14/(3*I14)))

employs the sum of squares regression (cell H14) and the sum of squares residual (cell I14) from the regression of the *covariate* on the coded group vectors. The degrees of freedom for the Group factor (in this case, 3) also appears in the fragment; it's represented as *k* in the symbolic representation of the formula, earlier in this section.

Compared to S, |D| is easy to calculate, but both values can require care when the number of observations per group changes from contrast to contrast, and when some contrasts are simple, two-group comparisons and others involve the averages of two or more groups. If each of your contrasts involves two groups only, as in the example just given, and if each of your groups has the same number of observations, then the value of S will be the same for all contrasts.

This is not the case in Figure 8.16. The values of S change from cell J25 to J26 to J28, because the contrast coefficients change from 1 to .5 to .33 as the comparison becomes a little more complex. However, the value of S in cell J29 is identical to that in cell J25 because each comparison involves a simple two-group contrast and all groups have the same number of observations.

Again, the approach described in this section is less powerful statistically than either planned orthogonal contrasts or planned nonorthogonal contrasts. But it does afford you more room to maneuver as long as you specify, before seeing the results, which groups or combinations of groups you want to contrast.

We've covered quite a bit of ground in the last few hundred pages, starting with measures of variation and covariation, moving through correlation and estimation to regression—both simple and multiple—and on to various coding methods that enable you to perform analysis of variance and covariance by adopting particular coding methods, to be followed in some situations by multiple comparisons. Along the way we've detoured briefly into more theoretical topics such as the robustness of these analyses to the violation of their assumptions, and the effects of choices you make on the statistical power of the tests you run.

You have now laid the groundwork for applying regression analysis to problems that interest you either intrinsically or professionally. Beyond that, you're in a position to work with more advanced topics such as logistic regression, factor analysis, and other techniques that study more than one outcome variable at a time. If you head in that direction, I think you'll find the going much smoother with concepts you've explored here, such as semipartial correlation, shared variance, and models comparison in your toolkit.

Index

Symbols

^ (exponentiation operator), 8

A

ABS( ) function, 7

adding

covariates, 298

sum of squares, 74–75

add-ins, Data Analysis, 203, 215

Correlation tool, 271, 276

dummy coding, 246

adjusted regression sum of squares, 312–313

alphas, 184

analysis

effect coding, 279–283

factorial, 272–277

LINEST() functions, 132

multiple regression, 166

outcome variables, 299

residuals, 92

results, comparing, 249

summarizing, 320–321

via proportions of variance, 133–149

via Sum of Squares, 132

with/without covariates, 308–312

analysis of covariance. ***See*** **ANCOVA**

analysis of variance. ***See*** **ANOVA**

ANCOVA (analysis of covariance), 295–297

adjusted regression sum of squares, 312–313

analysis
summarizing, 320–321
with/without covariates, 308–312
ANOVA, comparing results, 297–305
charts, 305–308
covariate sums of squares, 312
degrees of freedom, 301
F-ratios, 299
LINEST() function, 303
outcome measure sums of squares, 312
planned nonorthogonal contrasts, 330–332
planned orthogonal coding, 321–328
post hoc nonorthogonal contrasts, 332–336
regression
multiple comparisons, 328–330
structuring using, 315–316
residuals, 301–303
R^2 for covariates/outcomes, 312
structuring conventional, 308
within-cell product of deviations, 313–336
ANOVA (Analysis of Variance), 117
ANCOVA (analysis of covariance), comparing results, 297–305
dummy coding, 215–217
f-ratios, 129–132, 136–140
General Linear Models, 146–149
planned orthogonal contrasts with, 268
Single Factor tool, 75, 140, 247–248
Sum of Squares Within, 81–82
Two-Factor Without Replication tool, 273
arguments
const, 82
new x's, 85–86
TREND() function, 86–88
array-entering, 249
LINEST() function, 103–104
TREND() functions, 84–85
arrays
formulas, 84
LINEST() function, 104
assumptions
distributions, 211–213
dummy coding, 215–217
equal spreads, 213–215
overview of, 199–202
robustness, 202–204
statistical inference, 204
straw man example, 204–211
t-tests, 217
auditing, Formula Auditing group, 325
AVERAGE() function, 7, 10–12, 36, 73
avoiding traps in charts, 48–53

B

betas, 286, 311
bias
correlation, 41–44
reduction functions, 305–308
binominal variables, 117
biserial correlation, 30. *See also* **correlation**
Bubble charts, 47

C

calculating
betas in factors, 311
correlation, 34–44
bias, 41–44
coefficients, 38–41

CORREL() function, 41
covariance, 34–36
deviations, 12
errors, 173–176
predicted values, 63
prediction, 61–62
probability, straw man example, 209
R^2, 312
residuals, 201
standard deviations, 6
standard errors, examples, 176–181
statistical power, 285–286
sum of squares, 169
t-ratios, 187
zero-constant regression, 88
z-scores, 18
z-values, 20

canonical regression, 65

category variables, 48

causation, correlation and, 53–54

cause, direction of, 54–55

cells, unequal cell sizes, 288–289

central F distributions, 209

charting

ANCOVA, 305–308
correlation, avoiding traps, 48–53
prediction, 70–71
regression, 63–75
regression lines, 317
Scatter, 50
types of, 47
variables, 46

checking for common regression lines, 316–320

coding

dummy, 246–250
effect, 259
factorial analysis, 279–283
multiple comparisons with, 264–267
orthogonal, 267–272
contrasts, 267
factorial analysis, 274–277
planned orthogonal contrasts with ANOVA, 268
using LINEST() function, 269–272
planned orthogonal, 321–328
rules, 248
with –1 instead of 0, 260–261

coefficients

common regression, 310
contrast, 264
correlation, 30, 38–41
regression, 154
errors, 109–110
measuring probability, 112–113
predictions, 65
standard errors, 217–244
straw man example, 208
zeros, 110–112
standard errors
calculating, 173–176
using of the, 181–186

collinearity, 114

common regression

coefficients, 310
lines, checking for, 316–320

comparing

Dunnett multiple comparison procedure, 253–259
f-ratios to R^2, 146
LINEST() function, 106–114
models, 103

models comparison approach, 192
probability, 195
results, 249
results, ANOVA/ANCOVA, 297–305
components, summing component effects, 278–293
composite variables, 155
computational formulas, 73
conservative tests, unequal spreads, 220–225
const argument, 82
TREND() function, 86–88
constants, 51, 69, 87, 104, 154
trendlines, 160
variables, 166–167
contrasts
coefficients, 264
orthogonal coding, 267
planned nonorthogonal, multiple comparisons, 330–332
post hoc nonorthogonal, multiple comparisons, 332–336
control groups, 246
conventional ANCOVA, structuring, 308
correlation, 29
bias, 41–44
calculating, 34–44
and causation, 53–54
charts, avoiding traps, 48–53
coefficients, 30, 38–41
CORREL() function, 41
covariance, 34–36
directions
of cause, 54–55
determining, 32–34
linearity, checking for, 44–48
measuring, 29–30
outliers, checking for, 44–48
partial, 90–95, 166
Pearson, 45
prediction, 60–61
predictors, 161
R^2, 117–120
restriction of range, 55–57
semipartial, 95–101, 166, 303
strength of, expressing, 30–32
two-predictor regression, 167–169
variables, 34, 55
Welch's, 237–243
z-scores, 62
Correlation tool, 271, 276
CORREL() function, 38, 41, 67, 168
covariance, 34. *See also* **ANCOVA**
deviations, 36–38
COVARIANCE.P() function, 36
covariates, 130
adding, 298
analysis with/without, 308–312
R^2 for, 312
sums of squares, 312
critical values, 187
cross-validation, 198
curvilinear relationships, 219

D

Data Analysis add-in, 203, 215
Correlation tool, 271, 276
dummy coding, 246
LINEST() function, using instead of, 230–231
degrees of freedom, 13, 112, 125, 188
ANCOVA, 301

residuals, F-ratios, 172–173
straw man example, 207
Welch's correlation, 241
dependent groups t-tests, 244
dependent variables, 130, 134
design
experimental, 53–54
experimental design approach, 288
repeated measures, 273
deviations
calculations, 6, 12
covariance, 36–38
from the mean, 11
MAD (mean absolute deviation), 7
squared, 38, 73, 118
standard, 14–15, 71–73
standard errors as residuals, 125–128
sums, 6–10
within-cell product of, 313–336
DEVSQ() function, 9–11, 73
dialog boxes, Solver, 159
differences, 245
coding with –1 instead of 0, 260–261
dummy coding, 246–250
Dunnett multiple comparison procedure, 253–259
effect coding, 259
factorial analysis, 272–277
General Linear Models, relationships, 261–264
mean, 229
multiple comparisons with effect coding, 264–267
orthogonal coding, 267–272
proportions of variance, 277–278
standard error of the mean, 217
summing component effects, 278–293
T.DIST() function, 231–237
vectors, populating, 250–253
direct correlation, 32
directional tests, 191
directionality of causation, 53–54
directions
of cause, 54–55
correlation, determining, 32–34
distributions
assumptions, 211–213
central F, 209
families of, 188
sampling, 113, 125
standard normal, 22
t, 26, 121–122
unit normal, 22
dummy coding, 135, 215–217, 246–250
Dunnett multiple comparison procedure, 253–259

E

effect coding, 259
factorial analysis, 279–283
multiple comparisons with, 264–267
effects, summing component, 278–293
equal sample sizes
unequal spreads, 226–230
equal spreads, 128–129, 200
assumptions, 213–215
Equal Variances tool, 239
equations, regression, 49
errors, 120–121, 262. *See also* **standard errors**
as standard deviation of residuals, 125–128
calculating, 173–176
examples, 176–181
LINEST() function, 105

mean square, 173, 262
regression coefficients, 109–110, 181–186
standard, 72
standard error of estimate, 72, 120–121
standard error of measurement, 72
standard error of the mean, 15–18, 72
sum of squares, 156–160
t distributions, 121–122
type I/type II, 283–288
variance error of the mean difference, 229

estimation, 60
standard error of estimate, 72

evaluating predictors, 192

experimental design, 53–54, 288

exponentiation operator (^), 8

F

factorial analysis, 272–277
effect coding, 279–283

factors, 246
betas, calculating, 311

families of distributions, 188

F.DIST() function, 187

F.DIST.RT() function, 299

finding sums of squares, 169–170

F.INV() function, 187

Formula Auditing group, 325

formulas
array-entering, 249
arrays, 84
computational, 73
LINEST() function, 104
shrinkage, 197–198
standard error of estimates, 126

F-ratios, 116, 129
ANCOVA, 299
ANOVA, 129–131, 136–140
R^2, comparing, 146
regression, 131–132, 140–146
residual degrees of freedom, 172–173
sum of squares, 173

F-tests
omnibus, 215
straw man example, 208

fudge factors, 202

functions
ABS(), 7
AVERAGE(), 7, 10–12, 36, 73
bias reduction, 305–308
CORREL(), 38, 41, 67, 168
COVARIANCE.P(), 36
DEVSQ(), 9–11, 73
F.DIST(), 187
F.DIST.RT(), 299
F.INV(), 187
INDEX(), 108
INTERCEPT(), 69–70, 89
LINEST(), 48, 50, 82, 84, 90–101, 103–107
analysis via, 132
analysis via proportions of variance, 133–149
ANCOVA, 303
array-entering, 103–104
array formulas, 104
comparing, 106–114
errors, 105
f-ratios, 129, 136–140

mapping results to worksheets, 163–166
orthogonal coding, 269–272
partial correlations, 90–95
semipartial correlations, 95–101
significance of, 122–132
statistics for, 149
straw man example, 209
structuring ANCOVA using regression, 315
using instead of Data Analysis Tool, 230–231
zeros, 114–122
NORM.S.DIST(), 20, 25
PEARSON(), 41
SLOPE(), 65–69, 89
standard deviations, 36
STDEV.P(), 36
STDEV.S(), 36
SUMDEVSQ(), 74
SUMPRODUCT(), 265–266
SUMSQ(), 263
T.DIST(), 187, 231–237
T.DIST.RT(), 25
T.INV(), 187
TRANSPOSE(), 103–104
TREND(), 82–89, 104
array-entering, 84–85
const argument, 86–88
new x's argument, 85–86
TTEST(), 243–244
VARA(), 12
VAR.P(), 10–14, 36
VARPA(), 12
VAR.S(), 11–14, 36
VLOOKUP(), 251, 263

G

General Linear Models, 146–149, 157–158
relationships, 261–264
generalizing predictions, 64–65
group means, 245, 324
coding with –1 instead of 0, 260–261
dummy coding, 246–250
Dunnett multiple comparison procedure, 253–259
effect coding, 259, 264–267
factorial analysis, 272–277
General Linear Models, relationships, 261–264
orthogonal coding, 267–272
proportions of variance, 277–278
summing component effects, 278–293
vectors, populating, 250–253
groups
control, 246
Formula Auditing, 325

H

homoscedasticity, 128–129, 200
assumptions, 213–215

I

increasing statistical power, 286–288
independent variables, 130, 134
INDEX() function, 108
inference
assumptions, 204
statistical, 113–114
interactions, 272
INTERCEPT() function, 69–70, 89
LINEST() function, comparing to, 106–114

intercepts, 69, 87. *See also* **INTERCEPT( ) function**
interval scales, 29, 116–117
interval variables, 245
inverse relationships, 33

J

joint effects, 272

K

known x's/y's, 83, 86

L

lambda values, 144
least squares, 157–158, 203
levels, 246
liberal tests, unequal spreads, 225–226
linear transformations, 153
linearity, checking for, 44–48
Line charts, 47
lines
- common regression, checking for, 316–320
- median regression, 157–158
- regression, 31, 61

LINEST( ) function, 48–50, 82–84, 90–101, 103–107
- analysis via, 132
- ANCOVA, 303, 315
- ANOVA, 136–140
- array-entering, 103–104
- array formulas, 104
- comparing, 106–114
- errors, 105
- f-ratios, 129
- mapping results to worksheets, 163–166
- orthogonal coding, 269–272
- partial correlations, 90–95
- proportions of variance, 133–149
- semipartial correlations, 95–101
- significance of, 122–132
- statistics for, 149
- straw man example, 209
- using instead of Data Analysis Tool, 230–231
- zeros, 114–122

M

MAD (mean absolute deviation), 7
mapping results to worksheets, 163–166
mean
- deviations from the, 11
- groups, 245, 324. *See also* group means
- regression toward the, 62–63
- standard error of the, 15–18, 72

mean absolute deviation. *See* **MAD**
mean difference, 229
mean square (MS), straw man example, 208
mean square between, 130
mean square errors, 173, 262
mean square regression, 173, 210
mean square residuals, 173, 210
mean square within, 130
measurements
- correlation, 29–30
- probability, 112–113
- standard error of, 72
- variations, 5–6

median regression lines, 157–158
methods, Scheffé, 264
models
- comparing, 103
- comparison approach, 192

General Linear Models, 146–149, 157–158
statistics, 192–196
multicollinearity, 114
multiple comparisons, 253. *See also* **comparing**
ANCOVA/regression, 328–330
effect coding, 264–267
planned nonorthogonal contrasts, 330–332
post hoc nonorthogonal contrasts, 332–336
multiple predictors, 153–156
multiple regression, 65, 108, 114
analysis, 166
errors
calculating, 173–176
examples, 176–181
using of the regression coefficient, 181–186
F-ratios, residual degrees of freedom, 172–173
LINEST() function, mapping results to worksheets, 163–166
model statistics, 192–196
multiple predictors, 153–156
one-tailed tests, 189–192
predictors
evaluating, 192
variables, 152–153
R^2
estimating shrinkage, 197–198
standard error of estimate, 170–172
using instead of sum of squares, 196–197
semipartial correlation in two-predictor regression, 167–169
sum of squares
errors, 156–160
finding, 169–170
using instead of R^2, 196–197
trendlines, 160–163
two-tailed tests, 186–189
variables, holding constant, 166–167

N

National Center for Health Statistics, 71
navigating charts, 46
negative relationships, 33
new x's argument, 85–86
nominal scales, 116–117
nominal variables, 117
dummy coding, 246
non-centrality parameters, 144
non-directional tests, 191
nonlinear distributions, 211–213
nonorthogonal contrasts
planned, multiple comparisons, 330–332
post hoc, multiple comparisons, 332–336
NORM.S.DIST() function, 20, 25

O

omnibus F-tests, 215
one-tailed tests, 189–192
operators, ^ (exponentiation operator), 8
orthogonal coding, 267–272
contrasts, 267
factorial analysis, 274–277
planned, 321–328
planned orthogonal contrasts
with ANOVA, 268
using LINEST() function, 269–272

outcomes
measure sums of squares, 312
R^2 for, 312
variables, 246, 299
outliers, checking for, 44–48

P

parameters, 12
non-centrality, 144
populations, 23, 185
partial correlations, 90–95, 166
partitions, 77
Sum of Squares, 133–136
Pearson correlation, 45
Pearson, Karl, 41
PEARSON() function, 41
planned nonorthogonal contrasts, 330–332
planned orthogonal coding, 321–328
planned orthogonal contrasts
with ANOVA, 268
using LINEST() function, 269–272
pooled regression lines, 306
pooled variance, 229
populations
parameters, 23, 185
vectors, 250–253
positive correlation, 32
post hoc nonorthogonal contrasts, 332–336
power, statistical, 235
predicted variables, 130
predictions, 60–61, 120
calculating, 61–62
charting, 70–71
generalizing, 64–65
regression, coefficients, 65
values, 63
predictors
correlations, 161
evaluating, 192
linear transformations, 153
multiple, 153–156
variables, 50, 130, 152–153, 245, 249
priori ordering approach, 288
probability
comparing, 195
measuring, 112–113
straw man example, calculating, 209
of t-ratio if null is true, 229
proportions of variance, 277–278
analysis via, 133–149

Q

quasi t-ratios, TTEST() function, 243–244
quasi t statistic, 237

R

R^2, 117–120
calculating, 312
for covariates/outcomes, 312
F-ratios, comparing, 146
in linear simple regression, 77–81
shrinkage, estimating, 197–198
standard error of estimate, 170–172
straw man example, 208
sum of squares, using instead of, 196–197
ranges
restriction of, 55–57
of values, 214
ratios
f-ratios, 116, 129. *See also* F-ratios
ANOVA, 129–132, 136–140
comparing to R^2, 146

regression, 140–146
sum of squares, 173
quasi t-ratios, TTEST() function, 243–244
scales, 29
t-ratios, 111, 116, 182, 229
calculating, 187
Dunnet multiple comparison procedure, 253
straw man example, 207

regression
adjusted regression sum of squares, 312–313
ANCOVA
multiple comparisons, 328–330
structuring using, 315–316
approach, 288–293
canonical, 65
coefficients, 154
errors, 109–110
measuring probability, 112–113
predictions, 65
standard errors, 217–244
straw man example, 208
zeros, 110–112
common coefficients, 310
differences, 245. *See also* differences
equations, 49
f-ratios, 131–132, 140–146
General Linear Models, 146–149
lines, 31, 61
mean square, 173, 210
multiple, 65, 108, 114. *See also* multiple regression
simple. *See* simple regression
Sum of Squares, partitioning, 133–136
toward the mean, 62–63
zero-constant, calculating, 88

Regression tool, 203

relationships, 31
curvilinear, 219
General Linear Models, 261–264
inverse, 33
negative, 33

repeated measures design, 273

residuals, 120–121, 166, 262
analysis, 92
ANCOVA, 301–303
calculating, 201
degrees of freedom, 188
F-ratios, 172–173
independence of, 201
mean square, 173, 210
standard errors as standard deviation of, 125–128
sum of squares, 81–82

restriction of range, 55–57

results
ANOVA/ANCOVA, comparing, 297–305
comparing, 249
worksheets, mapping, 163–166

returning variances, 11–14

robustness, 202–204

RSQ() function, 79

rules, coding, 248

S

sample sizes, 220, 226–230

sampling distributions, 113, 125

scales
interval, 29, 116–117
nominal, 116–117
ratio, 29

Scatter charts, 47, 50

Scheffé method, 264
scores
prediction, 60–61
z-scores, 8, 18–23
semipartial correlations, 95–101, 166, 303
in two-predictor regression, 167–169
sequential approach, 288
shared variance, 71
shrinkage, estimating R^2, 197–198
significance of LINEST( ) function, 122–132
simple regression, 59–60
charting, 63–75
INTERCEPT() function, 69–70, 89
LINEST() function, 90–101
predicted values, calculating, 63
prediction, 60–61
calculating, 61–62
charting, 70–71
R^2, 77–81
regression toward the mean, 62–63
shared variance, 71
SLOPE() function, 65–69, 89
standard deviations, 71–73
Sum of Squares Regression, 79
sums of squares, 73–77, 81–82
totals, 76–82
TREND() function, 82–88, 89
zero-constant regression, calculating, 88
z-scores, 62
Single Factor ANOVA tool, 140
sizes, unequal cell, 288–289
SLOPE( ) function, 65–69, 89
LINEST() function, comparing to, 106–114
Solver dialog box, 159
spreads, equal, 128–129, 200
squared correlations
R^2, 117–120
squared deviations, 38, 73
sums, 7–10
standard deviations, 14–15, 71–73
calculations, 6
functions, 36
standard errors as residuals, 125–128
standard errors, 72
calculating, 173–176
coefficients, using of the, 181–186
of estimate, 72, 120–121, 174
examples, 176–181
of the mean, 15–18, 72, 217
of the mean difference, 229
of measurement, 72
R^2, 170–172
of the regression coefficient, 217–244
as standard deviation of residuals, 125–128
standard normal distributions, 22
standard scores, 8. *See also* **scores**
prediction, 60–61
statistical inference, 113–114
assumptions, 204
statistical power, 235, 283–288
calculating, 285–286
increasing, 286–288
statistics
for LINEST() function, 149
models, 192–196
quasi t, 237
STDEV.P( ) function, 36
STDEV.S( ) function, 36
straw man example, 204–211
strength of correlations, expressing, 30–32
structuring
ANCOVA using regression, 315–316
conventional ANCOVA, 308

studies, robustness, 202–204

SUMDEVSQ( ) function, 74

Sum of Squares Regression, 79

Sum of Squares Within (ANOVA), 81–82

summarizing analysis, 320–321

summing component effects, 278–293

factorial analysis with effect coding, 279–283

regression approach, 289–293

statistical power, 283–288

unequal cell sizes, 288–289

SUMPRODUCT( ) function, 265–266

sums

deviations, 6–7

squared deviations, 7–10

sums of squares, 10–13, 73–77

adding, 74–75

adjusted regression, 312–313

analysis via, 132

Between, 75

covariates, 312

errors, 156–160

finding, 169–170

F-ratios, 173

outcome measure, 312

partitioning, 133–136

R^2, using instead of, 196–197

residuals, 81–82

totals, 76–82

Within, 75

SUMSQ( ) function, 263

T

T.DIST( ) function, 187

differences, 231–237

t distributions, 26, 121–122

T.DIST.RT( ) function, 25

tests

conservative tests, unequal spreads, 220–225

directional tests, 191

F-tests

omnibus, 215

straw man example, 208

liberal tests, unequal spreads, 225–226

non-directional tests, 191

one-tailed tests, 189–192

t-tests, 111

assumptions, 204

dependent groups, 244

two-tailed tests, 186–189

T.INV( ) function, 187

tools

ANOVA: Single Factor, 75

Correlation, 271, 276

Data Analysis add-in, 230–231. *See also* Data Analysis add-in

Equal Variances, 239

Regression, 203

Single Factor ANOVA, 140

Trace Dependents, 325

Trace Precedents, 325

t-test: Two-Sample Assuming Equal Variances, 221

Two-Factor Without Replication, 273

Unequal Variances, 239

totals, sum of squares, 76–82

Trace Dependents tool, 325

Trace Precedents tool, 325

transformations, linear predictors, 153

TRANSPOSE( ) function, 103–104

traps in charts, avoiding, 48–53

t-ratios, 111, 116, 182, 229

calculating, 187

degrees of freedom, 112
Dunnet multiple comparison procedure, 253
quasi, TTEST() function, 243–244
straw man example, 207
TREND() function, 82–89, 104
array-entering, 84–85
const argument, 86–88
new x's argument, 85–86
trendlines, 31, 48, 61. *See also* **regression, lines**
multiple regression, 160–163
TTEST() function, 243–244
t-tests, 111, 217
assumptions, 204
dependent groups, 244
t-test: Two-Sample Assuming Equal Variances tool, 221
t-values, 23–27
Two-Factor Without Replication tool, 273
two-predictor regression, semipartial correlations, 167–169
two-tailed tests, 186–189
type I/type II errors, 283–288
types of charts, 47

U

unequal cell sizes, 288–289
unequal group variances, Welch's correlation, 237–243
unequal spreads
conservative tests, 220–225
equal sample sizes, 226–230
liberal tests, 225–226
unequal variances, 220
Unequal Variances tool, 239
unique variance, 99
unit normal distributions, 22

V

validation, cross-validation, 198
values
critical, 187
lambda, 144
of covariance, 38
prediction, 62–63
ranges of, 214
R^2 in linear simple regression, 77–81
t-values, 23–28
y, 82–83
z-values, 18–23, 20
VARA() function, 12
variability, 60
variables
binominal, 117
category, 48
charting, 46
composite, 155
constants, 166–167
correlation, 34, 55
dependent, 130, 134
dummy coding, 246
independent, 130, 134
interval, 245
nominal, 117
outcome, 246
predicted, 130
predictor, 50, 130, 152–153, 245, 249
variance, 10–11, 214
analysis of variance. *See* ANOVA
error of the mean difference, 229
functions, 11–14
pooled, 229
proportions of, 277–278
shared, 71

unequal, 220
unique, 99
variation measurements, 5–6
VARPA() function, 12
VAR.P() function, 10–14, 36
VAR.S() function, 11–14, 36
vectors, populating, 250–253
violations of assumptions, 202
VLOOKUP() function, 251, 263

W

Welch's correlation, 237–243
within-cell product of deviations, 313–336
worksheets, mapping results, 163–166

Y

y values, 82–83

Z

zero-constant regression, calculating, 88
zeros
LINEST() function, 114–122
probability, measuring, 112–113
regression coefficients, 110–112
z-scores, 8, 18–23, 62
z-values, 18–23
calculations, 20

que